Perspectives on Geographical Marginality

Volume 2

Series editors

Walter Leimgruber, Fribourg, Switzerland
Etienne Nel, Dunedin, New Zealand
Stanko Pelc, Koper-Capodistria, Slovenia

This book series *Perspectives on Geographical Marginality* comprehensively overviews research, on areas and communities impacted by processes of marginalization as a result of globalization, economic, environmental, political and social change. This series seeks to discuss and determine what is geographical marginality by inviting leading international experts to publish theoretical and applied work. It also seeks to rigorously debate the degree to which local areas and communities are responding to these process of change and with what success.

The series stems from the International Geographical Union's (IGU), 'Commission on Globalization, Marginalization, and Regional and Local Response' (C12.29). As is suggested by its name, the commission researches the problem of geographical marginality offering a leading forum from which this series will be led. Marginality cannot be defined without putting it into a certain perspective: economic, political and social (including cultural). Marginality has to be clearly distinguished from peripherality. Marginal areas may be a part of periphery or even the centre, but "cannot really be attributed to them".

Proposed themes which will be covered include:

- Mountainous regions and globalization;
- Regional development and policy/or: Globalization and its impact on local and regional development;
- Theory of marginalization;
- Transformation of rural areas from the viewpoint of globalization and marginalization;
- Drivers of marginalization in border and peripheral areas.

This present book provides an overview of wide range of examples of societies fighting with poverty, social inequity and marginalization from many different viewpoints. It will be especially interesting for those who want to gain a better insight into situation and position of ethnic groups living in harsh mountainous conditions in Himalaya. Livelihood problems of this part of the world, outmigration and environmental threats are further interesting topics. Examples from other parts of the world such as from Kyrgyzstan, Israel, Switzerland and Finland provide an opportunity to compare and to see how marginality and marginalization differs around the world, considering the conditions, the scale and the angle of observation.

More information about this series at http://www.springer.com/series/15046

Raghubir Chand · Etienne Nel · Stanko Pelc
Editors

Societies, Social Inequalities and Marginalization

Marginal Regions in the 21st Century

Springer

Editors
Raghubir Chand
Department of Geography
Kumaun University
Nainital, Uttarakhand
India

Stanko Pelc
Faculty of Education
University of Primorska
Koper-Capodistria
Slovenia

Etienne Nel
Department of Geography
University of Otago
Dunedin, Otago
New Zealand

ISSN 2367-0002 ISSN 2367-0010 (electronic)
Perspectives on Geographical Marginality
ISBN 978-3-319-50997-6 ISBN 978-3-319-50998-3 (eBook)
DOI 10.1007/978-3-319-50998-3

Library of Congress Control Number: 2016960572

Reviewer: Walter Zsilinczar

Printed on acid-free paper

This Springer imprint is published by Springer Nature
The registered company is Springer International Publishing AG
The registered company address is: Gewerbestrasse 11, 6330 Cham, Switzerland

Preface

In a world characterized by persistent and increasing socio-spatial unevenness and inequality, concerns about geographical marginalization are all the more concerning and need to be effectively understood in order to develop appropriate and sustainable responses. Rather than living in a world of decreasing differences, sadly we live in a time and space, where differences and inequalities appear to be growing and where concerns over the marginalization of people and places have yet to be fully understood and addressed.

International research challenges us to rethink the concepts of differences between and within nations and societies. There is abundant evidence, from around the world, that while levels of extreme poverty appear to have been reduced, inequality levels within most societies are increasing. This reflects both the inability of all people to participate fully and to the same degree in the dominant economic system and complex overlays of social exclusion, discrimination, displacement and conflict which sadly remain features of the world in the twenty-first century.

It is in this spirit that this book series seeks to extend our collective human understanding of what marginalization is and how it impacts on people and places, from both evidential and theoretical perspectives. Over time, the gathered evidence will, hopefully, facilitate the better informed analysis and understanding of what marginality and marginalization are and how it might be responded to.

This, the second book in the series 'Perspectives of Geographical Marginality', seeks to extend the academic debate on these issues through a particular focus on, as the title suggests, societies, social inequalities and marginalization. This book is a continuation of themes explored in the first volume. The majority of its chapters have also been prepared from papers presented at the conference of the International Geographical Union's (IGU) Commission on Marginalization, Globalization and Regional and Local Responses C08.27 held in Nainital, India, from 1 May to 9 May 2011, jointly organized by the IGU's Commission and the Department of Geography, DSB Campus, Kumaun University, Nainital, India. The conference theme was 'Local and Regional Responses to Globalization in the Mountains and Marginal areas of the World'.

The concept of marginalization has been applied in a broad variety of disciplines and perspectives and remained the focus of academic debates. In the last two centuries, marginality was perceived as a subject of economic and social discrimination, physical remoteness and political exclusion, overburdened by colonial suppression both in developing and in developed countries. Marginality became a relevant topic of geographical research and was viewed as a response to the deep-seated socio-economic and political changes and increasing complexities created by the on-going process of globalization in the twentieth and twenty-first centuries. Marginal regions are undergoing structural and demographic changes with increasing social insecurity and cultural clashes depicting interconnected positive and negative consequences affecting individuals in different kinds of marginal societies in the twenty-first century.

Nainital, India — Raghubir Chand
Dunedin, New Zealand — Etienne Nel
Koper-Capodistria, Slovenia — Stanko Pelc

Contents

Contributors

Asylbek A. Aidaraliev National Academy of Sciences of Kyrgyzstan, International University of Kyrgyzstan, Bishkek, Kyrgyz Republic

Chhavi Arya Home Science Department, Kamoun University, Nainital, Uttarakhand, India

Aditi Chand Department of Literature, Faculty of Arts, Katholieke Universiteit, Leuven, Belgium

Raghubir Chand Department of Geography, DSB Campus, Kumaun University, Nainital, Uttarakhand, India

Devi Datt Chauniyal Department of Geography, H.N.B. Garhwal Central University, Srinagar, Uttarakhand, India

Savita Chauniyal Department of Economics, D.A.V. (P.G.) College, Dehradun, Uttarakhand, India

Moushumi Datta Nagindas Khandwala College, Mumbai, Maharashtra, India

Seth Frantzman The Jerusalem Post, Jerusalem, Israel

R.C. Joshi Department of Geography, Kumaun University, Nainital, Uttarakhand, India

Ruth Kark Department of Geography, The Hebrew University, Jerusalem, Israel

Olli Lehtonen Natural Resources Institute Finland (Luke), Helsinki, Finland

Walter Leimgruber Department of Geosciences, Geography, University of Fribourg, Fribourg, Switzerland

Chandra Singh Negi Department of Zoology, L S M Government Postgraduate College, Pithoragarh, Uttarakhand, India

Etienne Nel Department of Geography, University of Otago, Dunedin, New Zealand

Gibji Nimasow Department of Geography, Rajiv Gandhi University, Itanagar, Arunachal Pradesh, India

Surendra Pal G. I. C. Ramak, Champawat, Uttarakhand, India

Lata Pande Home Science Department, Kamoun University, Nainital, Uttarakhand, India

B.R. Pant Department of Geography, M. B. Govt. P. G. College, Haldwani, Nainital, Uttarakhand, India

Stanko Pelc Faculty of Education, University of Primorska, Koper-Capodistria, Slovenia

Vishwambhar Prasad Sati Department of Geography and Resource Management, Mizoram University, Aizawl, India

R.C.S. Taragi Ministry of Development Planning and Statistics, States of Qatar, Doha, Qatar

Jyoti Tiwari Home Science Department, HNB Garhwal University Srinagar Campus, Srinagar, Uttarakhand, India

Markku Tykkyläinen Department of Historical and Geographical Studies, University of Eastern Finland, Joensuu, Finland

Havazelet Yahel The Ben-Gurion Research Institute for the Study of israel & Zionism, Ben-Gurion University of the Negev, Sde-Boker, Israel

List of Figures

List of Tables

Chapter 1
Introduction

Raghubir Chand, Etienne Nel and Stanko Pelc

1.1 Setting the Frame

Globally there is abundant evidence that 'convergence' in terms of the livelihoods, quality of lives, opportunities and the earning capacity of people around the world is remaining stubbornly elusive. The United Nations's Development Programme's 2015 Human Development Report (UNDP 2015), makes for sobering reading in terms of the scale of the disparities which exist between nations and peoples around the world. Equally concerning is the degree to which social inequalities increase dramatically in most part of the world, when one factors in considerations of gender inequality and income inequality within societies. Given the very real constraints the global population faces in terms of resource constraints and the stark reality of climate change, there is a very real risk that disparities might increase in future.

While some progress was clearly made in terms of reducing overall levels of global poverty through the Millennium Development Goals (UN 2015), clearly the implementers of the new Sustainable Development Goals have enormous challenges to face in a world characterized by population growth, resource depletion, economic challenges, and political and ethnic tensions, to name the most apparent.

R. Chand
Department of Geography, DSB Campus, Kumaun University,
Nainital 263002, Uttarakhand, India
e-mail: raghubir.seri@gmail.com

E. Nel
Department of Geography, University of Otago,
PO Box 56, Dunedin 9054, New Zealand
e-mail: etienne.nel@otago.ac.nz

S. Pelc (✉)
Faculty of Education, University of Primorska, Cankarjeva 5,
6000 Koper-Capodistria, Slovenia
e-mail: stanko.pelc@pef.upr.si

R. Chand et al. (eds.), *Societies, Social Inequalities and Marginalization*,
Perspectives on Geographical Marginality,
DOI 10.1007/978-3-319-50998-3_1

It is in this context that the concept of 'marginalization', sadly remains a persistent reality of life on this planet. Marginality or marginalization, which is examined as a concept in this chapter of this book, characterizes a situation of impediments—social, political, economic, physical, environmental—which impact on the abilities of many people and societies to improve their human condition. It is in this context that this book seeks to advance our global understanding of what marginalization is, how it is manifested and what causes it, and also, in the final part, to consider development options.

It is through offering an informed and detailed critique of what marginalization is and it causes, that development agencies and society more broadly will be in a better position to enact meaningful changes which help to address these persistent challenges. The information contained in this book draws from field-evidence of the manifestation and cause of different parts of the world, but particularly from the mountainous regions of South Asia where physical considerations and isolation have exacerbated social and economic challenges which communities face.

1.2 The Structure of the Book

1.2.1 Part I: Poverty, Social Inequalities and Marginalization

This section of the book reviews issues of poverty, social inequality and marginalization. The commonly held view put forward by neoliberal economists is that globalization reduces poverty and social inequality and is potentially beneficial to all. This single-minded approach is being challenged by other academics all across the world. Over the last four decades, the gap between the poorest and the richest has been rising and there are more social tensions arising out and created due to the forces of rapid globalization operating in marginal areas. In marginal areas and mountain regions in particular, unskilled labourers who are unable to take advantage of new technology are losing their jobs because of low cost and high-tech imports. The isolated pockets of population often surviving on their traditional livelihood strategies become victims of a sudden intrusion of outside global firms taking away all of this in favour of the activities of large multinational companies. In fact access to better infrastructure and communication and sound household conditions is crucial to reap the benefits of globalization. In the absence of qualitative social value considerations, the real goal of global economies benefitting the poor and the marginal sections of the society can hardly be realized.

In the first chapter of this section Pelc continues the long standing discussion about the definition of marginality and discusses different approaches to this phenomenon as well as the process of (de)marginalization. Marginality and marginalization are present in at least some way in every chapter of this book, but the authors of the chapters are more or less exclusively focussed on the problem that they present. They do not try to put this problem into the context of marginality and marginalization. The exception is Leimgruber's chapter. Pelc starts his discussion with an explanation of the word marginal then he presents initiatives from the Atlantic coast countries that started with marginality studies and comes to IGU study group and commissions that were previously involved in the research of marginality from many different angles. Leimgruber's definitions and approaches are presented, as are Park's concept of marginal man and University of Bonn's Center for Development Research approaches to marginality. His final discussion is about the dilemma: 'geography of marginalization or geography of marginal regions' and proposes that "In order to achieve better research results in the field of marginality studies geographers should adopt a 'geography of marginal regions' approach."

In the chapter Poverty and Global Warming: an Example of the Kyrgyz Republic, A. A. Aidaraliev examines the current status of poverty in Kyrgyz Republic of Central Asia. The phenomena of poverty and marginalization, depopulation, degradation of resources and environment, backwardness of transport and social infrastructure, and, unemployment are typical for mountain regions of the Kyrgyz Republic, and create conditions for social insecurity and conflicts. The general level of poverty was recorded at 31.7% in 2008, with the level of extreme poverty at 6.1%. It is clearly time to change the territorial development strategy of the republic in response.

Pande, Tiwari and Arya in their chapter on "Socio-economic wellbeing and mental health profile of rural hill women of Uttarakhand, India" report on a study of 250 women respondents of reproductive age (15–49 years) in three developmental blocks of Almora district, in Uttarakhand, India, with the aim to examine the socio-economic conditions and mental health profile of rural hill women. The overall results of the study revealed the existence of unsatisfactory socioeconomic conditions and revealed the compromised state of mental health of rural hill women in the study area.

Similarly, "The Status of Marginalized Women Tea Garden Workers in the Mountain ecosystem of Darjeeling in a Globalised Village", presented by Moushumi Datta provides substantial evidence to understand the status of women in the tea industry in Darjeeling. The data used for this qualitative study was obtained from a primary survey conducted in the tea gardens of Darjeeling in May 2009 and May 2010. The survey covers a representative sample of 180 women workers. This study sought to highlight the disadvantaged position of the women workers in the tea industry and the exploitation that they face. Empirical evidence shows that the management violates most of the provisions of fair and compensatory employment. As one of the oldest industries still surviving and prospering it is imperative that the women work force's concerns need to be addressed immediately. The development of these marginalized tea garden women will ensure a future for the industry, which, to date has exploited their main pillar of sustenance and survival.

A micro-level study by Chauniyal and Chauniyal on socio-economic development in the mountain watershed of Uttarakhand State of India is a good example of disparities in socio-economic development. They base their arguments on a set of demographic (7), agricultural (3), social (9), infrastructural (3), and economic (2) indicators. The analysis indicates that at the utmost 51.51% of the villages in the study area are below a moderate level of development of 40–60 index points, comprising 48.94% of the population in the watershed. The analysis concluded that 73.23% of the villages are below the average level of development while only 26.78% percent villages are above that level. This indicates that the Bino watershed area is experiencing only a low to moderately low level of development.

The concluding chapter of this book part by Aditi Chand elucidates the principle that marginality is created and re-created after every major reconstruction of the society at large. The main purpose of this chapter is to provide a general understanding of the attributes that characterise the cultural landscapes in Bhutan, their diversity, the cultural, social and economic processes that have shaped these landscapes, their genesis and their associations. The tales of origin and the settling process of marginal societies in Bhutan as compiled and narrated by the author describe how these marginal societies have evolved and survived providing for a better understanding of their role in the context of globalization and how to facilitate and plan for their entry into the market economy. A significant amount of development has taken place in these villages of Bhutan, and the government is trying to pull them out of the oblivion which they experience living on the margins and introduce them to numerous modern facilities, bringing them closer to global awareness than they were before.

1.2.2 Part II: Indigenous Communities, Identity, Livelihood Practices and Biodiversity

The section opens with an important contribution made by Kark about the consequences of the Ottoman Land Law and the agrarian and privatization processes that took place in Palestine from 1858 to 1918. The main focus of this chapter is to critically examine the objectives of the Ottoman Land Code, the Tabu Law of 1858 and the 1867 Law that permitted foreign citizens to acquire urban and rural land. The chapter assesses their significance, impact, and success or failure in Palestine. It touchs upon issues such as the abolition of the musha'a, land surveys, systematic mapping, and land registration and land settlement. It promotes our understanding of these historical processes and conflicts in Palestine, including the wide scale Jewish settlement and the Arab-Jewish conflict.

Yahel, Kark and Frantzman in the following chapter discusss the problem of indigeniety in the example of Negev Bedouins. The UN Declaration on the Rights of Indigenous Peoples (DRIP) is not legally binding, but it draws international attention to the situation of indigenous populations and strengthens their demands for rights. In Israel the debate concerning this problem focuses mainly on the Bedouin's demand for recognition as the indigenous population of the Negev, in southern Israel. Authors present the roots of the indigenous concept and the development of the indigenous rights regime under international law and explore legal definitions and present a set of widely accepted characteristics. A brief history of the region, the various regimes, and the Islamic and Ottoman legal heritage are presented distinguishing between the ME's history and the *terra nullius* or "discovered" territories where the indigenous concept was first applied. The main topic are the Bedouin of the Negev, their Arabian Peninsula origin, and the late date of arrival in the Negev of their fore bearers. The conclusion based on the presented arguments is that Negev Bedouin claims are not compatible with the prevailing notions of indigeneity.

The study on "Changing identity, livelihood and biodiversity of indigenous communities in The Eastern Himalaya with special reference to Aka tribe" by Nimasow and Joshi demonstrates how the society, economy and culture of the Aka have undergone a number of changes while trying to accommodate the pace of globalisation. The penetration of the scientific knowledge and technology into the land of this forest people has had a deep impact on their lives. Due to the assimilation of knowledge and the early adaptation by the local people, there are lots of changes in the spheres of the religion and festivals of the people. They are converting to Christianity, Hinduism, Buddhism etc., which may be the outcome of either the local people's inclination towards new faiths or the perceived receipt of better facilities from recognised religions. However, they have not progressed significantly during the post-independence period due to the lack of efficient and effective transportation systems and the inadequate means of communication. The economic organisation, material culture, food habits, social, cultural and religious practices of the Aka are largely determined by the existing forest resources. Hence, there is an urgent need to formulate a sustainable forest management for the wellbeing of the Aka tribe.

The chapter by Negi confines itself to the summer habitations of the Bhotiya community residing within the valleys of Darma, Vyans and Johaar valleys in the Pithoragarh district of Kumaun Himalaya, India and delves into the factors that sustain the very livelihood of these people. Trade with Tibet and the factors associated with its maintenance (agriculture and livestock) were the prime occupation of Joharis and Darmis tribesman until 1962, when the Sino-Indian war brought an immediate end to this thriving lifeline. The loss of trade brought about drastic changes in the transhumance-based lifestyles, with fewer households arriving at their summer homes, a smaller livestock population, disappearance of traditional handicrafts, and increased exploitation of wild medicinal and aromatic plants (MAPs), with traditional crops being replaced by more remunerative crops such as chives (Allium stracheyi) and caraway (Carum carvi) which became the chief source of livelihood, as in the case of Johaar valley. The present study highlights the causal factors behind changes in the lifestyles of these traditional people, inclusive of the causal factors behind the decline in transhumance, livestock population, of traditional crop diversity, and concludes with proposals for possible alternatives for a sustainable future for these people.

The chapters on Brokpa Yak Herders of Bhutan by Chand examine one of the most significant semi-nomadic indigenous people of Bhutan. Surviving on transhumance, herding and trade, they represent the initial mode of human adjustment to nature. They live in a harsh mountain environment and survive by their ecologically adaptive strategies to many kinds of environmental constraints. Altitude creates vertical zones of resources that range from sub-tropical forest to sub-alpine forest and alpine pastures. The Brokpas have therefore developed a symbiotic relationship with these forest and pasturelands which provide them with a perennial source of livelihood and sustenance. Their occupational patterns as yak herders and sellers of milk products, their social characteristics and their persistence as distinct entities in Bhutan are immensely significant. The main focus of this chapters is to bring to light the ecological mode of adaptation and changing livelihood strategies of the Brokpas of Bhutan. The study is based on the primary information at household level for which data have been collected from 210 households from 13 sample villages of the Brokpa in January 2010. In the present sample, about 54.76% of the families are practicing transhumance, while the "drukor" or brukor (which roughly translates into moving around for grains to exchange with their dairy products during winter) is still an economic necessity. Occupational change is minimal for the Brokpas with only 4.72% changing occupations between 1999 and 2010 (comparing the first study by the author in 1999 to the sample survey of 2010). The author focused on the new generation Brokpas, how they can adapt their livelihood strategies to the new democratic setup within Bhutan and the changing global context. In the 2010 sample, about 39.8% families were yak herders followed by 18.3% engaged in agriculture and 15.4% in weaving. Occupational change is very small for the Brokpas being recorded only 9.19% from 1999 to 2010 (as compared with the first study). For the Brokpas cattle rearing seems to be still perceived and favoured as the best occupation.

The chapter based on the socio-cultural and nutritional environment of Tharu population living in the Tarai belt of Kumaun Himalaya, Uttarakhand, India was contributed by Pant and Pal. The primary data was gathered from 907 households of 22 sample villages through structured questionnaires. The data pertaining to cultural aspects, demographic composition, and nutrition status, have closely reflected the situation as it exists in Tharu society. Actual food intake taken by Tharu people been collected during field work in local units and then converted into standard units of measurement and these food items again converted into various nutrients such as protein, fat, energy, carbohydrates, minerals and vitamins. Finally the actual availability of nutrients per head per day has been compared with the Recommended Dietary Allowances in India.

1.2.3 Part III: Migration and Depopulation

The issue of rural-urban migration is a serious concern in all marginal and mountainous areas of the world as more people drift to urban areas in search of better living conditions. In developed economies, the long-standing spatial formation of unemployment and migration reflects the imbalance in the labour market as a result of structural changes in the economy. This section includes two chapters beginning with Lehtonen, and Tykkyläinen's chapter on similarities in the spatial patterns of net Migration and unemployment across Finland. This chapter investigates the association between unemployment and net migration at the municipal level. The concept of spatial formation, defined empirically as the clustering of areas of similar development, is apposite as we chase uniform spatial structures brought about by economic development and its economic externalities. The main focus of this research is on the question of how the spatial pattern of net migration is linked to unemployment and how spatially unbalanced development is reflected by the socio-economic conditions in Finland. The years of investigation were 2004, 2006 and 2008. Methodologically, this chapter is based on autocorrelation analyses and a new similarity index, which are used to analyze the spatial patterns of net migration and unemployment, and the residuals of mismatch of the responses to the differences in job opportunities associated with migration are explained by the Finnish socio-economic conditions. It is concluded that demographic and some socio-economic characteristics in part explain the differences in the spatial patterns of unemployment and migration in push-lose and pull-win municipalities.

The chapter contributed by Taragi and Chand deals with the rural-urban migration in the Himalayan state of Uttarakhand, India. The creation of this new state more than a decade ago seems not to have been effective in checking the continued tendency of migration from villages to towns and cities within and outside the state in search of sustenance. The migration of people remains the biggest challenge faced by the state in its social and economic development. It has threatened the very basis of the hill economy, i.e. agriculture, besides being a serious threat to the existence of the hill society. The figure released by the 2001 census reveal that about 36% (3 million) of the Uttarakhand population are reported as being migrants. Also 43% of the urban migrants from Uttarakhand go to Delhi NCR, followed by 22% going to other cities including Lucknow, Bareilly, Moradabad, Kanpur etc. Mumbai and Chandigarh which are also seen to be important destinations. Intra-district migrants constitute about 18% (1.5 million) of the total population (mainly females due to marriage) and inter-district migration within the state is about 6% (0.5 million) of the total population. An overwhelming majority of out-migrants (66%) are concentrated within the state in districts such as Nainital, Udhamsinghnagar, Dehradun and Haridwar, lying in the plains and having large cities and industrial activities. All hill districts are losing population. The situation is alarming, as there has been a decrease in the total population of the Almora and Pauri Garhwal districts as revealed in the comparison of the findings of the 2001 and 2011 censuses.

1.2.4 Part IV: Policies and Strategies

Policies and strategies to cope with the conflicting situations of marginal and mountain world are the topic of this section. Marginal communities across the world are starting to learn to live with the realities of globalization. Mountain communities are marginalised on the basis of their geographical remoteness, their ethnicity, and their livelihood. Furthermore, there has been a severe lack of either public or private investment in infrastructure and economic development in mountain areas, combined with poor access to markets. The alternative livelihood strategies such as tourism, cash cropping, rural-urban integration and improved road transport and resource recovery techniques are some of the considerations required to empower mountain communities.

The chapter by Leimgruber focuses on tourism in the Swiss Alps and the human factor in local development. Since the 19th century mountains, in particular the Alps, were opened up through improved transportation, emerging tourism, and the internationalization of the economy, which led to an increasing influx of people from the outside. The late 19th century saw the first tunnels dug through the Swiss Alps; further tunnels for both road and rail were constructed during the 20th and early 21st century. The chapter discusses three cases of tourism and recreation in the Swiss Alps that demonstrate the importance of the cooperation on all scales, from the local to international. It shows that inhabitants of mountain regions are not narrow-minded when it comes to taking decisions; decisions that are related to their survival. However, the risks are considerable, and they may also result in certain persons coming into conflict with the law. The three examples chosen demonstrate the many faces of the marginality concept. Geometric marginality need not be a handicap; it is what humans make out of a particular situation that counts. Factors contributing to marginality are both internal and external: internal if the social actors deviate from a common goal (resulting in systemic marginality), external if, through technological innovations and/or the general economic situation, the local framework is upset and contingent marginality will be the outcome. Both can be avoided, as the three examples have demonstrated.

Achieving food security through opting for cash crops is the focus of the case study of Nepali immigrants in the Khanda Gad sub watershed of the Garhwal Himalaya, India by Sati. It examines the potentials of subsistence and cash generating crops keeping food security in mind and suggests at best practice for cultivating suitable crops, which could lead the way for ensuring food security for the next generation. Based on a case study of 65 Nepali households in the Khanda Gad sub watershed, findings were also supported by participatory observation after a routine field visit. This study on off-season vegetable farming in the Garhwal Himalaya reveals that the region has ideal conditions for vegetable production to exceed food security needs and to help attain sustainable development. The farmers of the region have already initiated potato and onion farming along with other spices and green leafy vegetables as an alternative to help ensure sustainable livelihoods. The main constraints on the production of vegetables are lack of cold storages, means of transportation, improved seeds, and fertilizer.

References

United Nations. (2015). *The Millennium development goals report*. New York: United Nations.
United Nations Development Programme (UNDP). (2015). *Human development report*. UNDP: New York.

Part I
Poverty, Social Inequalities and Marginalization

Chapter 2
Marginality and Marginalization

Stanko Pelc

2.1 Introduction

Marginality was recognized a relevant topic of geographical research in early nineteen-nineties when the Study group on Development Issues in Marginal Regions started to work within International Geographical Union. The same issues have been studied long before that, but with the formation of a study group and the following commissions all addressing marginal regions or marginality, we focused our attention on areas and regions that all face similar problems due to their position at the edge (fringes, border, …)—i.e. on the margin of something. That something may be spatial, social, economic, political etc. A 'marginal' position is therefore the starting point of marginality research in geography. When we are talking about regions or certain areas being in such position we would normally use a notion marginal regions (areas) and the process of putting them in such position is therefore marginalization. Finally the basic characteristic of such regions (areas—social groups, societies) is marginality. It all seems pretty simple and logical as long as we do not try to precisely define all mentioned notions. Leimgruber first defined marginal regions (Leimgruber 1994) more than two decades ago. He came to a conclusion that marginality is so wide notion that we cannot define it in a simple clearly distinct way. Besides, he also emphasized the importance of scale. Altogether that means that the definition of marginality can only be a very complex one. That may be the reason that, despite Leimgruber's first definition from 1994, and his book *Between Global and Local* (Leimgruber 2004), that is devoted to marginality and marginal regions and to globalization, we still continue the debate about the meaning of marginality and about the forms in which it can be observed. In this chapter I would like to present some of the basic ideas about understanding

S. Pelc (✉)
Faculty of Education, University of Primorska,
Cankarjeva 5, 6000 Koper-Capodistria, Slovenia
e-mail: stanko.pelc@pef.upr.si

R. Chand et al. (eds.), *Societies, Social Inequalities and Marginalization*,
Perspectives on Geographical Marginality,
DOI 10.1007/978-3-319-50998-3_2

the notion of marginality from existing literature. I intentionally do not analyze the meaning and the definitions of globalization as my intention was to focus as much as possible on marginality and marginalization, although the relation between marginality and globalization is also very important.

Globalization per se is a very vague term. The use of the concept, as observed by Robertson almost a quarter of century ago, diffused enormously during 1980s "… *across large number of areas of contemporary life in different parts of the world. By now, even though the term is often used very loosely and, indeed, in contradictory ways, it has itself became part of 'global consciousness', an aspect of the remarkable proliferation of terms centered upon 'global'*" (Robertson 1992, p. 8). Globalization undoubtedly plays major role in the process of (de)marginalization. Marginal areas, regions, social groups are this way or another influenced by the effects of globalization. In this chapter we present the broad meaning of the term marginal (marginality, marginalization) as it is explained in dictionaries and then come to some of the definitions presented by different authors and finally try to compile them in a way that we believe that could be useful for geographical research of marginality and marginalization.

2.2 Marginality from Different Angles

2.2.1 Explanations of Words Marginal and Marginality

The first question that we need to ask, when we are talking about marginality is what kind of noun is that. There is no doubt that it derives from adjective marginal. So it should be in similar relation to the word as popular is to popularity or actual to actuality etc. Dictionaries explain this relation as "the state of being" or in case of marginality "the property of being marginal" (American Heritage® Dictionary of the English Language, Fifth Edition. S.v. "marginality." 2011). Therefore somebody who is popular is experiencing popularity and somebody who is marginal is experiencing marginality. It may also be something and that something is showing characteristics that can be seen as marginality. So when we are talking about marginality we are talking about the noun that includes things, actions, characteristics etc. that all together manifest the state or property of being marginal or in case of popularity the state or property of being popular. Popularity of a movie star is characterized by the numerous fans that watch that star's films, buy different products promoted by that star, follow the star on social networks, read media news about the star etc. On the other side the popularity brings movie star several opportunities for well-paid work, influence and power. Popularity is not limited to persons, it can also be attached to things and ideas. It is, however, always connected with a large number of people all liking the same person, group, thing or idea. Marginality on the other hand is much more complex because the adjective marginal has numerous meanings and therefore marginality may be understood in many more ways than popularity.

Let us see what are the common explanations for the adjective marginal in four different dictionaries cited at TheFreeDictionary.com.

The first and the second explanation in three out of four dictionaries (American Heritage® Dictionary of the English Language 2011; Collins English Dictionary—Complete and Unabridged 2014; Random House Kernerman Webster's College Dictionary 2010) explain marginal with the position at the margin or close to the margin. Some explanations are more detailed as well as wider like the first explanation in American heritage Dictionary: "*Of, relating to, located at, or constituting a margin, a border, or an edge*". The other two are "*of, in, on, or constituting a margin*" and simply "*pertaining to a margin*" in the third of the above cited dictionaries. The second explanations are "*being adjacent geographically: states marginal to Canada*", "*close to a limit, especially a lower limit: marginal legal ability*" and "*situated on a border, edge, or fringe*". All these explanations have in common that marginal designates position or location. From this point of view we can consider them as spatial or geometric. That is why these explanations are the closest to geography.

The next set of explanations that is similar in the above cited dictionaries is from the field of economy. Marginal from economic point of view is "… *enterprises that produce goods … at a rate that barely covers production costs*" and "… *commodities thus manufactured and sold*" (American Heritage® Dictionary of the English Language 2011), "… *produced and sold at the margin of profitability: marginal cost.*", "… *small change in something, such as total cost, revenue, or consumer satisfaction.*" as well as "… *agricultural land on the margin of cultivated zones*" (Collins English Dictionary—Complete and Unabridged 2014) or "*selling goods at a price that just equals the additional cost of producing the last unit supplied.*", "…*goods produced and marketed at margin: marginal profits.*" (Random House Kernerman Webster's College Dictionary 2010). What all these explanations have in common are costs and profitability. Margin in these cases is a line between profit and loss and marginal is the situation of being just above this line. The only case that has to do with objective space is the explanation from the field of agriculture—land on the margin of cultivated zones. In this case the margin is spatial as well as economic. The land that is so unproductive that it can provide only minimum resources for bare survival is usually at the edge of settled area, close to the line that delimits inhabited and uninhabited areas and far from densely populated areas.

In politics marginal can be "…*designating a constituency in which elections tend to be won by small margins: a marginal seat.*" according to Collins English Dictionary (2014) and "*A constituency in which elections tend to be won by a small margin and may therefore be likely to change hands.*" according to Dictionary of Unfamiliar Words by Diagram Group (2008).

From the field of sociology surprisingly only one out four dictionaries have single explanation "*having contact with two or more cultural groups but not fully accepted in any of them*" (Random House Kernerman Webster's College Dictionary

2010). One would expect many more explanations as the first association connected with the word marginal would probably be the image of persons and social groups at the edge of society being partly or completely excluded.

Marginal as a synonym for insignificant can have variety of uses such as "*barely within lower standard …*" (American Heritage® Dictionary of the English Language 2011) and "*at the lower limits; minimal for requirements: marginal ability*" (Random House Kernerman Webster's College Dictionary 2010). In both cases the margin is close to the line that is delimiting important from not important and acceptable from unacceptable (satisfying the requirements from not satisfying).

The rest of the explications are "… *written or printed in the margin of a page.*" (American Heritage® Dictionary of the English Language 2011), (Collins English Dictionary—Complete and Unabridged 2014), "*A plant that grows with its roots submerged in the shallows at the edge of a pond or stream.*" (Dictionary of Unfamiliar Words by Diagram Group 2008) and the last one from the field of psychology "*Relating to or located at the fringe of consciousness.*" (American Heritage® Dictionary of the English Language 2011).

As we can see from the above presented explanations, marginal is always "… at the edge (fringes, border)" of something, so it designates the position. In most cases this is not a position within physical world. Nevertheless this is not crucial for geographical interest in marginality, because geographic research should focus on spatial distribution of "marginal …" and spatial consequences of "marginal …". The question that we have to resolve next is what do the three dots stand for! With other words who or what are we interested in when we are using the adverb marginal. Are this marginal individuals or marginal social groups or maybe marginal areas (regions, nations)? Or is it maybe all of that?

2.2.2 ISSMR and PIMA—Initiatives Interested in Marginality in Developed Countries of North Atlantic Area

When reviewing the history of the IGU Commission Marginalization, Globalization and Regional and Local Responses it is apparent that its work stemmed from the research of physically marginal areas, mainly of northern Europe, North America and of remote mountain areas. However, it was not only geographers who focused on the problems of these areas. At the beginning of the nineteen-seventies when Great Britain and Ireland were joining the European Community, academics from different fields and government officials from Wales, Scotland, Ireland and Norway met to discuss the problems of marginal regions in North-Western Europe. They started biannual seminars in 1972 and formally established the International Society for the Study of Marginal Regions (ISSMR) at their ninth meeting in Scotland in 1987. As stated on the webpage of the 2007 conference "*The purpose of the Society is to ensure the continuity of …*" their "… *hitherto informal efforts to bring together people with a common interest in the problems of marginal areas in industrialised countries in the North Atlantic region, notably Canada, Norway, Ireland, Scotland, Finland, Wales and Sweden.*

The Society aims to encourage an informed and interdisciplinary approach to an understanding of the problems of marginal regions in these countries, to promote active research on these problems by professional and academic practitioners, to disseminate the work of the Society as widely as possible through seminars and publications, and to seek funds to enable these aims to be carried out." (OM Høgskulen 2007). The Society is open to any individual or organization that wishes to participate.

Another group of researchers and practitioners held annual meetings in nineties under the name PIMA (Consortium for the study of Perceived Planning Issues in Marginal Areas). It was established in 1989 and focused on "*various aspects of marginal areas defined either in locational or developmental terms. Members of the core group …*" represented "*… universities in the United states, Sweden and Ireland …* subgroup within PIMA"*… has focused attention on studies of areas located between urban centres and rural peripheries.*" They named these areas as "*Intermediate Socio-economic Regions—ISER*" (Persson and Wiberg 1995).

2.2.3 United Nations University Definition of Marginal Areas

The name marginal areas was also mentioned in United Nations University's publication from the beginning of eighties about natural resources for human development (Ruddle and Rondinelli 1983). Interesting enough authors did not consider the definition of marginal areas as a problematic one:

> *The term "marginal area" is a convenient shorthand that defies simple definition. The term has come to be used for convenience to refer to an area that is distinguished by some or all of the following characteristics:*
>
> - *geographical remoteness (peripheral to the most highly developed and populous areas of a country);*
> - *high levels of ecological fragility or extreme or recurrent man-made or natural hazards;*
> - *dispersed, heterogeneous populations of minority ethnic groups living at subsistence levels;*
> - *actual or perceived lack, or low levels, of physical and social infrastructure;*
> - *lack, or low levels, of access to the towns and cities where services, facilities and economic opportunities are usually concentrated;*
> - *economic backwardnessor depression;*
> - *"low productivity" of economic activities using current technologies and techniques, which fail to return the profits on investments that could be realized elsewhere; and*
> - *populations with little or no political influence on the decisions affecting their lives.*

Marginality from this point of view derives from physical remoteness (low accessibility to services and working places), ecological fragility, low population density, ethnic structure, having an underdeveloped economy, the unavailability of resources or inability to use them and isolation from political influence. The definition is simple and covers the main characteristics of marginal areas. However, in geography we use the concept of the region from the micro to macro level. So we definitely have to consider the scale. That is one of the important things pointed out by Leimgruber in his book: *Between Local and Global* (2004, p. 15). He gives an example of Saar-Lor-Lux transborder region (Saar, Lorraine, Luxemburg) stating that "*Looking at its location within the European Union, the Saar-Lor-Lux region is doubtlessly central. From an economic and traditional resource perspective (coal and iron ore, steelmaking), on the other hand, it is rather marginal. Similarly, the politico-administrative heterogeneity of the three countries involved constitutes significant drawback. However, the existence of peaceful and intensive transborder contacts since World War II speaks in favor of centrality.*" Marginal or not is therefore dependent from the "spatial" scale. However, the time dimension is also very important and Leimgruber considers "time" scale as equally important. There are numerous examples of once prosperous regions that have fallen into decline because the resources that their success based upon were depleted or lost importance because of the change in technology.

2.2.4 Leimgruber's Understanding of Marginality

When Leimgruber first defined marginal regions (1994), he proposed four different approaches:

- ***Geometrical****. In this sense, marginal regions would be those at the geometric periphery of a larger area (e.g., a State, a Continent or an otherwise territory)*
- ***Ecological****. This view is ambiguous: it can either be taken as the natural potential of an area for human survival, or as the state of the environment.*
- ***Economic****. Marginality in this case would be defined by the production potential, accessibility, infrastructure, attraction within a space economy.*
- ***Social****. In this case we would focus on minorities and socially marginal groups, according to various criteria (ethnicity, language, religion etc.).*

These are not all of the possible approaches, he also mentioned political and cultural approach, but he further elaborated only on the above-mentioned four. A decade later, in a more recent work, he only kept the *geometrical* aspect in combination with newly included *systems* and *processual* one (Leimgruber 2004, p. 56). Systems aspects could be simply explained as being outside the system (economic, political, social). The system does not need the marginal "part" and it has to survive on its own at a subsistence level. Of course only being outside the system from one point of view does not necessary mean marginality. Leimgruber gives the example of Switzerland. This country is not a part of EU and is therefore, in a way, outside the system, but it is impossible to say that it is marginal in any aspect (Ibid., p. 61). Processual aspect of marginality encompasses types of marginality that derive from different processes: (un)intentional human activities. After Mehretu and his co-authors Leimgruber quotes four types of processual marginality:

- contingent,
- systemic,
- collateral and
- leveraged.

The first one is an outcome of market competition, where some actors are uncompetitive and thus marginalized. Systemic marginality is produced by hegemonic forces within political and economic system that generate inequities trough the distribution of social, political and economic benefits. Collateral is a kind of a byproduct of certain process or simply: somebody living in a marginal neighborhood otherwise not marginal may be considered marginal due to the fact that he lives among marginal neighbours. Leveraged marginality is more intentional and can be described as a pressure on the labor force in developed countries because of the competition of low paid workforce in the countries of the global South. The first are thus becoming marginalized (Leimgruber 2004, pp. 61–62).

Leimgruber's understanding of marginality, as presented above, is a very broad one and requires aspects and scale to be considered in order to approach marginality as a geographical research topic. In their 2012 article Leimgruber with co-authors (Dery et al. 2012) states that geographers did not pay much attention to what sociologists have to say about marginality and vice versa. We devote some more attention to social aspect of marginality that has been mentioned when we were presenting the definitions of marginal.

2.2.5 Park's 'Marginal Man' as an Example of a Sociological Viewpoint of Marginality

Probably the most common use of the adjective marginal: "to characterize a person's or social group's position in society" is often overlooked in geographical research. In sociology marginality is more or less a synonym for social exclusion. The concept of marginality as constructed by Robert E. Park a sociologist from Chicago school of Sociology is closely related to migration and immigrants. He is writing about the "marginal man" (Park 1928, p. 881):

> One of the consequences of migration is to create a situation in which the same individual-who may or may not be a mixed blood - finds himself striving to live in two diverse cultural groups. The effect is to produce an unstable character - a personality type with characteristic forms of behavior. This is the "marginal man." It is in the mind of the marginal man that the conflicting cultures meet and fuse.

Six decades later Adam Weisberger reconstructed this concept "*in order to yield a more complex general theory of marginality*" (Weisberger 1992, p. 425). Marginality was in this case defined by Park as "*a state of limbo between at least two cultural life worlds*". Marginality is therefore within this theoretical framework determined by socio-cultural factors. Weinberger's reconstruction tends to produce an improved structural analysis of marginality that would be valid for investigation of different empirical cases. His argument is that marginality is constituted of different social coordinates. A man that no longer fully belongs to his old culture (i.e. of the area of his origin) and is not integrated into the new one constructs the responses to his situation within these new coordinates in order to alleviate or resolve his double ambivalence. Weisberger's typology is designed of four divergent responses:

- assimilation,
- return,
- poise and
- transcendence.

Marginal man may use one or more of these possible answers in his attempt to resolve the ambivalence and crosscutting pressures that he experiences.

Weisberger states that Simmel's "*Stranger*" influenced Park when he was constructing his image of a "*marginal man*". Simmel views on spiritual conflict and instability, according to Weisberger, embodied in many ways what the conventional Park's marginal man should be. "*The Stranger*" is one that is near and far at the same time, the one who comes today and stays tomorrow. These characteristics are showing basic instability, unsecure social position that makes life complicated and unsatisfactory.

The difference between Park's marginal man and Simmel's stranger, according to Weisberger, derives from different situation in Chicago and in Germany. Park was confronted with tremendous ethnic heterogeneity of Italian, Polish, Irish, Ukrainian and other ethnic minorities that were in transition from the old world to the new one. That is why Park's marginal man "*... appears to be a bewildered foreigner who is on the road to successful assimilation, joining the dominant culture and sacrificing his ethnic peculiarities to the melting pot.*" (Weisberger 1992, p. 428). Simmel on the other hand was a Jew in ethnically homogenous Germany and as such was "*in society but not of it, unable to enter society in the way, many ethnic groups in America did.*" (Weisberger 1992, p. 428).

Weisberger claims that Park does not capture the full extent of the ambivalence that characterizes marginality and that he understands marginality as a single social position. While Park observes that the marginal man is unable to cut off from his original culture and to merge into the new one and therefore remains on the margin of both, Weisberger states that he is also unable to return to his original culture or at least he is not able to do that without the influence of the new one. So if Park considers the marginal man's position between two cultures unidirectional, Weisberger speaks about a structure of double ambivalence. Within this marginal man wants to return, but cannot, wants to stay, but cannot, wants to assimilate, but cannot, wants to reject the new culture, but cannot. From this double ambivalence Weisberger extracts his four possible theoretically pure responses that we have cited above.

Assimilation means the absorption of the cultural standards of the host society. Of course that means that the former practices originating in his old culture have to be abandoned. Return means that the marginal man goes back where he came from and has to reconstruct what he has lost while he was confronting with the new culture in the host society. His interpretation of the culture of his origin can never be the same as before because of his new experiences. Transcendence is actually an attempt to find a third way that is supposed to overcome the opposition of both cultures. Usually this means that marginal man adopts radical political, philosophical or cultural ideas. Finally poise is nonresponse—meaning the abiding in ambivalence often causing loneliness and anxiety.

Within this concept marginality is the consequence of a process in which migrants from one culture encounter another. The products of such collisions are conflicts that may result in new combinations of elements of each culture, created by the most sensitive and creative among immigrants. However, there are many different factors influencing particular marginal situations that, change through time and space.

In graphic terms responses to marginality can be presented graphically as shown on Fig. 2.1. Of course ideal responses (those on both axes) are not common in real life. Usually the individual's response would be somewhere between the two axes closer to one or another within four possible combinations (four quadrants).

Presented revised sociological concept of marginality is without any doubt an interesting tool for the analysis of immigrant groups and their confrontation with the ambivalence situation that they are faced with in the host society. With the worsening political situation, instability and armed conflicts the urge for migration is growing not to mention the economic reasons.

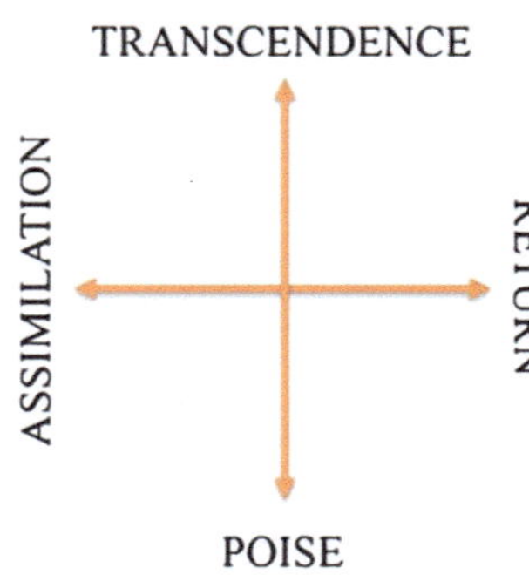

Fig. 2.1 Directions of marginality [drawing according to Weisberger (1992, p. 432)]

2.2.6 University of Bonn's Center for Development Research Approaches to Marginality

Another concept that often tends to associate with marginality is poverty. Gatzweiler with co-authors from Center for Development Research, University of Bonn researched the roots of extreme poverty in connection with marginality (Gatzweiler et al. 2011). They start the presentation of their research with a statement that the number of poor globally grew despite economic growth quoting Grant and Shepherd saying that: "Policies that redistribute wealth from growth are often not in place." It is a statement similar to simplified definition of systemic marginality. "Proximate reasons for the exclusion can be that the extreme poor:

- live in unfavorable areas (poor agricultural asset base, poor or no transport infrastructure),
- can (for various reasons) only make minimal use of their labor and lack opportunities to acquire skills,
- spend most of their effort to achieve a calorific and nutritional minimum,
- are socially or ethnically excluded or lost their role or status in society, e.g. because the income earning household head passes away, because of disease, or as a result of remaining childless, or because of being born into a class of extreme poor,
- are excluded from public services or poverty reduction programs." (Ibid., p. 1)

These reasons are in a way showing different characteristics of marginality as can be seen from different perspectives (spatial, social, economic). The above cited authors are emphasizing the correlation between extreme poverty and physical remoteness (rural areas), low accessibility to all kind of services and the status of (ethnic) minorities. They define marginality as "*an involuntary position and condition of an individual or group at the margins of social, political, economic, ecological and biophysical systems, preventing them from access to resources, assets, services, restraining freedom of choice, preventing the development of capabilities, and eventually causing extreme poverty.*" The emphasis is on an individual or a group, but the reason for marginality (which is an involuntary position)—is not (practically) outside the system, as discussed by Leimgruber under the systems aspect of marginality—in this case it is the position at the margin of different systems. Gatzweiler et al. (2011, p. 3) are also emphasizing the role of spatial and environmental dimension of the concept of marginality. Their approach is multidisciplinary and based on complex adaptive system thinking, complexity and sustainability science in order to "*… shift from predominantly economic views toward examining multiple factors and their evidence as informed by systems-based analyses*". They also include ecological systems because they believe that the patterns of marginality emerge from the relations between casual variables from ecological and social systems (Ibid., p. 4). They observe marginality from the perspective of complexity and dynamic systems and therefore, according to Kay, they claim that they "*… no longer see systems as complicated and principally* *predictable, with answers to problems which can be correct with some degree of probability, given sufficient information is available. Instead, complex systems are characterized by irreducible uncertainty and scientists can arrive at possibly correct answers, at best. In a complexity setting, 'being correct' changes its value. When decisions directly affect people's lives and their survival, being roughly right is better than being precisely wrong*" (Ibid., p. 4). According to this they find different marginality patterns and claim that poverty may be explained by different types of marginality that cause it. Their categorization encompasses types with societal dimensions such as social, political, cultural economic as well as bio-physical categories with dimensions such as geographical and agro-ecological. However, they consider this only as one of several possible categorizations. They adopted another one from livelihood approach that consider identification of different kind of capital (social, financial, human, manmade and natural) as a starting point. They believe that different causal networks produce different types or patterns of marginality (Fig. 2.2).

Marginality according to Gatzweiler and co-authors can only be explained by networks of causal factors, not just by one of them alone. Different factors combined together lead to marginality. "Having a low income alone, for instance, is not a sufficient cause for qualifying as marginalized, as someone with no income could be cared for within a family or social group. That means, in combination with underlying causes of being excluded, experiencing discrimination or not having access to services and facilities, causality crystalizes to specific causal networks of marginality."(Ibid., p. 7). Marginality is understood by cited authors as the situation in which a person or a (social) group is trapped. It is also involuntary and whoever finds himself in such situation tries to escape from it aiming to achieve better

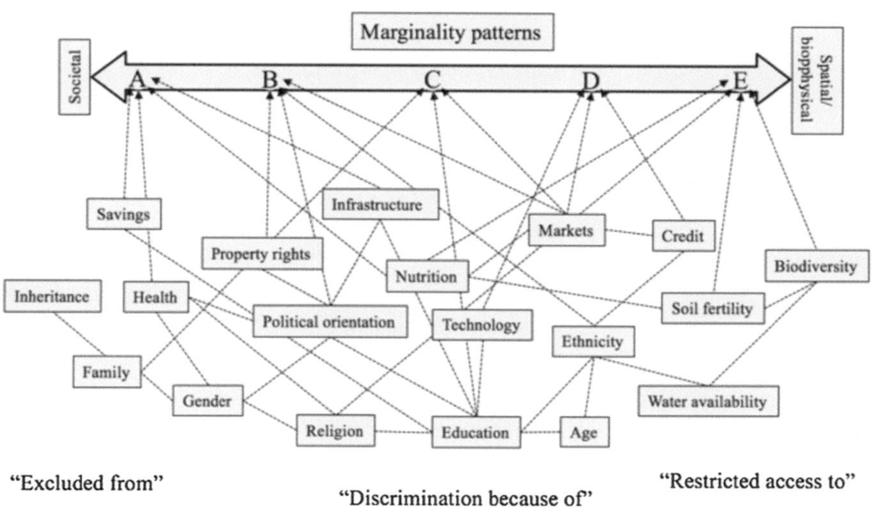

Fig. 2.2 Examples of causal factors underlying different marginality patterns (adapted after Gatzweiler et al. 2011, p. 6)

quality of life. They believe that every individual or social group is a part of multiple systems and marginalized are at the edge of more of them. To change their marginal situation they need to move from the edge in order to gain the access to the benefits that the systems provide (access to goods and services). Even though they consider spatial dimension as very important within the complexity of causal networks, they do not operate with the term marginal areas or marginal regions. The reason may be that their primary focus is on poverty.

2.3 Geography of Marginalization or the Geography of Marginal Regions

After presentation of the above considerably different notions of marginality, we need to consider how are they related to geographical research of marginality where the spatial aspect should play the most important role. The question is: "Do we practice geography of marginalization or geography of marginal regions when we research different aspects of marginality and marginalization from geographical point of view?".

If we chose the later then we obviously practice a sub-discipline of human geography with strong emphasis on the regional aspect. It can be thus compared with rural or urban geography. In order to do our job we need to define the vital characteristics and the extent of marginal areas. That means that we need to delimit them from the rest of the territory. As in case of urban-rural division this is a more or less impossible task. No single definition of marginal areas can be so accurate and precise that we could simply draw a line between marginal and non-marginal areas. As we presented above, marginality is too complex and approaches so different and always at least partly subjective. Marginality is therefore, by my opinion, not objectively measurable. We can state that a certain area or region is (at least partly) marginal using appropriate set of indicators that can show the presence of certain characteristics of marginality in the area (region). However, it is not very common that researchers of marginality would first try to objectively define the case study area's marginality. Rather they choose it on the basis of their subjective judgment. Very often this includes only one or two aspects of marginality, such as remoteness, usually connected with lower economic status of the case study area (low economic output or high share of people living under the threshold of poverty etc.) . The majority of chapters in this book present the problems of case study areas that have been chosen in this manner. This may seem contradictory to the emphasized complexity of the topic and consequential need for an interdisciplinary approach to it. However, this is a subjective choice made by experts, knowing the topic is usually not problematic. Besides, through the research of the problem, their focus on the characteristics of marginality usually become more and more obvious.

The problem is that we do not have broader research enquiries focused on marginal areas, such as involving large team of experts from different fields. For now we have a lot of fragments revealing different faces of marginality in geographical space, but no overall, general study of the phenomenon of marginal areas. To do such a study we need to provide a basic framework in order to know how and what to research. We can do the research at different scales, but we need to choose the approach and to come to an agreement how are we going to define what is a marginal region/area in different regional/scale context.

If we consider marginality as the state of being marginal then we could say that a marginal region (country) experiences certain kinds of marginality. In the case of a region this could be correct statement as region may be understood as a system formed by society and environment. 'Region' has certain common characteristics and encompasses certain part of Earth's surface (i.e. a certain territory). In the case of an area, it would not be completely wrong, but we believe that marginality of a territory in a sense that it is far from something, can be equally well described with the adverbs like remote or distant. Use of the notion of marginal with an area or region should be limited to areas where marginality is experienced by individuals or social groups or society in general to such an extent that it can be considered as an important and noticeable issue. From that point of view we could classify regions according to their degree of marginality into regions with:

- no marginality,
- traces of marginality,
- clearly observable marginality issues,
- severe marginality problems and
- marginalized regional society.

Considering scale from a global perspective there is no such macro-region (continent, nation), where there would be no marginality. Even if we go downward it is very hard to say that there are regions at the mezzo-regional level where there are no individuals or social groups that are experiencing at least some kind of marginality. Therefore it is possible to state that, potentially, regions without marginality can only be found at micro-level. They need to be small and homogenous enough to be without any marginal individual or group.

Regions with traces of marginality are those where marginality is not an important issue, but there are some individuals or social groups that are marginal. Again we have to consider the scale and, similar to what has been stated above, we can say for this regions that smaller scale means a higher probability that some regions can meet the criteria of having only a few individuals or social groups that find themselves in marginal position, despite the fact that the society tries to prevent marginalization.

Regions with clearly observable marginality issues cover all scales. Europe as a continent could be classified as this kind of region at continental level. It is a continent with the lowest level of inequality, but even in the most developed European nations we have social groups that are obviously marginalized. This may be immigrants, jobless, and ethnic minorities such as Roma population in many European countries. At the level of single nations the situation may be similar or worse and in case that we consider Turkey as a European nation (being partly in Europe) then we could probably classify it even in the lower class as their challenges with the Kurd ethnic group can be considered as a severe marginality problem. With the rising problem of refugees and illegal immigrants Greece and Italy can also become nations with severe marginality problems, not to mention the regions within these countries with the highest numbers of these marginalized social groups. These regions are undoubtedly regions with severe marginality problems. All the humanitarian aid cannot change the fact that immigrants trying to enter EU are marginalized on their way to their final destination and many of them after they reach it too.

The last and the lowest class is again one that can be observed at smaller scale. It can be the type of region that is described by Leimgruber (1994, p. 6, 2004, p. 60) as Reynaud's 'angle mort' and 'isolat'. It is an area or region existing on its own, almost completely out of the system, existing at the margin of survival. That is of course the extreme example of completely marginalized society from economic, political, cultural and social aspects. However, a less extreme situation is also possible when only from some of this aspects the regional society as whole is trapped into a marginal position.

Marginality research can therefore take place almost everywhere, but it can focus on many different topics, considering the scale and the type of marginal region that the research is dealing with. The regions with traces of marginality, with clearly observable marginality issues and with severe marginality problems will probably attract the research interest for the following:

- who is marginal (identification of marginal individuals or social groups),
- what are the manifestations of this marginality (identification of the type of marginality),
- what are the consequences of this marginality (identification of consequences),
- what is causing this marginality (identification of marginalizing factors),
- what is the role of geographical (geo-spatial) factors in marginalization process (identifying the role of geographical factors).

Ideally a geographical research of marginality in certain marginal region would involve all of the above, but in practice, researchers usually limit themselves to just some of the listed topics as can be seen throughout this book too. In this sense we should probably talk about the geography of marginalization rather than the geography of marginal regions. In cases where the researcher focuses only on one problem, no matter how important its role in the process of marginalization is, we cannot talk about geography of marginal regions, it may only be considered as a partial study within geography of marginalization.

2.4 Conclusion

In this text we presented different views of marginality and the common denominator of most of them is that marginality is very complex and hard to define. It needs to be approached from different viewpoints observed from different perspectives considering the scale. Better co-operation between different disciplines that are interested in marginality would certainly bring better understanding of this phenomenon and of drivers of the process of marginalization. Marginality can be observed as an important cause of extreme poverty or at least as a reason for the lower quality of life of those that are trapped in marginal situation (that are marginalized by the rest of the society).

In order to achieve better research results in the field of marginality studies geographers should adopt a 'geography of marginal regions' approach. At present their efforts are atomized in numerous partial studies of different problems either connected with marginalization, either taking place in remote and underdeveloped regions or both. The examples of such studies are in the following chapters of this book. They reveal some important new findings, but we need to go further in future, and synthesize findings such as these into a broader theoretical framework of marginal regions, which draws on the insights discussed above. Hopefully we can overcome this insufficiency with a more holistic regional approach performed by an international team of experts. In this sense this book series may play important motivating factor.

References

American Heritage® Dictionary of the English Language. (2011). TheFreeDictionary. Accessed May 9, 2016. http://www.thefreedictionary.com/marginal

American Heritage® Dictionary of the English Language, Fifth Edition. S.v. "marginality." (2011). Accessed May 9, 2016. http://www.thefreedictionary.com/marginality

Collins English Dictionary—Complete and Unabridged. (2014). TheFreeDictionary. Accessed May 9, 2016. http://www.thefreedictionary.com/marginal

Dery, S., Leimgruber, W., & Zsilincsar, W. (2012). Understanding marginality: Recent insights from a geographical perspective. *Hrvatski geografski glasnik, 74*(1), 5–18.

Dictionary of Unfamiliar Words by Diagram Group. (2008). TheFreeDictionary. Accessed May 11, 2016. http://www.thefreedictionary.com/marginal

Gatzweiler, F. W., Baumüller, H., Ladenburger, C., & von Braun, J. (2011). *Marginality: Addressing the root causes of extreme poverty*. Bonn: Zentrum für Entwicklungsforschung/Center for Development Research, University of Bonn.

Leimgruber, W. (1994). Marginality and marginal regions: Problems of definition. Marginality and development issues in marginal regions (pp. 1–18). In *Proceedings of the IGU Study Group 'Development Issues in Marginal Regions'*. Taipei: National Taiwan University.

Leimgruber, W. (2004). *Between global and local. Marginality and marginal regions in the context of globalization and deregulation*. Aldershot: Ashgate.

OM Høgskulen. (2007). ISSMR|ISMR—The international seminar on marginal regions. Accessed May 15, 2016. http://www.hivolda.no/ismr

Park, R. E. (1928). Human migration and a marginal man. *The American Journal of Sociology, 33* (6), 881–893.
Persson, L. O., & Wiberg, U. (1995). *Microregional fragmentation—Contrasts between a welfare state and a market economy*. Heidelberg: Physica-Verlag.
Random House Kernerman Webster's College Dictionary. (2010). TheFreeDictionary. Accessed May 9, 2016. http://www.thefreedictionary.com/marginal
Robertson, R. (1992). *Globalisation. Social theory and global culture*. London: Sage.
Ruddle, K., & Rondinelli, D. A. (1983). *Transforming natural resources for human development: A resource systems framework for development policy*. Resource Systems Theory and Methodology Series, (1). United Nations University. Accessed May 19, 2016. http://archive.unu.edu/unupress/unupbooks/80469e/80469E05.htm#1.1.The%20problem%20of%20marginal%20areas
Weisberger, A. (1992). Marginality and its directions. *Sociological Forum, 7*(3), 425–446.

Chapter 3
Poverty and Global Warming: An Example of the Kyrgyz Republic

Asylbek A. Aidaraliev

3.1 Introduction

The report on human development annually published by the United Nations testifies that the mountain states (especially those with mountains covering over 75% of their territory), as a rule, take places in second hundred among 194 states under income per capita. About 80% of the territory of the Kyrgyz Republic is mountainous, and 93% of the KR territory is situated on altitudes higher than 1000 m.

Such phenomena as poverty and marginalization, depopulation, degradation of resources and environment, backwardness of transport and social infrastructure and unemployment are typical for mountain regions and create conditions for social insecurity and conflicts. During the last 10 years the poor mountain states have faced with the growth of special challenges and threats: terrorism, extremism, drug-dealing, criminality, social stratification of a society, intensive degradation, and an exhaustion of vital natural resources. Large-scale emigration of marginal mountain inhabitants to the safe countries has led to globalization of above-mentioned social anomalies, political and social confrontation. The last year's tragic events on the south of the Kyrgyz Republic can be one of sad examples of such appearing conflicts (Libishevski and Bakhler 1999).

The analysis has shown that a principal cause of the above-named social anomalies inherent in highlands is the poverty of mountain inhabitants. From 1 billion of the poorest world's population more than 800 million live in mountain regions. The poverty of mountain inhabitants is appreciably affected by the extreme nature-climate conditions of life in mountains, inaccessibility and isolation from economically and technologically developed centers of a civilization (Aidaraliev et al. 2008).

A.A. Aidaraliev (✉)
National Academy of Sciences of Kyrgyzstan, International University of Kyrgyzstan, Chui Avenue, 255, Bishkek 720001, Kyrgyz Republic
e-mail: iuk@mail.elcat.kg; board-iuk@mail.ru

R. Chand et al. (eds.), *Societies, Social Inequalities and Marginalization*, Perspectives on Geographical Marginality,
DOI 10.1007/978-3-319-50998-3_3

In general, all the mountain states have essential distinctions between their low- and high-mountainous areas. The high mountains inhabitants are characterized by having higher levels of poverty. The analysis and generalization of actual materials convincingly testifies that mountain territories are areas of economic risk. Life-support of mountain populations, in comparison with flat territories, demands essentially big biological and physical power expenses as shown below:

- The work of people in high mountains is correlated to high physiological expenses, and working capacity of the person decreases on average by 10–20%;
- Energy losses in electric systems because of remoteness and severity of a climate is from 10 to 40%;
- The power of internal combustion engines decreases by 12–20%, but fuel expenses increases on 10–12%, because of this the operational period of technical devices decreases on average by 50–60%;
- It is several times more expensive to manage the infrastructure—roads, electric systems, water-pipes, inhabited and industrial buildings;
- The productivity of grain, potato, other cultures, and the efficiency of agricultural animals is 8–17% lower (Aidaraliev et al. 2001).

Owing to these reasons the prices of the basic foodstuff in mountains are from 8 to 49% higher depending on their remoteness from the plain agro-industrial centers.

Mountain ecosystems are especially sensitive to natural and anthropogenic influences. Meanwhile there is a process of large-scale degradation and an exhaustion of the vital natural resources (ground, water, flora and fauna) because of compelled and intensive use by the population that in turn leads to a poverty and socio-political instability. Progressing degradation of land resources (desertification, salinization, waterlogging, and loss of humus), environmental pollution, decrease of forests and biodiversity, and drug dealing, religious and national extremism, migration and social disintegration of local societies and general crime growth are the characteristic features of Kyrgyzstan and the countries of the Central Asia nowadays. The reasons for the growth of the aforementioned social anomalies in the countries of the Central Asia are well-known: economy decline in most of these countries, as well as its effects unemployment and poverty of the major parts of the population, especially in mountain regions. The states of Central Asia have made great efforts to overcome the crisis by developing various, mainly economic, as a rule, short-term strategies and programmes. However, in the medium-term and especially in long-term perspective Nature can install very difficult obstacles for these strategies and programmes. The name of the obstacle is Global warming. Unfortunately, this key factor is not given the appropriate value in the countries of the Central Asia when designing development strategies. Global warming causes not only climate change but also the correlated chain reaction of transformation of all systems of human life support: economic, ecological, social, political and cultural. The anxiety of world society is heightened by the consequences of global climate change; that is clear. The climate is a key factor defining environment conditions of living organisms: temperature regime, humidity, pressure, soil composition, water content, biodiversity etc. It is obvious that environmental changes will entail a number of cardinal, and for some regions of the Earth, disastrous consequences. It should be remembered that natural climate changes occurred in geology history of our planet more than once. Associated transformations of environment resulted in deep reconstruction of the bio-physical structure, disappearance of whole species and classes of animals and plants; and in human society to the great migration of people, wars, change of farming forms and life style, and ruin of civilizations (Anonymous 2009a).

Across the whole world, despite official acknowledgment of the fact of global climate warming, investigations continue in forecasting of ecological-economic and social consequences of the climate change, and developing strategies and programmes of reaction to these new challenges and threats. Here, it is necessary to make one important remark. Seventeen centuries ago, or even earlier, there were vast regions on the Earth that were free of human activities and, therefore, ecological and economic systems functioned separately and independently from each other. At present, these two systems combine into one new united integrated ecological-economic system where ecological and economic parameters are interconnected and interdependent. Notably, destruction of the ecological component or degradation or exhaustion of natural resources inevitably will entail the destruction of the economic component of the integral ecological and economical system.

One of the severest consequences of global climate change will be the acute deficiency of sweet water, as most researchers think. In early 1984 the International Geological Congress acknowledged that sweet water is the number one mineral resource and it is forecasted that in future the world will enter a period of global crisis of sweet water with all associated consequences and the fight for water will be crueler than the present fight for oil and gas. Actually, carbons can be exchanged by alternative energy resources: like hydropower, sun, wind, biological, atomic, but life on the Earth is impossible without sweet water (Aidaraliev et al. 2001).

3.2 The Global Warming and Poverty

The World Bank came to the conclusion in its report "Adaptation to climate changes in the countries of Europe and Central Asia" (2009, p. 37) that the countries of Europe and Central Asia are exposed to considerable threat due to climate change, taking into account that some of the more serious factors have already become reality. Rise of average temperature by 0.5° is observed in Central Asia and by 1.6° in the south part of Siberia (Fay et al. 2009).

45% of the glaciers of Central Asia are concentrated and located in the territory of Kyrgyzstan. Among the 8200 glaciers, with a total area 8.17 thousand km^2, more than 2000 glaciers melted for the last time in Tien-Shan and Pamir-Alai. The average speed of glacier recession is 8 m/year and reduction speed is 0.6–0.8% in area and 0.8–1% in volume per year. According to instrumental observation, the glaciers of Alai and Pamir lost 25% of sweet water stock for 43 years in the period from 1957 to 2000. It should be emphasized that since the 1970s glacier melting speed has tended to increase. For example, according to the data of "the second national message of the Kyrgyz Republic on the framework convention of the U.N. O. of climate change" (2009a, p. 135) a decrease of the glaciated area of about 64–95% since 2000–2100 is predicted for the Kyrgyz Republic depending on variants of climate scenarios.

Therefore, in mountain regions global climate change will result in the disappearance of mountain glaciers and snowfields—the main water resources, and as a consequence, will entail serious social-economic and political consequences in the whole Central-Asia region.

For the last 40 years the mountain area covered by forests decreased, e.g. by half in the Kyrgyz Republic (Kolov 2001). Natural restoration of forests does not go on. Grasslands around villages are trampled down due to overgrazing and faraway summer grasslands are used not enough or are not used at all and are covered by weeds. The area of cultivated land (about 7% of the total territory) catastrophically decreased due to desertification, bogging or construction.

Soil erosion affects 70% of mountainous areas. The area of glaciers and perpetual snow has decreased. The number of surface and underground water sources is decreasing as is the quality of the water. Landslips, snow-slips, floods and other disasters are becoming more frequent (over the last 50 years) on the tops of mountains because of soil erosion and man-caused impact. Destruction of nature and mountain resources might be irreversible. In many respects the level of poverty of Kyrgyzstan's population has a tendency to be lowering but still it is high on the base of the before mentioned factors. For example, according to the information of the National Statistic Committee of the KR the general level of the poverty was 31.7% in 2008, and level of extreme poverty was 6.1% (Table 3.1).

Although the problem of poverty remains at the same level in rural districts, most of them are situated in mountainous areas, and it is still acute. The number of the population who are living in poverty in the cities is 22.6%, but in rural and mountainous areas the number of poor people is more than 1/3rd of the population (36.8%) (Table 3.2).

Table 3.1 Level of poverty (%) of population in Kyrgyz Republic (KR) (calculated on consumer's expenses)

	2004	2005	2006	2007	2008
General level of poverty	45.9	43.1	39.9	35.0	31.7
Level of extreme poverty	13.4	11.1	9.1	6.6	6.1
Depth of poverty	13.2	10.5	9.1	6.6	7.5
Acuteness of poverty	5.2	3.6	3.1	1.9	2.6

Source The National Human Development Report of the Kyrgyz Republic in 2012, UNDP, Bishkek

Table 3.2 The number of the population under feature of the poverty with their consumers expenses in Kyrgyz Republic (in %)

Years	1996	1997	1998	1999	2000	2001	2002	2003
Total	43.5	43.0	54.9	55	62.6	56.4	54.8	49.9
City	36.3	23.1	42.2	42	53.3	45.4	44.5	35.7
Village	49.6	55.3	62.4	60	67.6	62.3	60.3	57.4

Years	2004	2005	2006	2007	2008
Total	45.9	43.1	39.9	35.0	31.7
City	28.3	29.8	26.7	23.2	22.6
Village	55.5	50.8	47.7	41.7	36.8

Source The National Human Development Report of the Kyrgyz Republic in 2012, UNDP, Bishkek

The reason of the low level of the investment in mountainous areas is low income, unemployment, the shutdown of the many governmental industries and agricultural enterprises, low harvest and so on. This has led to mass migration of the population to other countries or areas. The system of professional education is weak and unpopular in the country, only 10% of the youth study at professional schools. The quality of teaching is becoming lower and the schools are not available for the majority of the families because of the poverty.

The possibility of self-actualization of the citizens in Kyrgyz Republic is reflected in the index of human development. It is considered as the main integrated index that includes the level of the lifetime, education and the level of income and that is based on the gross domestic product per person at par for purchase power in US dollars.

According to the data of the National Report on Human Development (UNDP in Kyrgyz Republic 2010, p. 7) index of human development and its main features and constituents are presented in Table 3.3. According to the research, the index of human development in 2007 increased by 5.7% in comparison with 1993. In 1994–1998, the human development index decreased; that is conditioned with the growth impairment of personal consumption of the gross domestic products. The ultimate point of fall of the human development index is seen in 1995, when it was 0.642 and in 2007 it reached 0.704. In comparison with 1995 it is observed the stable growth of the index was the outcome on the base of the gross domestic product: in 1995 it was 0.384; in 2007 it was 0.498. The education index level for these periods was at the same level and in 1995 at 0.859; in 2007 it was 0.897. The index of lifetime decreased from 0.725 in 2000 to 0.715 in 2007. The index of the expected lifetime decreased from 0.725 in 2000, to 0.715 in 2007. Some decrease in the indexes is connected with the transition of the country on the live birth criteria that was recommended by the World Public Health Organization and children's death is included here. Generally it is necessary to focus on the main components of the human development index—index of incomes and index of the lifetime stayed unchanged or at a very low level and the index of education is low too.

Table 3.3 Components of the human development in Kyrgyz Republic

Index	1993	1995	2000	2003	2004	2005	2006	2007
Expected lifetime after birth	67.3	66.0	68.5	68.2	68.2	67.9	67.7	67.9
The level of literacy of the adults in percentage (%)	97.3	97.3	98.7	98.7	98.7	98.7	98.7	98.7
Cumulative part of the students of the junior, high, higher schools in % (from 7 to 24 years old youth among the population)	66	63	71	71	71	71	71	72
Real gross domestic products per person among population (GDP in 2005, $ США)	1282	1000	1332	1558	1697	1728	1813	1980
Index of the expected lifetime	0.705	0.683	0.725	0.720	0.720	0.715	0.712	0.715
Index of the education level	0.867	0.859	0.895	0.896	0.895	0.896	0.895	0.897
Index of the income	0.426	0.384	0.432	0.458	0.473	0.476	0.484	0.498
Human development index	0.666	0.642	0.684	0.692	0.696	0.696	0.697	0.704

Source The National Human Development Report of the Kyrgyz Republic in 2012, UNDP, Bishkek

Experts of the International University of Kyrgyzstan, as well as International Institute of Mountains and National Center on the development of the mountainous countries of the Kyrgyz Republic with the support of the Asian Development Bank developed the programme "National and regional strategy and acting plans on the sustainable development of the mountainous regions like Kyrgyz Republic, Kazakhstan, Uzbekistan, Tajikistan, Turkmenistan, and China." (Tashybekov 2002).

Comprehensive information was presented in the documents about the damage to nature, damage to plants as well as animals of the mountainous countries. In the document ecological and social problems are not only described but also they are analyzed deeply, in detail and systematically. Their correlations, the level of their subordination with each other, the level of the dependence on each other, from the economic, social, and political conditions are presented in detail. Different schemes that are called the tree of the problems are worked out by the experts in different fields like agriculture, environment, forests, mineral resources, tourism, and water (Fig. 3.1).

These schemes can show us all the connections and dependence with each other and the level and the reaching of the problem decision. The decision of the problems of the lower level is available to decide the problems of the next level, including the highest level. This scheme "the tree of the problems" may give us the chance to understand what problems we must begin with, how to identify them, and what we must use for the resolution of these problems. Beside these "trees of the problems", we worked out "the trees of purposes". The main and sub dominate goal

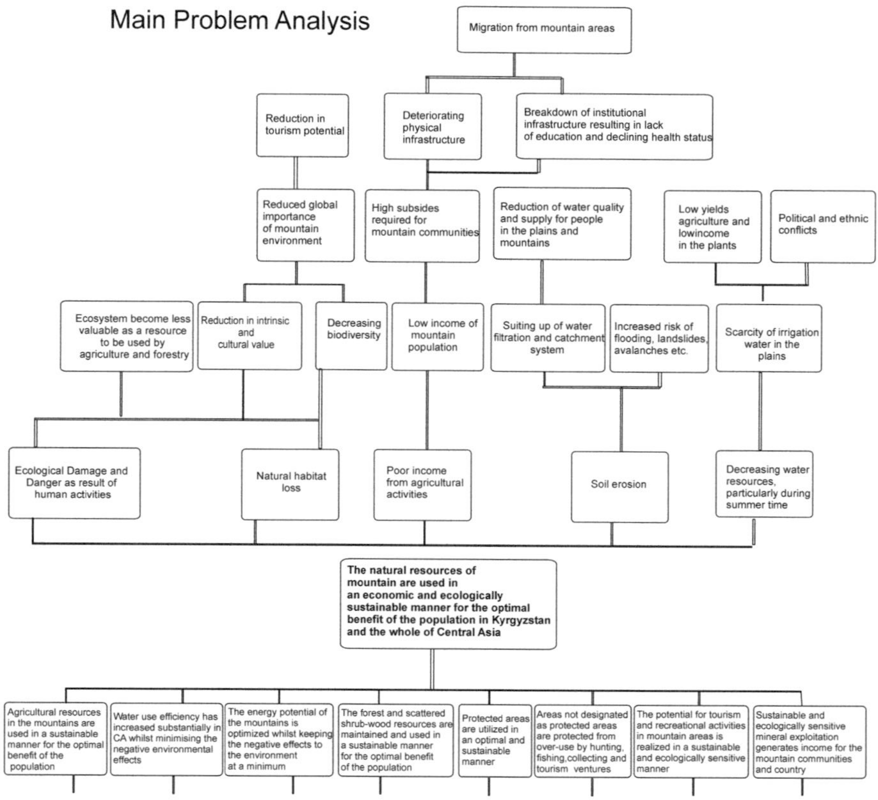

Fig. 3.1 Scheme 1—Analysis of the main problem

of the problems are formulated, they are necessary for the defense, saving, and correct exploitation of the natural resources, for the elimination of the poverty, for the saving of health, development of culture and national traditions (Fig. 3.2).

The strategy and detailed plan of the activities are described in these documents for the sustainable development of the mountainous countries and their territories.

As a result of the National Strategy and Action Plan for sustainable development of the mountainous territories of Kyrgyzstan, a special guide on the development of the mountainous territory was stated as the base document in 2002 by the Government of the Kyrgyz Republic.

The main aspects of the document were presented at the international forum on the problems of the sustainable development in China and were also presented in the world congress of the Mountainous People's Association in Ecuador (2002).

Thus systematic approach to the research on ecological and social problems of the mountainous territories which was worked out in Kyrgyzstan entered into the planetary information field of the concepts on Sustainable Development. The existing situation demands immediate and drastic changes concerning the mountain

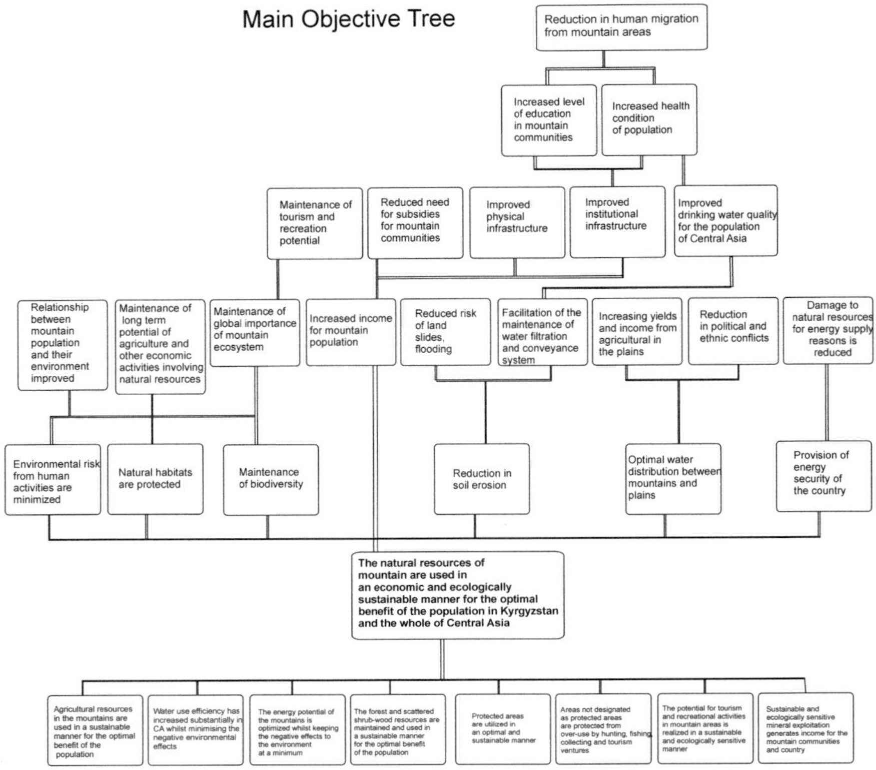

Fig. 3.2 Scheme 2—The main tree of purposes

regions, as in the 21st century together with the growth of the mountains accessibility, there appears degradation of the mountain resources which will increase and will be connected with the intensive economic learning in the mountainous areas.

One of the preventive measures to reduce the impact of global climate changes in the Kyrgyz Mountains is the increment of the forest areas by means of planting of different types of rapid growing trees. During all the existing periods in the Republic, more than 50–60% of the trees were cut out, and nowadays forest areas are about 4.6% of the land cover. Increment of the forest areas in the republic to 8% helps address some problems in a complex way:

- increment of the carbohydrate drain: annual carbohydrate drain in forest areas reaches to 784 Gg (1 Gg = 1,000,000,000 g) CO_2;
- the process of desertification of the flat area ecosystem will be stopped; and there will be new ecosystems;
- soil moisture will be increased and water resources will be kept and they will be restored;
- dependence on imported wooden material will be decreased to a considerable level.

In the Republic solar energy is practically not used, nevertheless its potential is great. The location of parts of the Kyrgyz Republic in low flat desert areas on the borders of the subtropical belt makes for the availability of intensive sun radiation. Sun heat receipt is 120–160 kcal/cm^2, which is twice that of the other two regions of the CIS countries. There is a long time solar aurora that is typical for these areas. It changes from 17.00 in lowlands and till 29.65 kcal/cm^2 in the highlands. The usage of solo aurora scales back the usage of the electric energy in up to 60% of houses (Anonymous 2009b).

The population of the country burns organic matter (wire wood, trash, chaff etc.) for the preparation of meals and heating of houses. The total amount of dry organic matter that is burned by one family may be 10.5 tons a year. In burning all this organic matter about 2% unhealthy carbon monoxide may be extracted and according to the expert's evaluation the total amount of the emissions of CO_2 will be 444.57 tons a year. By converting all of this matter in biogas treaters we can receive 5.5 million tons of the highest effective liquid petroleum gas in the form of fertilizer that gives a chance of growing the productive capacity for 1.43 million hectares of arable and mowing acres (Anonymous 2009c).

In addition during the treatment of dung in the biogas machines there will be energy production (biogas) for the provision of the electric needs of the agricultural population. There is some other technology for pure energy production (wind power, geothermal energy); the usage of these types depends on the economic effectiveness.

By decree of the President of the Kyrgyz Republic National Committee on the consequences of the climate change, a programme on the sustainable development of the country was established and it will represent a national instrument in the mechanisms of pure development. This interagency committee will be a coordinating body (organ) including representatives of all key partners in the republic. For the effective work of the committee necessary legal documents define the criteria of the project sampling and the approval of project procedures. The main tasks of this committee are:

- the establishment of a national system for the evaluation of anthropogenic emissions and absorption of all greenhouse gas scrubbers;
- the formation of national registration for the accounting of produced anthropogenic emissions and absorption of the all greenhouse gas scrubbers;
- the provision of regular presentations of the National Information about climate changes;
- the establishment of interagency coordinating activities in preparation and realization of the projects on the mechanisms of pure development;
- monitoring the project activities on the base of pure development mechanisms.

It is necessary to note that from the point of monitoring the situation concerning climate change, the committee fulfills its functions very effectively. For example, in 2009 with the financial support of the Global Ecological Fund and the support of UN Development Programme committee, the Second National Announcement of the Kyrgyz Republic on the base of the convention of the UN concerning climate changes was published. It is considered the base document for the definition of strategic directions of the republic's development, taking into consideration all the consequences of global climate changes.

On the base of this research, backbone tasks of the Economic Republic Strategy until 2030 years were defined. It is devoted to the distant high mountainous areas in the context of the transformation of environmental habitation concerning with global climate changes, including:

- Communication network reanimation and development
- Energy provision: building of medium and small hydroelectric power stations
- Use of non-traditional kinds of energy
- Governmental support of the high mountainous villages.

The importance of this moment is to change the traditional approaches in the realization of the backbone tasks.

For example, governmental support of the remote mountain villages should be realized not by giving them privileges (now in our country we have more than 550 thousands privileged people, and the number of the pensioners is growing day by day compared with the reduction of the working people—who produce products). It should be done through a self-support and self-sufficient economy taking into consideration all the natural peculiarities of every village and region. Establishment of cooperative societies in every region will assist in the development of the economy and the development of interregional cooperation in commodity producing (agricultural regional complexes).

3.3 Concluding Remarks

It is high time to change the territorial strategy of the republic's development. Today the strategy of social and economic development of the republic should be include the complex of measures of learning in medium and high mountainous areas and flat areas where only 5% of the population of the country live. It is necessary to research the zone of compensated discomfort. It is about 35% of the territory of the country. These lands are the strategic reserve for next generations.

We must begin to work out our long term governmental programme of the economic development of high mountainous and far away countries on the same basis as the National Committee on Climate Change Consequences. At the very beginning of activities the programme should be agreed on and connected with the plans of the necessary offices directly or indirectly influencing the development of the mountainous areas. The goals should be clearly available, scientifically justified, and practically probated. The programme should be serviced from the budget of the Kyrgyz Republic.

References

Aidaraliev, A. A., Musakozhoev, Sh. M., & Suerkulov, E. A. (2001). *Problems of exploitation of natural resources of the mountains of Kyrgyzstan, National Center of development of mountain regions of Kyrgyzstan*. Bishkek: International University of Kyrgyzstan, Research Institute of Economy of Development of Mountain Resources.

Aidaraliev, A. A., Bokonbaev, K. J., Japarov, A. U., Sultanov, M. A., & Shanazarov, A. S. (2008). "Mountain initiatives": Justification of the external debt transmission of the mountainous countries for the sustainable development (economy, geopolitics, ecology), Bishkek.

Anonymous. (2009a). *The main social-economic indexes of the level of the population life, statistical annual of the Kyrgyz Republic 2004–2008*. Bishkek, KR: National Statistic Committee. www.stat.kg

Anonymous. (2009b). The second national information of the Kyrgyz Republic on framework convention of the UN about the climate change—2009, Bishkek.

Anonymous. (2009c). *Inventory of anthropogenous emissions from sources and absorption by absorbers of greenhouse gases. Second national communication of the Kyrgyz Republic on the UN framework convention on climate change*. Bishkek: UNDP.

Anonymous. (2010). *National report on human development "Kyrgyzstan: Prosperous youth—Prosperous country"*. Kyrgyzstan, Bishkek: UNDP.

Anonymous. (2012). *The national human development report of the Kyrgyz Republic in 2012*. Bishkek: UNDP.

Fay, M., Block, R. I., & Ebinger, J. (2009). *World Bank Report "Adapting to the climate change in Eastern Europe and Central Asia", World Bank*. © World Bank. https://openknowledge.worldbank.org/handle/10986/2407. License: CC BY 3.0 IGO.

Kolov, O. V., Musuraliev, T. S., & Bikirov, S. B. (2001). forest and forest management in the mountains, Mountains of Kyrgyzstan, National Center for Mountain Development Kyrgyzstan, National Academy of Sciences of the Kyrgyz Republic, International Institute of Mountains, Bishkek.

Libishevski, S., & Bakhler, G. (1999). *The conflicts in mountainous areas—The factor interfering a sustainable development*. Moscow: Mountains of the World.

Tashybekov, T. A. (2002). *National centre in action*. International year of mountains—An important event of the new century. National Organizing Committee on preparation and holding of an international year of mountains, National Center for Mountain Regions Development of the Kyrgyz Republic.

Chapter 4
Socio-economic Wellbeing and Mental Health Profile of Rural Hill Women of Uttarakhand, India

Lata Pande, Jyoti Tiwari and Chhavi Arya

4.1 Introduction

Bounded in the north by Tibetan autonomous region of China and in the east bounded by Nepal, Uttarakhand is located at the northern margin of India. It is marginal not only in its physical setting but economically it is underdeveloped. It is characterized by small and fragmented land holdings. Fragile ecosystem, rain-fed agriculture poor productivity of the crop. Meager means of transport and communication, women-centered agriculture and economic migration of males in search of employment. The state also faces several challenges due to climate change and deforestation. As a result of which water and agricultural produce are adversely affected. The impact of development is witnessed at a very slow pace in the hilly regions of the state. Majority of the rural people in Uttarakhand belong low economic status and marginalized families.

The women belonging to this region bear the brunt of all these conditions and problems. Here the role of women is multi-dimensional and they shoulder the responsibility of agricultural and livestock production, making arrangements for water and fuel and also managing the chores of the entire household. Women in hilly region are the most disadvantaged people in terms of their health and nutritional status. They have less access to basic resources such as health care facilities,

L. Pande (✉) · C. Arya
Home Science Department, Kamoun University, DSB Campus, Nainital 263002, Uttarakhand, India
e-mail: latanutrition@yahoo.co.in

C. Arya
e-mail: chhavi_arya@hotmail.com

J. Tiwari
Home Science Department, HNB Garhwal University Srinagar Campus, Srinagar, Uttarakhand, India
e-mail: j.tiwari_7@yahoo.co.in

R. Chand et al. (eds.), *Societies, Social Inequalities and Marginalization*, Perspectives on Geographical Marginality,
DOI 10.1007/978-3-319-50998-3_4

education and employment opportunities and even fewer opportunities to enhance their social circle compared to their male counterparts. It is generally recognized that work burden of women in Uttarakhand is very high. However, they have limited say with regard to important decisions pertaining to family and the outside world. Financial decision is an aspect handled by male family members. Due to lack of independent economic status and lack of control over income generating resources of women they are marginalized both inside and outside the household in terms of social and economic status (Metha 1999). The role of women in the developmental process assumes great significance, as they constitute a substantial portion of the population. They play a variety of roles both within the household as well as outside the household. Woman‘s ability to perform these roles, and her quality of life, is determined by the health status she enjoys. Women‘s health matters not only to women themselves but it is also crucial to the health of the children they will bear. Women's health involves their emotional, social and physical well-being and is determined by the social, political and economic context of their lives, as well as by biology. However, women are unable to reach their full potential because of persistent health, nutrition, social and gender inequalities and health system inadequacies (García-Moreno et al. 2005). The Overall health status of women diminish by the harsh realities of malnutrition, illness injury, demanding physical labor, strains of child-birth, fatigue, and unhygienic conditions the scarcity of food and gender discrimination towards women.

Women are victim to stress at various levels in India that attribute their stressed condition to a large number of factors including poverty, ill health, traditional expectations of women role, lack of affection, conflict within the family, widowhood and divorce. Post-natal depression in India affects 11–23% of women, and is linked to poverty, antenatal psychiatric morbidity, poor marital and family relationships, lack of support, marital violence and the birth of a girl child (Chandran et al. 2002; Rodrigues et al. 2003). Domestic violence against women has a variety of health consequences. The health consequences of violence against women, includes non-fatal outcomes such as injury, unwanted pregnancy, alcohol/drug abuse, HIV/AIDS etc. and long term outcomes include organ damages, chronic disabilities, mental disorders, depression and adverse pregnancy outcomes. In rural areas, majority of women live in poor socio economic condition, wife beating is common affair in many families. Women in rural area never consider it as a serious matter and they are accustomed to it (Prakasam 2008). Domestic violence (physical, emotional and sexual violence) is reported in married women of Uttarakhand, which is one among many reasons leading to poor mental health status or stress. Rural women face number of conflicting situation in family and society. Rural hill women of Kumaun have been neglected and vulnerable since long time. Very few studies focus on their socioeconomic environment and absolutely no study was available on the mental health of these women. Therefore in view of these considerations the present study was planned to study the socioeconomic environment and mental health profile of rural Kumauni women.

4.1.1 Material and Methods

The study was conducted in the historically important district of Almora which had a population of 621,927 as per 2011 census. Out of the total population 62,332 (10.02%) were urban and the remaining lived in rural areas. In district Almora, there are 3 tehsils,[1] 11 development blocks with a total of 2247 villages. On the basis of easy accessibility and availability of some basic data three developmental blocks viz. Dwarahat, Hawalbagh and Takula comprising seventy two villages were included in the case study.

4.1.2 Selection of Sample

The selection of the present sample for the study is based on the idea that each segment of the society with varying social and educational categories and the potential age group of the women should be represented well in order to derive the meaningful conclusions. The sample was therefore selected by using multistage proportionate stratified sampling technique; 1428 households were chosen from 14 villages of Takula, 28 villages of Dwarahat and 30 villages of Hawalbagh. On total 250 respondent units (married women of reproductive age from 15 to 49 years) were taken for the present study (Fig. 4.1). An attempt was also made to get a representative sample from each group of respondents (Table 4.1).

The study was carried out from December 2005 to February 2008. An interview schedule was formulated and tested for data collection prior to survey. General information regarding religion and caste, sex ratio, educational level, occupational activities, household income, income sources, total working hours per day and participation in decision making and information regarding mental health status of respondents was collected using interview schedule. Information regarding household environmental conditions was collected through questionnaire cum observational method. Statistical analysis was used to interpret the data. Percentage, arithmetic mean and standard deviation were used to analyze data. Chi square test was used to test for differences in categorical variables (decision making power and mental health) between selected villages.

[1]An administrative division of some countries of South Asia. It is an area of land with a city or town that serves as its administrative center, with possible additional towns, and usually a number of villages.

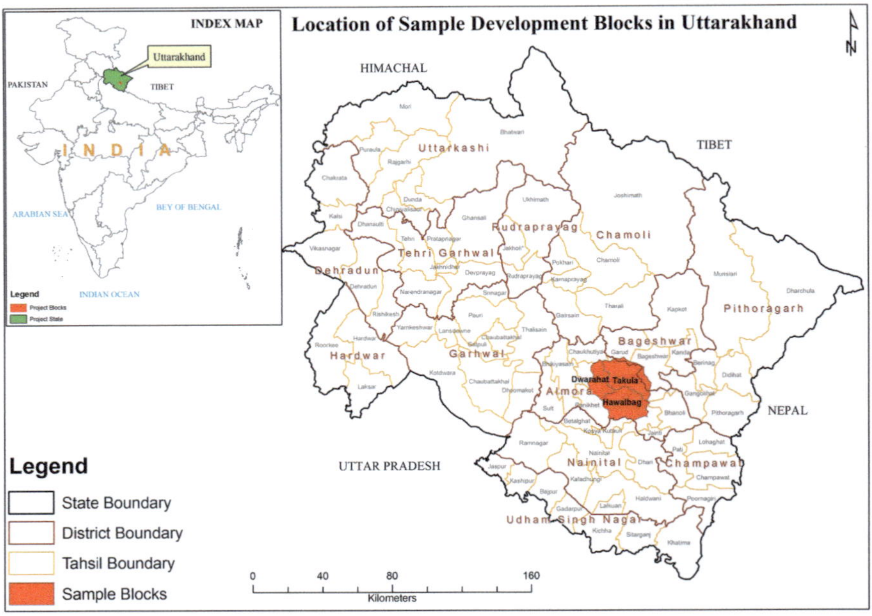

Fig. 4.1 Location of sample development blocks in Uttarakhand, India

Table 4.1 Block wise distribution of respondents in district Almora, Uttarakhand, India

Categories	Respondent from Takula	Respondent from Dwarahat	Respondent from Hawalbagh	Total
General category	24 (30%)	28 (31.11%)	21 (26.25%)	73
Pregnant category	31 (38.75%)	28 (31.11%)	24 (30%)	83
Lactating category	25 (31.25)	34 (37.77%)	35 (43.75%)	94
Total	80	90	80	250
	32.00%	36.00%	32.00%	100.00%

4.2 Results and Discussion

4.2.1 Socio-economic Scenario

The Socio-economic status (SES) is an important determinant of health and nutritional status as well as of mortality and morbidity. It influences the accessibility, affordability, acceptability and actual utilization of various available facilities (Aggarwal et al. 2005). The Socio-economic status (SES) of a community is influenced by several variables such as the composition of types of people, resources, their age, family size, health, literacy, occupation, income, wealth, possession of materials, life style (Park and Park 1983).

4.2.2 Categorization of Respondents on the Basis of Caste System

Results of the study showed that within the selected study area caste system is a dominant feature within the village society. Several caste groups (Brahmins, Khatriyas (Thakur, Rajput), Vaishya (Shah), Scheduled caste (Silpkar—Craftsmen), Scheduled tribe, other backward classes (Goswami etc.), Muslim and Christians etc.) are generally found in all the villages. Out of total 250 respondents, Muslims and Christians were found to be in minority. The area was dominated by the upper caste Brahmins and Thakurs constituting more than half (50.77%) of all respondents (Fig. 4.2).

4.2.3 Sex Ratio

The investigation of sex ratio discloses the numerical relationship between women and men which can be articulated as a number of females per 1000 males. Results showed that in the study area 46.80% households had high sex ratio i.e. surplus women. While remaining 53.20% households had low sex ratio. The important factors for the low sex ratio in this region were a preference for the son, unequal treatment given to boys and girls, female infanticide, neglect of female infants, early marriage, death during pregnancy, bad treatment of women and hard work (Fig. 4.3).

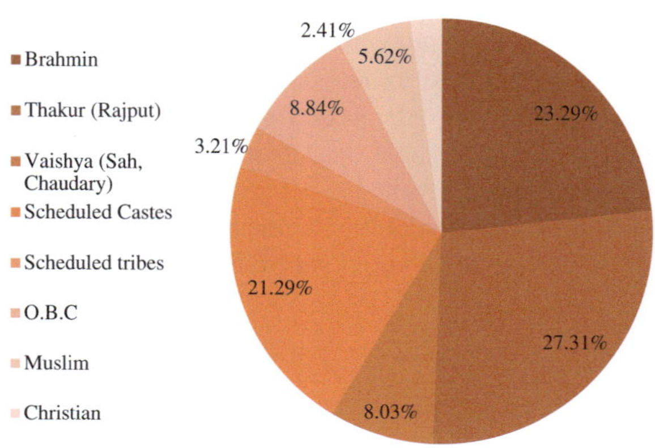

Fig. 4.2 Caste-wise distribution of respondents in the study area

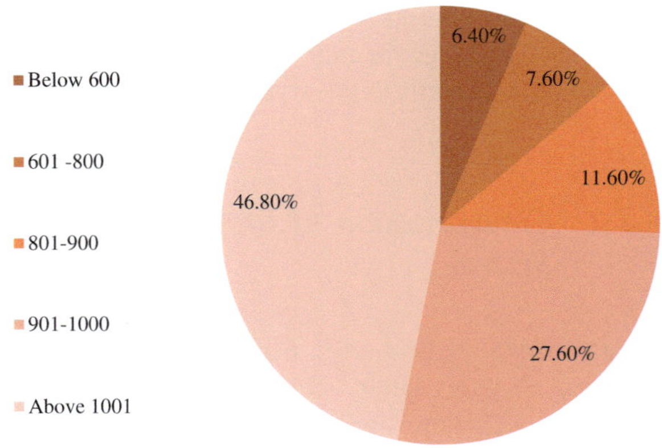

Fig. 4.3 Sex-ratio of respondent households in study area

4.2.4 Age of Respondents at the Time of Marriage

The majority of respondents (46.74%) got married at the age of 18–21 years followed by 27.73% married at the age of 21–25, 17.02% at 15–18 and 8.62% respondents who got married at 25 and above years respectively. It shows that the women in study area do not tend to marry at an early age. Only one-sixth of women are married at the age under 18 years. It reflects a high education level in the study area.

4.2.5 Educational Level

The educational level of the selected respondents is presented in Fig. 4.4. Firstly respondents were divided into two groups i.e. literate and illiterate. Results showed that there were almost two-thirds literate and a bit more than one-third illiterate respondent. Out of the total selected respondents about one fourth had received education up to primary level, and a bit more than one-fourth middle and High school. Respondents with Intermediate and graduate and above education represented the smallest share. The shares presented in Fig. 4.4 confirm the high educational status of Uttarakhand and justify its eighteenth rank in female literacy in India. This remarkable data in female literacy shows the heavy emphasis in girl education by parents and government. However, the dropout rate of girls in Primary level to High school level is remarkably high which can be attributed to the load of domestic chores and working on the land.

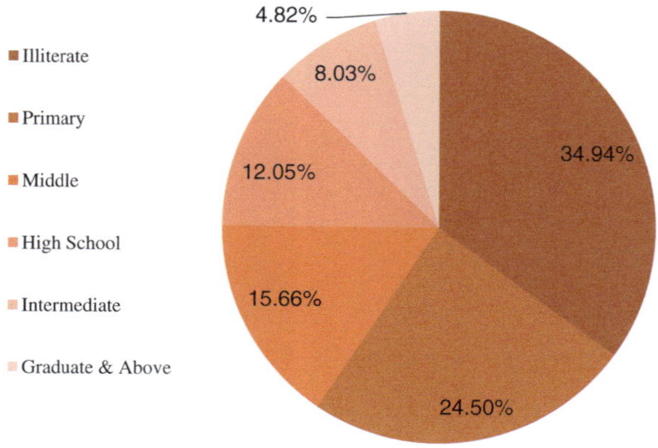

Fig. 4.4 Education level of respondents in study area

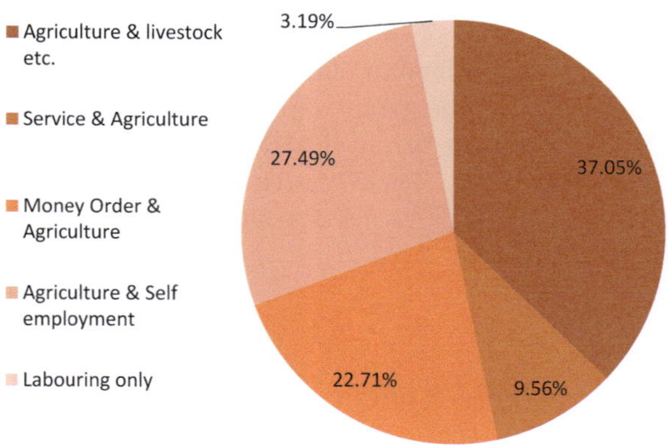

Fig. 4.5 Occupational status of respondents in study area

4.2.6 Occupation

Respondents were engaged in three main occupational activities i.e. in Agriculture and livestock etc., Service and Laboring. The vast majority of respondents were engaged in agricultural activities including animal husbandry, less than 5% were engaged in services like Angan Wadis, School Ayahs or teaching, whereas only a few were laborers. The Angan wadis are village level small institutions to help educate the rural people and school Ayahs are those women workers serving the children providing them mid-day meal etc. These data confirms that the hill economy is entirely based on agriculture and livestock (Fig. 4.5).

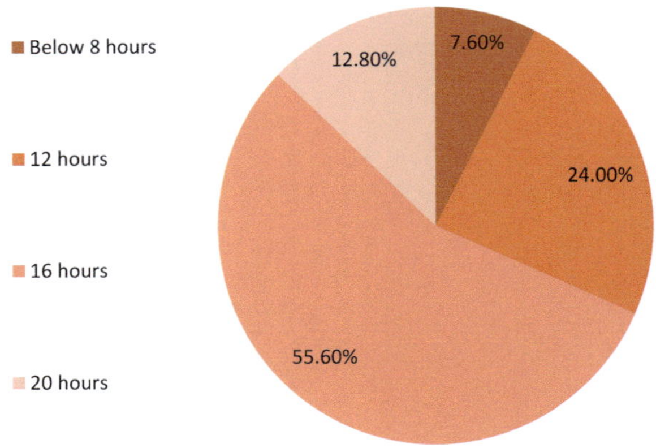

Fig. 4.6 Working hours of respondents in study area

4.2.7 *Working Hours*

The rural women of the mountains are overburdened with the work load. Women participation is found in almost all the rural occupations. Total respondents were classified into four groups on the basis of their average per day working hours (Fig. 4.6) amount of physical labor that rural women folk have to perform in various activities. This excessive burden of work, in the absence of proper nutritious diet and health care, is bound to generate serious health problems for both mother and child.

4.2.8 *Total Income*

Respondents were classified into five income groups on the basis of their family income.

Figure 4.7 represents that the share of respondents in the higher economic class with family income more than 15,000.00 rupees per month was the smallest while the share of respondent‘s with the lowest income was the highest and exceeded one-third of all respondents. Their family income was less than 2000.00 rupees per month; it shows that mass of respondents was living in below poverty line.

4.2.9 *Decision-Making Power*

Respondents were classified into six groups on the basis of their freedom to take decisions regarding Agriculture and livestock, Family Affairs, Social affairs, Child Education, Economic matters and Health and Reproductive issues (Table 4.2). Maximum (80%) respondents participated in the decision related to agriculture and

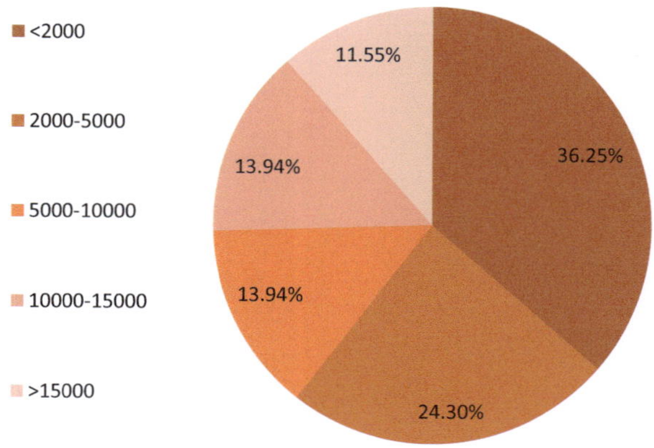

Fig. 4.7 Total income of respondents in study area (values are in percentage)

livestock activities. This may be because agriculture and livestock activities are major income generating activities in rural Uttarakhand and usually performed by women members of households. About 62.8% respondents took decisions regarding their reproductive health; this is due to the improved education level in the study area. Economic decisions were taken by more than half (54.8%) of respondents. Family decisions were taken by about 32.4% respondents whereas social decisions were taken by only 31.4% respondents. Minimum (15.6%) respondents were involved in decisions regarding their child's education. No significant differences were found in decision making pattern of women selected from the three developmental blocks ($p \leq 0.05$). Overall picture about decision-making power of the woman in Uttarakhand is better than the rest of the states in India. A study carried out by Jan and Akther (2008) on hundred married and unmarried belonging to Jammu and Kashmir state of India showed that women generally possess low decision-making power and are mainly dependent on masculine and/or familial decision-making. Participation of the women in decision making is likely to be affected by socio-economic status (Bano 2014).

4.2.10 Factors Affecting Mental Health of Respondents

Factors affecting Mental health of respondents can be classified into four groups, namely Physical Violence, Social Insult, Antepartum depression and Postpartum depression, all causing mental stress. Figure 4.8 shows that respondents that are mentally stressed with the trauma of social insult represent the highest share followed by those experiencing physical violence (a bit above two-fifths); this is because relatively large share of male members in rural areas are addicted to drinking and playing cards, in spite of their poor economic condition. Their low economic status, unavailability of jobs and feeble will for hard work lead them to frustration which is poured on their female counterpart in the form of insults and

Table 4.2 Decision-making power of respondents in three developmental blocks of district Almora, Uttarakhand, India

Type of decisions	Takula	Dwarahat	Hawalbagh	Total	*P* value
Agriculture and livestock decision	60 (75.00%)	75 (83.33%)	65 (81.25%)	200 (80.00%)	0.46
Family decision	25 (31.25%)	30 (33.33%)	26 (32.50%)	81 (32.40%)	0.77
Social decision	28 (35.00%)	26 (28.89%)	24 (30.00%)	78 (31.20%)	0.86
Child education	10 (12.50%)	17 (18.89%)	12 (15.00%)	39 (15.60%)	0.37
Health and reproductive decision	52 (65.00%)	54 (60.00%)	51 (63.75%)	157 (62.80%)	0.15
Economic decision	41 (51.25%)	47 (52.22%)	49 (61.25%)	137 (54.80%)	0.63

Fig. 4.8 Factors affecting mental health of respondents in the study area

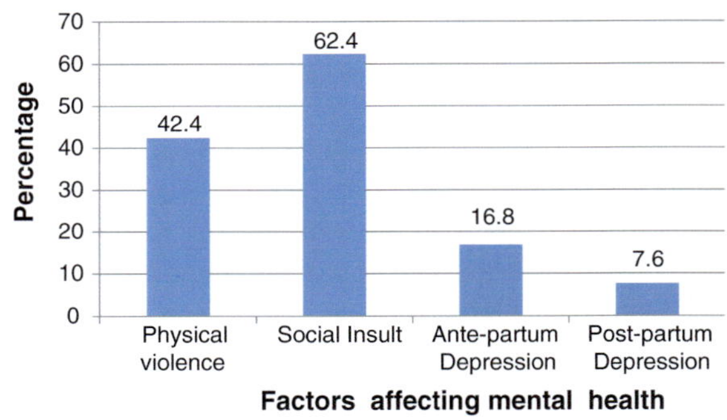

physical violence. Pregnant respondents were also stressed by ante-partum and postpartum depression due to social and cultural malevolence. About one-sixth of respondents faced ante partum depression whereas 7.60% respondents faced post-partum depression. The respondents from the three developmental blocks showed no significant difference on parameters of mental health.

4.3 Conclusion

Present findings conclude that the community in the selected villages was divided and stratified on the basis of caste system. The majority of the women respondents were literate however, only a few were graduates. A gradual decline was observed in the number of respondents with the increase in educational level. In some areas like education, agricultural work participation, sex ratio and decision making they were in a better position as compared to rural women elsewhere in India. The respondents from the three different developmental blocks did not differ significantly on decision-making aspects and mental health. However, most of the respondents were living below poverty line and were overburdened with work and suffered from various kind of mental stress. The rural society needs to change its attitude towards women.

References

Aggarwal, O. P., Bhasin, S. K., Sharma, A. K., Chhabra, P., Aggarwal, K., & Rajoura, O. P. (2005). A new instrument (scale) for measuring the socioeconomic status of a family: Preliminary study. *Indian Journal of Community Medicine, 30*(4).

Bano, S. (2014). Women and decision making in urban India: A gender geographical study of Varanasi city, Uttar Pradesh, India. *IJES, 3*(4).

Chandran, M., Tharyan, P., Muliyil, J., & Abraham, S. (2002). Post-partum depression in a cohort of women from a rural area of Tamil Nadu, India: Incidence and risk factors. *British Journal of Psychiatry, 181,* 499–504.

García-Moreno, C., Jansen, H. A. F. M., Ellsburg, M., Heise, L., & Watts, C. (2005). *WHO multicountry study on women's health and domestic violence against women*. Geneva: WHO.

Jan, M., Akthar, S. (2008). An analysis of decision-making power among married and unmarried women. *Studies on Home and Community Science, 2*(1), 43–50.

Metha, G. S. (1999). *Development of Uttarakhand: Issues and perspectives, status and empowerment of women* (pp. 166–191). New Delhi: APH Publishing Corporation.

Park, J. E., Park, K. (1983). Textbook of preventive and social medicine (pp. 72–75). Jabalpur: Banarsidas Bhanot.

Prakasam, C. P. (2008). *National seminar on population, environment and development in Uttarakhand.* International Institute for Population Sciences, Deonar, Mumbai, NFHS-3 data analysis.

Rodrigues, M., Patel, V., Jaswal, S., & De Souza, N. (2003). Listening to mothers: Qualitative studies on motherhood and depression from Goa, India. *Social Science and Medicine, 57,* 1797–1806.

Chapter 5
The Status of Marginalized Women Tea Garden Workers in the Mountain Ecosystem of Darjeeling in a Globalised Village

Moushumi Datta

5.1 Introduction

Tea industry is one of the oldest industries in India and enjoys the status of one of the best organised industries in the country. It occupies an important place and plays a significant role in the national economy. The Indian organised tea sector currently employs over 1.1 million workers throughout the tea growing states. Women constitute more than 50% of the total labour force. This is the only industry in the organised sector that employs such a high proportion of female labour. The industry boasts as an organised sector with various legislations in place specially the Plantation Labour Act of 1951. Owing to certain specific soil and climatic requirements tea cultivation was confined to certain parts of the country only. Major tea growing areas of the country are in Assam, West Bengal, Tamil Nadu and Kerala. The other areas where tea is grown on a lesser scale are Karnataka, Tripura, Himachal Pradesh, Uttaranchal, Arunachal Pradesh, Manipur, Sikkim, Nagaland, Meghalaya a, Mizoram, and Bihar. Tea plantation women are one of the most marginalized in this globalized village. Socially, economically and culturally they are excluded from the mainstream. They are the prime target of deprivation and exploitation.

5.1.1 *Research Objectives*

The study that we present in this chapter was a micro level study. It aimed to highlight the marginal position of the women worker in the tea industry and the

---

M. Datta (✉)
Nagindas Khandwala College, Road No.1, Off S V Road, Bhadran Nagar, Malad West, Mumbai, Maharashtra 400064, India
e-mail: moushumi@nkc.ac.in

R. Chand et al. (eds.), *Societies, Social Inequalities and Marginalization*,
Perspectives on Geographical Marginality,
DOI 10.1007/978-3-319-50998-3_5

exploitation that they have to face with. Several precise and rational objectives have been followed to explore the problems faced by the women working in the tea industry of the study area:

- To assess the socio-economic conditions of women workers in the tea industry in the study area.
- To review the living condition of workers in the tea industry.
- To identify their major health problems.
- To ascertain the knowledge, awareness and perception of the workers regarding the health hazards.
- To find out the nutritional status of the tea workers.
- To analyze the causes and consequences of psychological stress of the women workers.
- To analyze the status of marginalized women tea garden workers in the mountain ecosystem of Darjeeling in a globalised village.

We anticipated that health problems and education of women workers are not independent of different locations.

5.1.2 Study Area and Research Methods

Darjeeling is known as City of tea. It is located in northern West Bengal. It is bound by latitude 26° 31′N–27° 13′N and longitude 87° 59′E–88° 53′E. It is the head-quarter of Darjeeling district, in the Shiwalik Hills on the lower range of the Himalaya, at an average elevation of 2134 m (6982 ft.). Darjeeling is hemmed in by Sikkim in the north, Nepal in the west and Bhutan in the east (Fig. 5.1). Dar-jeelingtown, Kalimpong and Kurseong are the major urban centers.

Research method refers to the operational technique of data collection. In order to study the "Quality of Life of Tea Garden Women: Selected Case Studies in Tea Gardens of Darjeeling District, West- Bengal" the methodology adopted by the researcher is a rationalistic one. To get a comprehensive knowledge of the study area, different libraries like the Departmental (Geography) Library of the Mumbai University and National Library of Kolkata were consulted. In this study, four tea gardens—Makaibari, Jungapana, Sourenee and Gaybaree were selected for detail study (Fig. 5.2). Forty five women workers were chosen from each tea garden with the help of simple random sampling method and primary survey was conducted with the help of structured and unstructured questionnaire. Interview schedule; transect walk, group discussions and observation methods were mostly used to collect the primary data for the study. The data was collected during the period of May 2009 and May 2010. Facts were also recorded after direct observations of the work place, working methods, housing conditions of labourers, health facilities,

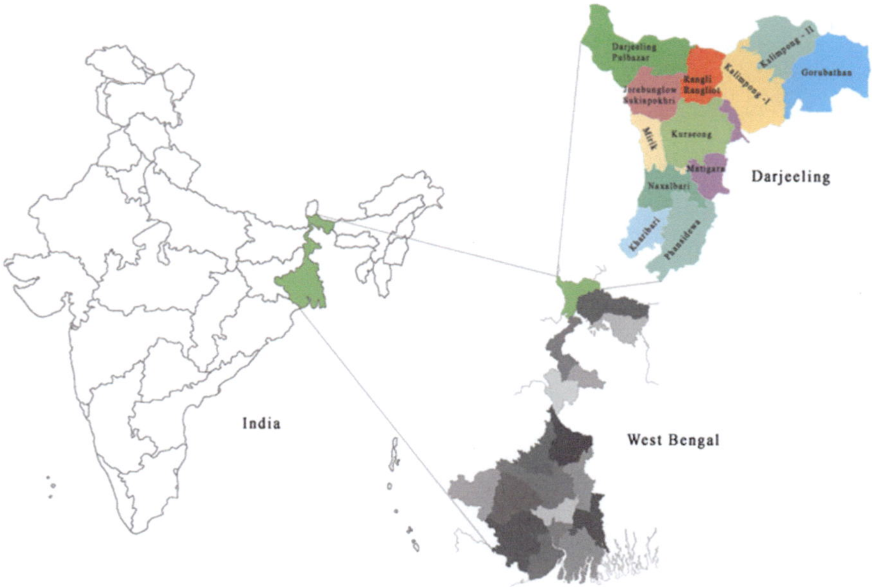

Fig. 5.1 Location of the study area

education and other public services. The Likert type of 3 point, 5 point and 7 point scale were also used to get the responses of the respondents. The attitude and opinions of the respondents were assessed at every point. The collected data, both primary and secondary, were processed, assimilated and analyzed. Editing, coding and decoding of collected data were also done simultaneously, according to the aims and objectives of the study, using simple statistical techniques. A suitable bibliography was prepared to show the references used. As a whole, the research design is qualitative in nature.

5.1.3 Review of Literature

Studies related to the plantations in India began in the early 1930s. However, most of the published works on tea plantations in India are about tea in North India. Griffith (1967) has done a commendable work on the history of the South Indian tea. He stated that the plantation sector emerged in India in the early 1830s when the East Indian Company was seeking alternative sources for procuring tea to Europe. The importance of tea plantations as the largest employer of women

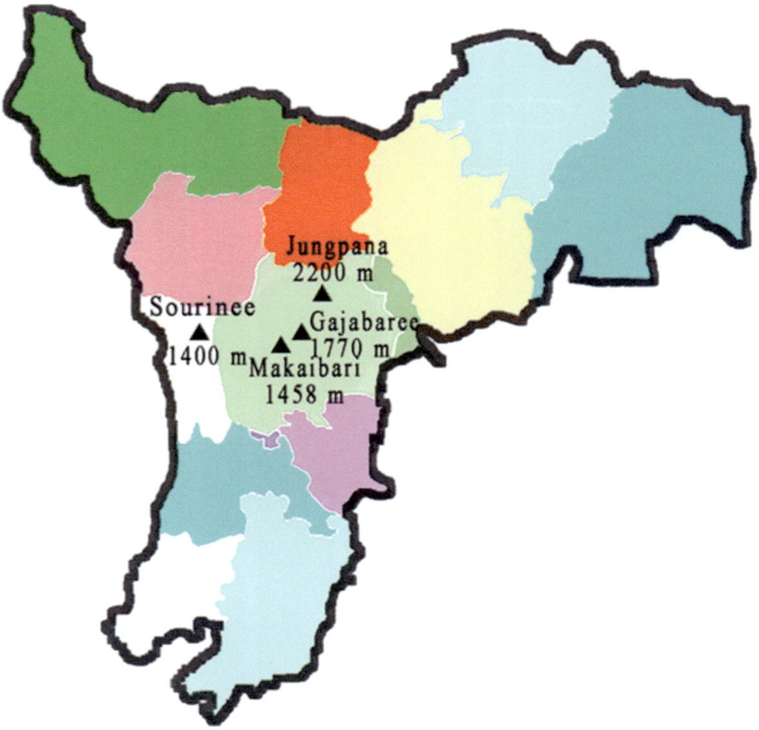

Fig. 5.2 The location of selected tea gardens

workers is stressed by Sen Gupta (1954), Bhowmik (1981a, b) studied the concept of class formation among tribal workers in West Bengal. Bhowmik (1992) and Bhowmik et al. (1996) suggested that 20% of the tea garden workers houses do not meet the specifications against the housing provision of the Plantation Act.

It is to be remembered that Darjeeling is a place of cultural and economic heritage and that of academic interest. This region has encouraged intensive study and investigation by various research groups and government authorities. In several publications like Malley (1970), Banerjee (1980), Das (1947), Khawas (2006), and the annual reports tea statistics of Tea Board of India (2001, 2002, 2008, 2009, 2009) many interesting aspects like quality of life of tea garden women workers have been studied. The study of quality of life became an important aspect of geographical literature in the past three decades. Some Indian geographers like Bhat et al. (1976), also took interest in the study of social well-being. A review of the above factors reveals that often the conditions that contribute to motivation will also contribute to quality of life. In the case of women employees special measures for their protection, availability of child care facilities like crèche, no discrimination in recruitment and equal treatment on the job, etc. assume additional importance.

5.2 Development of Tea Gardens in Darjeeling Introduction

The first thought of the possibility of producing tea in India came from the Britishers themselves, when the practical mind of Warren Hastings clearly realized that if the Company was to prosper, it must promote industrial development. In 1778, Sir Joseph Barks was asked to prepare a series of notes for the Company on the cultivation of new crops, and in them he advocated the cultivation of tea in India. In 1793, Barks was sent to China for obtaining detailed information on the cultivation and manufacture of tea. The issue was again put forward when Lord William Bentinck appointed the Tea Committee on 1st February 1834. In 1841 Dr. Campbell brought China tea seeds from the Kumaon and planted them in his residence at Beechwood in Darjeeling at a height of nearly 2150 m (7000 ft.). As Dr. Campbell had been very successful in raising the plant, the Government in 1847 decided to create tea nurseries in this area and also distributed seeds to those who decided to cultivate tea. In 1852 Dr. Jackson came to tour Darjeeling and reported that tea plantations were very healthy and luxuriant in the area showing that the soil is ideally suited for the cultivation. According to Dr. Campbell's report, it is clear that though experimental planting of tea has started in 1841 in the Darjeeling region, much headway was not made till 1853. Hunter (1876), however, pointed out that the real date of commencement of the industry may be taken as 1856–57 and this era marked the beginning of tea industry in Darjeeling. In Darjeeling tea gardens the plucking of the leaves is mainly done by the women workers. Plucking is a highly specialized work and requires a great deal of care. The woman workers consider tea bushes extremely sensitive and perform the job very efficiently. The technique is a traditional skill, handed down from generation to generation. The picking of two-leaves-and-one-bud shoot has to be followed here, unlike other tea growing areas in India.

5.3 Socio Economic Profile of the Respondents

The women workers of the four selected tea gardens are of Nepalese origin. They are short stature, have mongoloid features with a yellowish complexion, flat nose, small eyes and black hair. Though, originally from Nepalese origin, these early settlers of Darjeeling prefer to be known as "Gorkhas". The majority of the women workers belong to "Rai" and "Thapa" ethnic groups. There are small numbers of "Chettris" here who belong to the upper caste. Most of the respondents speak Nepali language. Majority of the women respondents belong to the age group of 26–35. About 30% of the women workers fall in the age group of 16–35 and only 12% are in the age group of above 53 years. 63% of the women tea garden respondents are married. 32% of the women respondents have the family size of less than 3 members. 38% have the family size ranging from 4 to 6 members and

30% of the women workers belong to the family size of 7 members and more. In Darjeeling, the daily wages of a tea plantation worker was Rs. 54 (less than 1 USD) in 2008. The wages increased to Rs. 67.00 (approximately 1 USD) in 2011. A women worker has to pluck 6 to 8 kilograms of leaf, which is usually a day's labour. The labourers are also entitled to rations of food grains at a highly subsidised rate. This forms a part of their wages. According to the Plantation Labour Act, 1951, a labourer is entitled to get slipper, umbrella, raincoat, and blanket. It is observed that some of the basic requirements namely provident fund, Durga Puja bonus,[1] medical and ration are provided on a reasonable scale by all the tea gardens. Overall it can be said that the awareness about economic entitlements is low. The household property of the women workers includes earthen pitcher, aluminum utensils and limited furniture. It has been noticed during survey that male-headed families are predominant in tea garden areas. In spite of their contribution towards the families needs, the participation of women in the decision making process of the family is virtually non-existent.

The expenditure of women workers on food is maximum in all the four tea gardens. Though the workers get ration on subsidized rate, they purchase food materials from outside and their expenditure on food goes from 48.89 in Sourenee to the extent of 62.22% in Makaibari (Fig. 5.3). They also spend a substantial amount of money in clothing. The women workers have to spend some amount of money for fuel because the amount of fuel given by the management is not enough for the full month. They collect fuel from forest or nearby places. Overall their expenditure on food is the maximum and that in the education sector is the minimum. This also implies that education is not very important to them.

From the data collected it is found that the number of illiterate women in tea gardens is 8.89% in Makaibari followed by Jungpana with 28.89%, Gayabaree 28.90% and Sourinee with 33.34%. Women workers who have secured primary education (up to IV standard schooling) in Sourinee and Makaibari are 35.56% each as against 40.00% in Gayabaree and 57.78% in Jungpana. It can be noted that only a few women workers have completed their secondary education. The share is 4.43% in Gayabaree followed by 4.44% in Sourinee. It is observed that none of the workers in Jungpana and Makaibari have secured secondary education.

5.4 Living Conditions

The housing facilities available to the women workers of the tea garden are not at all satisfactory. Most of the workers are living in kucchha[2] houses. Most of the houses have wooden floor. Tin and plastic are the main materials used for the roof.

[1]Durga Puja is the main festival in West Bengal. Durga Puja Bonus is an extra amount paid to the workers to celebrate the occasion.

[2]Temporary houses constructed using natural materials like mud, stones and thatched roofs.

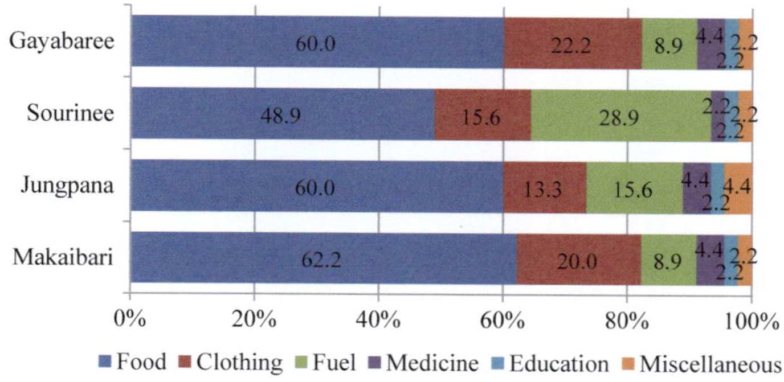

Fig. 5.3 Expenditure pattern of women workers in the four plantations

Under clause 8 of the Plantation Labour Act, in every plantation the labour houses should have effective arrangement of drinking water. But this is totally absent in most of the tea gardens of Darjeeling. Most of the houses lack minimum light and proper ventilation and this gives rise to several diseases. The researcher found there were several elderly women who are suffering from eye troubles and the reason may be absence of suitable exit arrangement of smoke. Non availability of fresh air during night may be contributing towards respiratory troubles. Even though the houses are situated in steep hill slopes, the paths are not maintained. As there are no garbage bins, the garbage lie scattered all over. When it rains, the paths are extremely slippery. There is no housing facility for casual workers. They live in the front portion of the houses of permanent workers. The Plantation Labour Act, 1951 lays down that there should be separate latrines and bathroom in the labour line. It is said that there should be at least one bathroom per 25 people. It is also stated that there should be separate bathrooms for women. Our observations states that the bathrooms are very badly maintained in the labour colony. There is a common bathroom and sometimes workers do not use them at all. Open field defecation is widely practiced in this area. In fact most of the provisions with regards to sanitation and hygiene have not found to be implemented properly in most of the tea gardens.

Women workers have to travel long distance to fetch water for drinking and other domestic uses. In Makaibari tea garden the main source of water is the tap. The highest share of women spending 5 min or less is in Sourinee and more than half women from this plantation need ten minutes or less. At Gayabaree plantation situation is quite different and almost 30% women there need 25 min or more. In Jungpana most women have to travel 15 min to collect drinking water followed by 20 min. In Makaibari the situation is almost the same as Jungpana (Fig. 5.4).

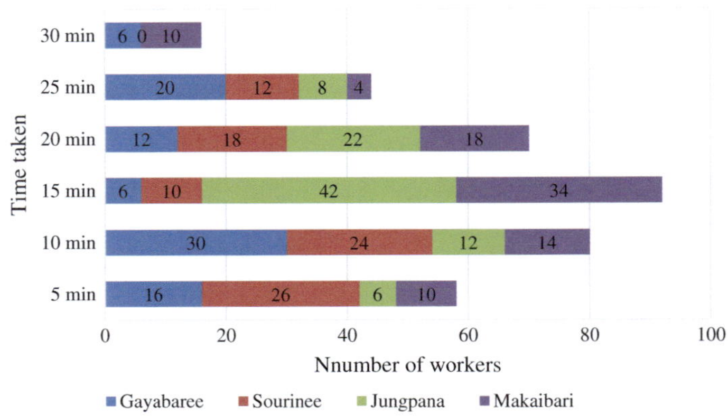

Fig. 5.4 Time taken to collect drinking water

5.5 Health Provisions

The statutory provisions as per the Plantation Labour Act, 1951 require every plantation to provide and maintain medical facilities to its employees. Most of the tea gardens do not have their own hospitals. The Panchayat Health care supply medicines but they are also limited. Generally cold and fever, gastroenteritis, anemia, dysentery, skin diseases, eye diseases, rheumatism and respiratory problems are the most common diseases. Some of the other frequent diseases are boils, urinary infections, jaundice, and malaria (Fig. 5.5). In fact, malnutrition coupled with overall unhygienic living conditions and poor sanitation makes the women workers prone to various diseases. Some women have the habit of drinking rice beer and these results in several respiratory problems and complications during pregnancies.

The women labourers often get infected with anemia and gynecological disorders. In all the tea gardens it was noticed that skin problem among the women workers is the major health problem. Besides skin problem a great proportion of the workers reported of constant back and neck pain. In all the tea gardens under study (Makaibari 13.21%, Jungpana 15.83%, Sourenee 25.74% and Gayabaree 36.26% have workers facing such a chronic problem. The workers do not take any effort to preserve the cleanliness of drinking water. This may be one of the reasons for which they suffer from waterborne diseases. Diarrhea, Cholera and Intestinal problems are common in all the four tea gardens. According to the World Health Organisation, individuals with a BMI[3] of less than 16.0 are considered at "high risk of mortality from starvation," while a population with more than 40% of its adults with a BMI of

[3]BMI stands for Body Mass Index which helps to indicate fat based on height and weight. A healthy BMI score is between 20 and 25. It helps in denoting the health conditions of an individual.

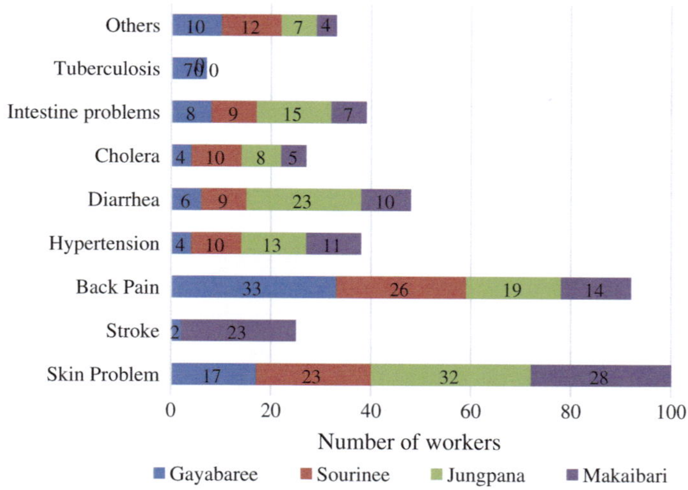

Fig. 5.5 Health problems of women—tea garden workers

less than 18.5 may be termed "at critical risk for mortality from starvation" or a "starving community." First of all, taken as a whole, the entire population of 180 adults surveyed can be labeled as a "starving community" or "at critical risk for mortality from starvation" (because the percentage of individuals with BMI less than 18.5 is above 40%). 45% of the total women workers belong to normal category and a very small %age fall in overweight and obese category.

5.5.1 Nutritional Status and Dietary Pattern

The level of nutrition attained by women workers in the Darjeeling tea gardens is very low. Milk, egg and animal meat are occasionally present. Pulses do not form a part of the meal every day. Poor nutrition makes them more vulnerable to infectious diseases. In all the four tea gardens it is found that the food habits are more or less similar. The staple food is rice. Generally, the women tea labourers take meal three times a day. Women workers must be made aware to take food which is needed for health and their body. Their notion is that nutrients are found in costly food items only. This idea should be changed and there is a need to make them realize that even low cost items can provide them nutrition.

5.6 Job Satisfaction

Most of the workers are not satisfied with the existing medical facilities (Fig. 5.6). Also most of them are not satisfied with the leave facilities, but the permanent workers are satisfied with the job security. The workers are not satisfied with the facilities provided to them. They want more facilities as far as their housing and sanitation is concerned. During interview 82% of the women workers said that they are not happy with their present job but as their education qualification is low, they are not in the position to change to a better job. The women workers are not satisfied with present hours of work. 56% of the respondents feel that the wages are very low. The wages and facilities provided to the women workers as against the hours of work are much lower which affects their performance level. Infact, all estate workers need to work only the legally stipulated eight hours and for additional work, double wage is to be paid. The women workers are not happy with their management. Most of the women workers feel that there is a lack of motivation and they do not have any chances of promotion. This is another major cause of dissatisfaction with work. On the whole it was found that the women workers are fully committed to their work.

5.7 Leisure and Recreation Activities

Recreational activities are available only in the form of radio, TV, video tape, etc. Most of the women workers said that they spend their leisure time by watching television and listening to songs in radio. The younger women during weekends go to Darjeeling town to watch movies.

5.8 Domestic Work Responsibility

Primary Survey revealed that most of the domestic work responsibilities are done by the women. This hectic work schedule makes them physically and mentally tired. Out of 180 women workers surveyed 82% of the women said that they do the work of cooking single handedly. Only 12% said that even their husbands who are mostly in the tea plantation help them. Rest 6% did not provide any answer clearly. 71% of the women said that all the cleaning and washing of clothes is done by them. Only 15% of the respondents said their husbands help them. The rest said that it was done by other members of the family. Fuel collection for cooking is done by the women in the family. Here 53% women workers are responsible for buying provisions from the market. 20% said their husband share the responsibility and the

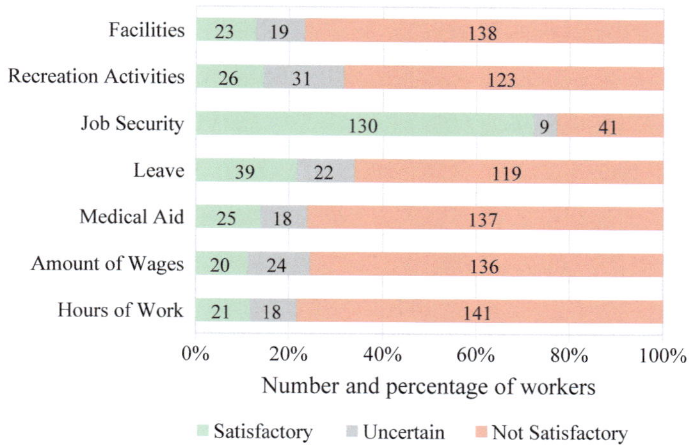

Fig. 5.6 Job satisfaction of women—tea garden workers

rest said that it is done by other members of the family. Fetching water from the nearby jhora is mostly done by women. 54.2% of the women said it is solely the responsibility of the women. However, 25.5% said reluctantly that the responsibility is shared by both. The responsibility of taking care of aged in most of the cases is shared by both male and female. Taking care of sick includes taking the family members to the doctors and hospitals and giving medicines to the sick at home. Here most of the women said that the husband and other members of the family take equal responsibility.

5.9 Stress Evaluation

For stress evaluation we have taken 6 questions in the scale of 1–5. It was found out that most of the women workers faced pronounced stress with regards to all six categories of causes. The lowest was the share for general wellbeing, but still almost exactly 50% and the highest for health issues (55.0%). Health issues were also the main factor for extensive stress reaching 36.6% of all respondents. That means that more than 90% of women workers experiences pronounced or extensive stress considering health issues. Family and financial issues are two factors that have the highest shares for causing moderate stress and the lowest shares for causing extensive stress, therefore we can consider them as the least stressful even though the differences are relatively small (Fig. 5.7).

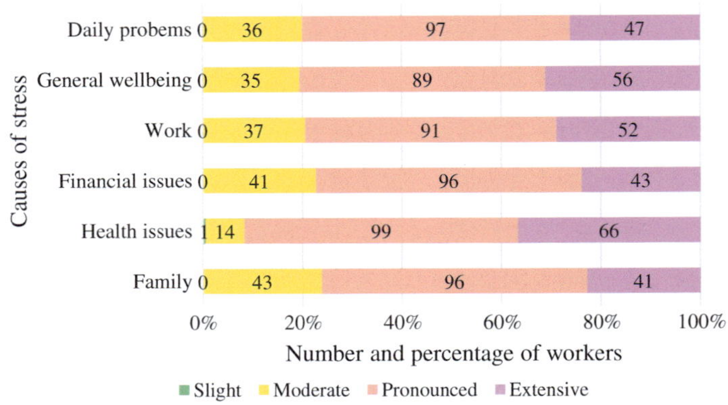

Fig. 5.7 Stress evaluation among women workers

5.10 Overall Quality of Life

In the light of above scenario, an attempt has been made here to examine and analyze the quality of life of the tea garden women workers in the mountain ecosystem of Darjeeling. To evaluate the overall quality of life seven questions were framed in a 1–7 scale. It has been found out that as far as personal life is concerned most of the women are dissatisfied or have mixed feelings about it. This may be because of their personal problems. Some may not be happy with their husbands and some have job related problems. Many of the women are not happy with their job structure as well. More than half of the respondents chose the values that are showing mixed feelings towards all categories of the overall quality of their life (from 52.8%—co workers to 61.1—personal life and household). The highest shares of satisfied and some even pleased is with co-workers (almost one fifth), while the shares of more satisfied at handling with life problems, adjustment to changes and life as a whole are closer to one quarter than to one third. The shares of more dissatisfied with their personal life and husband are the highest reaching 28.9% (10.0% age points above the following share—job structure: 18.9% of the 180 respondents are not very happy with the present job structure (Fig. 5.8). Most of them also believe that they acquire an ability to adjust to changes in their lives. When enquired about their life as a whole 55% said that their lives are a combination of happiness and sadness. They have a mixed feeling towards their lives. However, 28.9% are satisfied with their present life structure. The women prefer plantation work due to various socio-economic compulsions prevailing in labour society and thus they do not prefer to go outside for any other job. The facilities of residential accommodation, firewood, ration facilities, bonus, sick leave, provident-fund; possibility to get jobs to their sons and daughters in near future perhaps keeps them permanently tied to the plantations. The tea workers are tied up in the vicious cycle of poverty. The women workers do not see much hope in their future and seem to be resigned to their destiny.

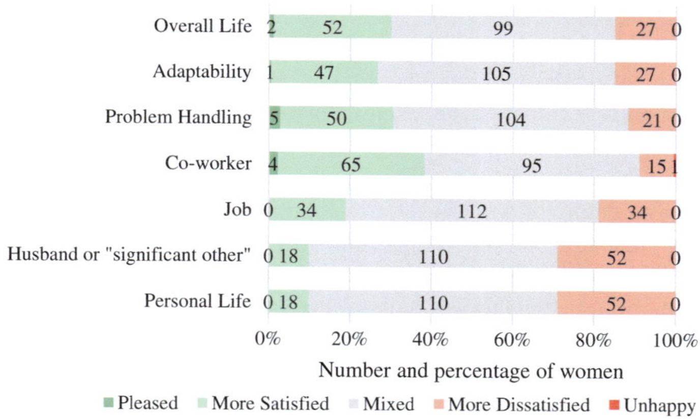

Fig. 5.8 Response of women workers considering overall quality of life

Most of the women workers are of Nepalese origin. They speak Nepali language and majority of them are Hindus. The houses in the labour lines are given by the employers. One worker gets one house, which is supposed to be maintained by the employer. However, generally, the workers themselves do the repairing and maintenance. Living conditions in these houses are unsatisfactory. Typically a single room is crowded with a whole family. The researcher noticed that there are both nuclear as well as joint families. The sources of drinking water are mostly the jhoras. Collection of fuel is also the job of the women workers. Depending on the economic background some of the women workers possess television, radio, furniture and kitchen utensils. The researcher found that most of the houses of the women workers have electricity. Some of the women workers in Makaibari and Jungpana tea estate also use a mobile phone.

According to the Plantation Labour Act 1951, there is a provision of medical care through garden hospitals. When the facilities in the garden hospitals are not adequate the women workers have to go to the hospitals in Darjeeling town. But the fact is that most of the gardens do not have their own hospitals. Makaibari has a dispensary with the necessary equipment and arrangement for visiting doctors who visits the dispensary at least thrice a week. The continuous heavy work in the field and home make women workers physically weak. Moreover, most women workers view that their health problems as a necessary evil and do not think much about how their nature of work and lack of rest affects their health. They also do not take adequate recovery time off as in the absence of clearly enforceable guidelines most workers land up with no pay in the days of their absence. The women workers are ignorant about health care.

Most of the women workers do not read newspapers or any other magazines but some of them are interested in movies. The women workers are mostly unaware of their basic rights especially they are ignorant regarding the importance of education. The awareness about economic entitlements is low among the women workers. They have very low level of awareness about the concept of gratuity.

In most of the families, they take food after satisfying all other family members. So there is the possibility of malnutrition and under nutrition among female workers. In most of the families, the women folk are the last to take food after ensuring all family members have taken their food. Malnutrition is rampant among female workers.

Plucking is the main job in the tea plantations and women are preferred to men for this job. The nature of work is either on permanent basis or on temporary basis. The wage of plantation workers is determined by the government, management and trade union. The women workers in the four plantations studied have to go to work early in the morning. Hours of work are limited to 8 h per day. The workers have to report for duty in the field before 8.00 a.m. They have to work continuously up to 5.00 p.m., except one hour of lunch break. The place of work is generally some distance away from home. The workers mostly have to leave their home before 7.30 a.m. It has been found that stress among the women workers is mainly caused due to health concern, work related problems, general wellbeing and difficulties in coping with daily problems.

5.11 Remedial Measures to Improve the Status of Marginalized Women Tea Garden Workers in a Globalised Village

Most plantation industries although mechanized are still dependent on the skills of workers who pluck the leaves which form the ingredient for the final output. No matter how big or modernized the industry is, the women workers and their skills form the backbone. The future of this industry is dependent on the quantity and quality of the plucking which in turn is directly related to the quality of life of the women tea garden worker.

In the light of the above study and findings a number of recommendations are made to improve the status of the women workers in the tea gardens of Darjeeling District. All estate workers need to work only the legally stipulated 8 h and for additional work double wage is to be paid. The houses should be maintained satisfactorily. Facilities for sewage and drinking water should be satisfactory. Pathways and street lights should be made available. Sewage drains should be constructed and maintained properly. Electricity should be extended to every house. Safe and potable drinking water should be made available. Drinking water which is collected from jhoras in the plantations may contain high amount pesticides due to the use of pesticides in the plantation. During the rainy season pesticides are spread all over and pollute the underground water, therefore, the levels of pesticides should be checked. There must be continuous chlorination of water, maintenance of water filters, pipe, taps, etc. There is also a need to educate the community on safe domestic storage and use of water. There is an urgent need to identify the health care system and requisite medical facilities should be made available to every worker without any discrimination. The medical facilities should be extended to the family members of the workers. Doctor, lady doctor and a qualified nurse should be appointed in the garden hospitals. For health promotion, chlorination of water, spraying of pesticide for mosquito control and the control of leaches etc. at appropriate intervals should be carried out. Health awareness is also very low among the women workers due to their illiteracy. Hence, it is necessary to educate the workers on the need for keeping the household clean, the importance of good sanitation to health and the kind of food which provide better nutrition. It is also essential to develop a data base of the sick workers. The data should contain the causes of their illness, frequency of the illness, number of man days lost due to illness, amount of absenteeism, etc. Specialised treatment should be available in the garden hospitals. Women workers must be educated to take food which is needed for health and their body. Their notion is that nutrients are found in costly food items only. This idea should be changed and there is a need to make them realize that even low cost items can provide them nutrition. The system of family employment in the plantation has made this sector the largest employer of women workers. Majority of these women workers are married and in their reproductive age group. There is a possibility of higher birth rate in this sector as most of the women are illiterate. Family planning programmes should be widely spread in this sector. There should be formation of recreational clubs and both indoor and outdoor games can be arranged in these clubs. Tea gardeners must be made aware of their basic rights and needs, governance, labour law of the country and gender equity through motivation. There is an urgent need to educate the community about personal hygienic practices and environmental sanitation, taking preventing measures against commonly prevalent diseases, health risk associated with alcohol and tobacco, appropriate child-rearing practices etc. All in all, there is a crying need of the involvement by the government in ensuring the welfare facilities and rules are in place. Active information campaigns by NGO's and the government on the rights of the tea garden worker will also help in creating awareness amongst the workers and ensure that they are not easily exploited.

5.12 Testing of Hypothesis

To see if our prediction that there are no major differences in health status of women workers among four plantations we tested the following hypothesis:

H0: Health problems of women workers are independent of different locations.
H1: Health problems of women workers are not independent of different locations.

We used Chi-Square test of independence to test the above hypothesis (Table 5.1).

Table 5.1 Chi square test results and measures of the degree of association for health status

103.90	Chi-square
24	df
6.45E−12	*p*-value
0.499	Phi coefficient
0.446	Coefficient of contingency
0.288	Cramér's V

Table 5.2 Chi-square test results and phi coefficient for education

4.29	Chi-square
12	df
1.36E−05	*p*-value
0.496	Phi coefficient

The above test (Table 5.1) almost undoubtedly implies that the differences among locations are statistically important and according to Phi coefficient we can say that there is weak correlation between the location and health status of woman workers.

The second hypothesis that we tested was:

H0: Education of women workers is independent of different locations.
H1: Education of women workers is not independent of different locations (Table 5.2).

We can consult the same as in previous case. The education is in weak correlation with location. In Makaibari two thirds of interviewees had secondary or higher secondary education while in other location only one third or even less in case of Sourinee (only one fourth) Table 5.2).

5.13 Conclusion

To understand the status of the marginalized women workers in the tea gardens, two most important indicators viz. **health** and **education** have been assessed using chi-square test.

In this study we have documented the major diseases suffered by women in the tea gardens. Almost 96% of the total sample population suffered from one or the other health problems in the last one year. Most of the sufferers consulted doctors or some medicinal practitioner in this connection. The issue of health in Darjeeling Hills has to be seen in the context of geo-environmental set up of the region. The geographical locations of human habitations and climatic constraints in the region have a deep bearing on the overall health of people in the region. To make the situation bad, irresponsibility on the part of health system with respect to proper health planning often exasperates conditions of health. Majority of the women workers are devoid of proper health centres and medical facilities. Tea companies and other legal custodians of tea gardens/estates are legally responsible to protect the health of the workers as per the Plantation Act 1951. Many of the health problems as listed in the above table are the results of the improper health planning and medical facilities in the area. Although people across the garden villages are quite aware of medical advice to be taken in recent times, poor accessibility of health centres often proves to be costly. Moreover, tea management in the pretext of falling production, productivity and market value of Darjeeling tea, in recent times, has withdrawn many of the health problems.

The Medical Officer of the Makaibari garden is of the opinion that female workers are more prone to different kind of diseases than man. Addiction to rice beer is the major cause of the diseases like influenza, sinusitis and various other kind of skin diseases for which they take treatment from garden hospital. Unhygienic living conditions and ignorance further aggravate the health problems. Cases of deficiency of vitamin, night blindness and anaemia are acute.

The second most important indicator is educational level. This may retard or accelerate the growth of individuals. Here, it has been found to be very low. And different for different tea gardens which may be due to the difference in the economic background of their paternal families which enabled some and disabled some to gain education.

From the study it can be concluded that the women workers in the tea gardens are socially and culturally cut off from the mainstream of national life. On the whole the study clearly throws out the fact that a lot needs to be done to improve the quality of work and life of the tea garden workers. On every parameter lie a scope for progress and the need for immediate improvement. What has been observed are the facts, that a commitment to the workers from the tea garden owners in particular and the government at large will bring about the requisite change without a major increase in the cost. Frequent communication between the women workers of tea gardens and mainstream society should be arranged, so that the mainstream people do not undermine the status of the women of tea garden workers.

Tea is a product which is there to stay and probably keep growing. This study had highlighted the disadvantaged position of the women worker in the tea industry and the exploitation faced by them. Empirical evidence shows that the management violates most of the provisions of fair and compensatory employment t. With one of the oldest industries still surviving and prospering it is imperative that the women work force concerns need to be addressed immediately. The development of this marginalized tea garden women will ensure a future to the industry which till date has exploited their main pillar of sustenance and survival.

It is recommended that there is a need to increase productivity, reduce wastage and increase the output per day per acre. At present the production of tea in this part of the globe has reduced and if the production does not increase the industry cannot flourish. The tea industry which is heavily dependent on the women tea workers need to realise the fact in the bright and peaceful future of these women workers lies the future of the tea industry.

We can assume that the better the industry takes care of its women workers, the better will be the growth and profits of the tea industry. Mergers of weaker tea gardens into the larger tea companies, clear marketing plans and productivity based remuneration will go a long way in improving the conditions of these workers.

Moreover, the involvement in creating common infrastructure facilities like hospitals, crèches, schools and colleges will not only create investments but also generate alternative jobs thus, reducing pressure on the tea gardens.

Innovative methods like promotion of tourism in tea gardens and such other measures will also bring in revenues which can be used to fund developmental projects for the welfare of the marginalized women tea garden workers. Therein lies the challenge and the opportunity.

Acknowledgements This research paper is dedicated to the people involved with the tea industry, specially to the women folk of the tea gardens who toil day and night through the good and the bad times with only one hope and dream- that of a bright and brilliant future for themselves and their industry.

References

Banerjee, A. K. (1980). West Bengal District Gazetteer, Darjeeling, Govt. of West Bengal.

Bhat, L. S., et al. (1976). *Micro-level planning: A case study of Karnal area, Haryana*. New Delhi: K.B. Publications.

Bhowmik, S. (1981a). *Class formation in the plantation system*. Delhi: People's Publishing House.

Bhowmik, S. K. (1981b). Labour welfare in tea plantation: An assessment of the plantation labour act. In S. Karotemporal & B. Dutta Roy (Eds.), *Tea garden labourers in North East India: A multidimensional study on the Adivasis of North East India*. Shillong: Vendrame Institute.

Bhowmik, S. K. (1992). Plantation labour act and child labour. *Economic and Political Weekly, 27*.

Bhowmik, S., Xaxa, V., & Kalam, M. A. (1996). Tea plantation labour in India, New Delhi.

Khawas, V. (2006, August 15). Not brewing right: Tea in the Darjeeling hills is losing its flavour. *Down to Earth*.

Malley, L. S. S. O. (1970). *Darjeeling District Gazetteer*. New Delhi: Gyan Publishing House.

Sen Gupta, P. (1954). Women in plantations. *Social Welfare*, *1*(6).

Tea Board of India. (2001). *Techno-economic survey of Darjeeling tea industry*. Tea Board of India: Kolkata.

Tea Board of India. (2002). *Tea statistics, 2004–2005*. Kolkata: Bina Enterprise.

Tea Board of India. (2008). Tea digest, 2006 & 2007, Bina Enterprise, Kolkata.

Tea Board of India (2009). Tea statistics. Bina Enterprises. Kolkata.

Tea Board of India. (2009). 54th annual report, 2007–08, Kolkata.

Chapter 6
Level of Socio-economic Development in the Mountain Watershed of Uttarakhand State (India)

Savita Chauniyal and Devi Datt Chauniyal

6.1 Introduction

The watershed approach has been adopted in the present study for the estimation of level of development within a catchment area. Currently $1000 million is invested yearly in watershed development programmes (WSDP) that are implemented by a range of Departments at the Central Government and State Government level. The Ministry of Rural Development implements the Integrated Wasteland Development Programme (IWDP), the Drought Prone Area Program (DPAP) and the Desert Development Program (DDP). There are also a number of donor funded and research oriented watershed development projects. The goal of most watershed projects is the proper utilization of resources, increase productivity through soil and water conservation, rainwater harvesting, and sustainable development for the upliftment of rural people at the micro watershed scale. Therefore, Bino watershed has been taken as a case study for the assessment of level of development.

Development is a continuous process which never ends. The growing needs of the human population require availability of natural resources, resource utilization measures and level of their utilization. The term development is variously used. Development means quantitative growth and qualitative changes in the social and economic reality of any region. Basically development refers quality of regional system in terms of economic progress, social advancement, political maturity and environmental development (Krishan 1981). This can be identified by the increase in the real production, availability of amenities and services, infrastructural facilities, modernization, better living conditions and increasing employment opportu-

S. Chauniyal
Department of Economics, D.A.V. (P.G.) College, Dehradun, Uttarakhand, India
e-mail: schauniyal@gmail.com

D.D. Chauniyal (✉)
Department of Geography, H.N.B. Garhwal Central University, Srinagar, Uttarakhand, India
e-mail: devidattchauniyal@gmail.com

R. Chand et al. (eds.), *Societies, Social Inequalities and Marginalization*,
Perspectives on Geographical Marginality,
DOI 10.1007/978-3-319-50998-3_6

nities (Babita 2006). The level of development in a region is conceived as the extent of people's command over material and immaterial wealth which optimally used in optimal level of economic welfare (Mishra and Chopra 1979). First hand development is depending upon the existing available resource of any area. In mountainous area, topography is the main factor for retarding utilization of resources and imbalance development (Mamgain 2007a, b, p. 15). In spite of difficult terrain conditions, people adopt the ways which are suitable to environment and utilize available resources for their livelihood. However, the causes of backwardness at the village level or block level are not the same and also the lagging sector in each village is not the same. This requires identification of lagging sectors in each village or village level or regional level planning. In this chapter we present the study that is an attempt in this direction.

The usable resources on the mountain are extremely limited. Only 14% of geographical area in Uttarakhand is available for cultivation. Food insecurity is very common feature in all over the remote area of the region. Water, forest and mineral resources are rich but their utilization is not accessible on mountain people. Besides this, the access to natural resources, though limited, is constrained by lack of access to markets and traditional techniques. Local people face problems mostly restricted by legal, commercial and environmental ones. Most of the income flows out from the hill areas (Papola 2002).

There are evidences to show that in mountain regions a higher degree of diversification among rural households may purely be a distress syndrome 'with little impact on the improvement of household income (Sharma et al. 2001). Mountain areas in Uttaranchal are virtually devoid of any major industry. As a result, the share of workforce employed in manufacturing is abysmally low at 2.5% in the mountain districts of the state. The situation in rural areas is worst.

There are few micro level studies on out migration and rural employment in the hilly region of Uttarakhand. Most of the studies focussed on rural employment for the house hold which also includes out migration. Some studies have focussed on backwardness of the district includes the study of agriculture, labour use, productivity, income etc. Hardly any study has been initiated in the contest of a level of development of Uttarakhand that could exclusively focus on the disparities and diversification of level of development in the rural areas. Surprisingly there is no study based on the primary data on the structure of level of development and its diversification within the state in general and district and village level particular during past or so on. Thus the present study is a modest attempt to fill this gap.

6.1.1 *Objectives of the Study*

The presented study outlines the following objectives:

1. To identify the spatial structure of socio-economic level of development of the study area.

2. To find out the extent of regional disparities in the level of socio-economic development of the study area.

6.1.2 Research Methodology

In an operational sense, there are four basic steps for the identification of level of socio-economic development:

- selection of judicious indicators,
- basic unit of investigation,
- data base generation and
- techniques of data analysis.

Selection of Indicators

Quantitative and qualitative indicators of developments were used in the presented study. The quantitative indicators are numerical indications of development while qualitative indicators include descriptions of living conditions and people's quality of life. The quantitative indicators are economic (GNP/per capita, unemployment rates, energy consumption, percentage of GNP), social, demographic and environmental. Composite or qualitative indicators combine several quantitative indicators into one figure and generally provide a more balanced view of a study area. Usually they include one economic, one social and one demographic indicator.

The selection and the validity of the indicators are crucial in all research problems (Gupta 1978). The selection of indicators depends upon the thinking level of individuals (Raza 1975) and nature of study. Before selecting the indicators it was decided that which indicator will be adequate for the level of development in rural area on the hill. It was considered that indicators should be meaningful in terms of basic need, aspiration and welfare of the people. Haq (1976) only applied gross national product as an indicator for the level of development. United Nations Research Institute for Social Development employed 18 main indicators pertaining to demographic characteristics, consumption, school enrolment, housing, and utilization of services, production and trade for the same purpose. Recently, Babita (2006) also applied social and economic indicators for the analysis of regional disparities in socio-economic development of Upper Ganga-Yamuna Doab. It is therefore, essential that indicators should be chosen in the light of the geographical location of the area. Indicators should encompass the maximum possible domain of the factors that are relevant to the study. Their choice leads to the approximation of the study.

In our study, it was intended to analyze the level of development in the Bino watershed. The development has been studied in five segments i.e. the demography (population density, literacy, female literacy, workers, employed persons, BPL families and unemployment), agriculture (percentages of agricultural land, irrigated land, agricultural density and production), modern means of development (distances of PO, bank, hospitals, water sources, veterinary center, market centre, link road and block head quarter), economic (income pattern) and social development (percentages of electric connection, LPG connection and telephone and mobile sets per house hold) (Singh 1985).

Basic Unit of Investigation

The selection of study unit for socio-economic investigation is depending upon the intensity of the study and nature of work. Such a unit should meet at least three requirements: unit should be homogeneous in term of development level, areal or regional differentiation must be found among the units, and it should have requisite data on developmental indicators. It starts from regional scale to the micro level (household). It is a macro level study so that revenue village units have been taken for the analysis. For the purpose of case study of village, household has been taken as a unit of study. Each aspect of population and economic activities has been studied in 198 villages.

Data Base

Basically, Census (2001) data of each village have been taken as secondary sources. Some aspects of the study like out migration, production, income etc. are based on the primary sources through personal enquiry. Other demographic and economic data are collected from their concerning villages through questionnaire, scheduled, interview and case studies. District Census Handbooks of Almora and Pauri districts (1981, 91, and 2001) provide Tehsil maps with village boundaries and location code numbers. There are 64 villages in Pauri district and 134 villages in Almora district. There are numerous techniques of data analysis for the disparities of socio-economic development such as standard score technique for the data analysis. Another technique uses ranking method for making the data of various indicators comparable. One frequently used method is to categorise data into mean values. Another available method is to arrange the enumeration units in a rank order.

Keeping in view the nature of problem and nature of study area a refined form of rank order technique is used in the present study. All the scores in particular parameter arranged in ascending or descending order and divided into a reasonable number of equal size groups. These categories are subjectively divided into high, moderate and low categories. Techniques of data analysis. Many techniques have been used by researchers in describing the regional disparities in India. Most of them calculate ratio or percentage of indicators and compare them subjectively. Standard scoring technique was applied by Pal (1966), and many others. Gosal and Krishnan (1979) have used ranking method to compare data of various indicators. Another frequently used technique is to categorize score with reference to national or regional mean values, with the units having a value around the mean fall in moderate category, above which falls in high, and below which falls in low.

In view of the limitations of above mentioned techniques and the nature of the problem, ranking technique was applied in the present study. The scores (percentage) of each of the indicators for the 198 villages were arranged in ascending and descending order separately. These were divided into classes and subjectively nomenclature as low, moderate and high. The same method was also applied for describing spatial pattern of composite index. For example, literacy percentage was ranked in descending order because higher percentages of literacy reflect the employment pattern which is also treated as the index of development. Higher percentage of literacy was ranked first and so on. In the case of distances of social development services from the villages like post office, bank, hospital, market, link road etc. were ranked in ascending order. Lowest distances were ranked first which indicates the level development in comparison to longer distances. Finally all indicators' ranks were combined and average value was found for each village. After which an ascending order was assigned to each average value. This composite level of development of villages has been grouped into five classes, ranging in 20 intervals.

6.2 Profile of the Study Area

There are 8 large river basins and about 173 small watersheds and catchments in the Uttarakhand Himalaya (watershed management, Dehradun). Bino watershed is one of the major watersheds of Ramganga river of Lesser Himalaya. It is located in the centre of the Garhwal and Kumaun region reflecting socio-economic characteristics of both. The area is tree shaped covering an area of 298.10 km^2 within 29° 47′ to 30° 2′N latitude and 79° 7′ to 79° 18′E longitude (Fig. 6.1). The average height of the watershed is 2113 m above sea level (Table 6.1).

It is one of the most fascinating agriculturally fertile valleys of the Uttarakhand which is the best representation of agricultural watershed in the region The Bino watershed comprises of altitudinal variations between 800 and 2913 m. The entire study area is divided into six altitudinal zones. The watershed is well linked by motorable roads from all sides. The major roads are Ramnagar-Deghat-Bungidhar, Marchula-Champanagar, Rmnagar-Saraikhet-Uphainkhal Bungidhar, Mehalchori-Nagchula-Bungidhar, Baijro-Uphrainkhal and Pirsain-Masoun-Bungidhar roads. All kinds of infrastructural development are influenced by the means of transportation. The construction of road is very difficult on the mountain slopes and high cost of maintenance prohibits construction of roads in the mountains. Wherever in study area where roads were constructed in early stage, these parts are more developed and prosperous as compared to those where roads were constructed later or are still without any approach road. Education, health, market, employment etc. are other essential factors which enhance the levels of socio-economic development of the region. Besides, the location of central places plays an important role in local

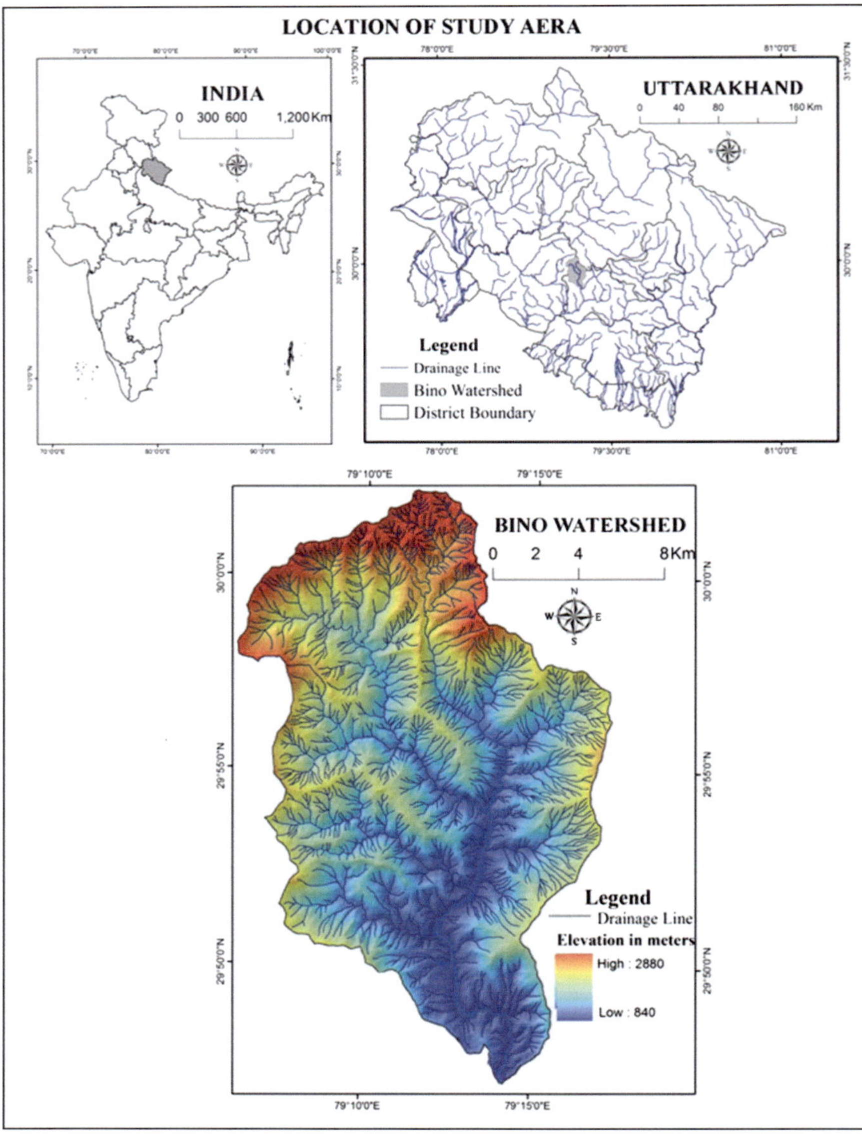

Fig. 6.1 The location of Bino watershed, the study area

development. The service centres are usually found located at the confluence of rivers, junction of two roads, hill gaps (khal) at religious place and on river terraces. The infrastructural activities found concentrated in such central places so that maximum local people can interact and derive benefits. Keeping in view the above situation, following analysis is performed in the study area.

Table 6.1 Some characteristic features of Bino watershed as compared with Uttarakhand state, India

S.no.	Characteristic features	Bino watershed	Uttarakhand state
1	Total geographical area (km^2)	298.10	53,483
2	Total number of inhabited villages	198	15,761
3	Percentage of forest area	26.43	61.14
4	Percentage of cultivated area	45.6	14.19
5	Percentage of irrigated area	5.17	44.91
6	Total population (2001)	45,438	8,489,349
7	Density of population (km^2)	152	159
8	Agricultural density (km^2)	211	
9	Percentage of male	44.44	50.96
10	Percentage of female	55.56	49.04
11	Sex ratio	1250	962
12	Literacy percent (+7 years) %	65.59	71.6
13	Female literacy (%)	51.86	59.6
14	Population growth (1971–2001) (%)	33.56	20.26
15	Birth rate	NA	17.0
16	Occupational structure (%)	51.77	36.9
17	Percentage of scheduled caste[a]	13.25	17.87
18	Land man ratio (per hectare)	0.007	11.92
19	Per capita income (Rs.)	19,657	42,000
20	Districts	2 (Pauri and Almora)	13
21	Block	2 (Thalisain and Syalde)	49

[a]The Constitution (Scheduled Castes) Order, 1950

6.3 Demographic Development

Demographic parameters play significant role in the level of development. The study area falls under two districts (Almora and Pauri) of two regions (Garhwal and Kumaun) in Uttarakhand (Fig. 6.2). It is the most ideal valley which integrates culture of Garhwal and Kumaun region. Being a remotest zone from Almora and Pauri, it is the most backward zone. The developmental processes have started here very late i.e. after 1980. Topographically, lower valley is flat with river terraces which are fertile and irrigated. It is the prosperous valley in the Kumaun region. Beside this upper valley is most backward because of steep slope, deep valleys, unirrigated land and forested land. Animal husbandry is the predominant occupation in this zone. Local peoples are called "Rathi" with primitive life style. The lower valley falling within Almora district is comparatively more developed than the upper part of the valley which falls under Pauri district of Garhwal. Thus, socially and culturally, upper part of the Bino is backward in comparison to lower valley. In the assessment of demographic characteristics, parameters like Arithmetic

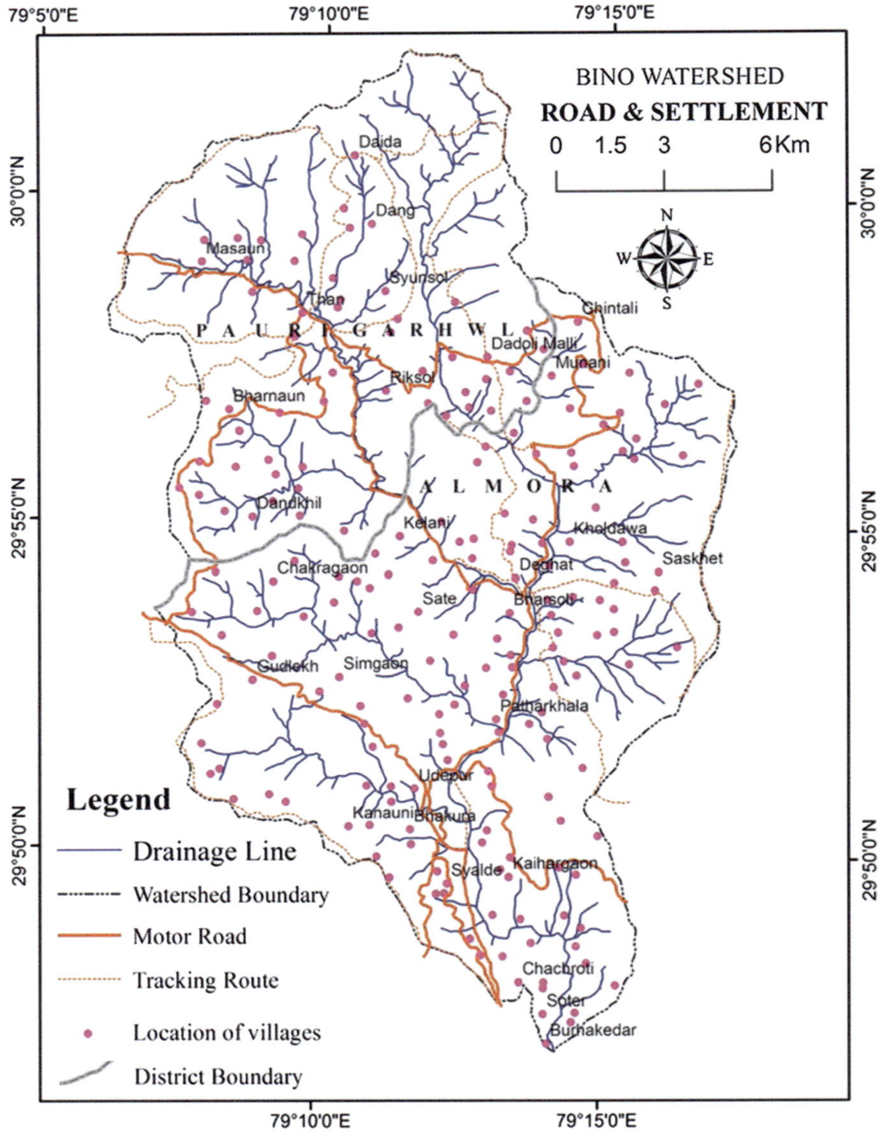

Fig. 6.2 Bino watershed road system and the location of villages

density, male and female literacy, Percentage of workers, employed persons, below poverty line (BPL) families have been taken into consideration to study the level of demographical development. All 198 inhabited villages are allotted ranks in their respective order of the value of different variables. Finally average value has been found out for indicating the level of development.

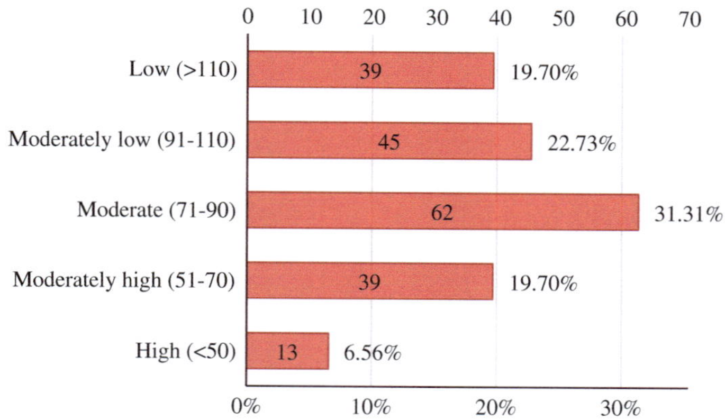

Fig. 6.3 Number of villages and their share in different levels of demographical development 2008–2009

Level of demographical development for all 198 villages is shown in Fig. 6.3. It shows that the smallest share of all villages is under high level of development. High demographic development is in villages Pipaora, Jaspur, Paithana, Udepur, Bungidhar, Bhakura Talla, Timili, Bhanariya etc. Main localities are Syalde near block headquarter, middle valley between Ulmara to Deghat and around Deghat about one fifth of all villages of the study area are under moderately high demographic development which includes 39 villages. Most of these villages are located near the motor road or near the rural service centres. There is a close association between the density of population and other demographical parameters. High population density is associated with high percentage of total literacy and female literacy, high percentage of tertiary workers, and low percentage of workers. Developed infrastructure and moderately gentle topography are primary factors of high demographic development. The area of high demographical development is found in good fertile cultivated areas with easy accessibility.

6.4 Parameters of Agricultural Development

Agriculture is the main source of living in the mountain region. In the assessment of agricultural development, four parameters like Percentage of agricultural land and irrigated land, Agricultural density and Per capita agricultural production have been taken into consideration.

Figure 6.4 shows the agricultural potential of the Bino watershed. Table reveals that maximum share of all villages are under agricultural rank group denoted by moderate level of agricultural development followed by moderately high level of development. Less than one tenth of all villages is under high level of agricultural development. Prominent villages are Lalnagari, Palpur, Basnalgaon, Bharsoli, Bhakura Talla, Paithana, Jaspur, Udepur, Tamadhon etc. Except Lalnagari all

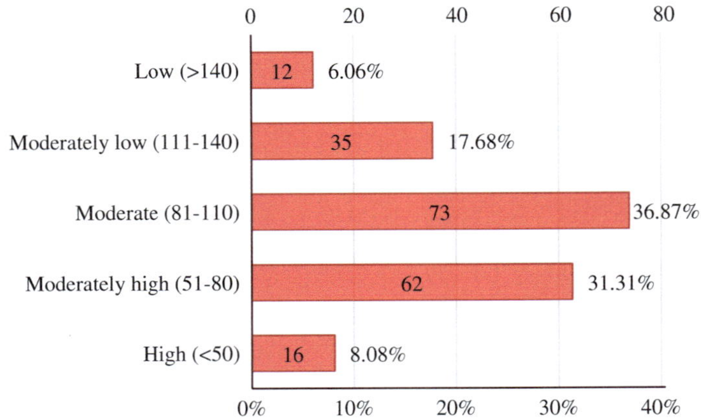

Fig. 6.4 Number of villages and their share in different levels of agricultural development 2008–2009

villages are located in the valley bottoms. Large amount of flat, fertile agricultural land with well irrigational facilities is the main cause of high agricultural potentialities. Low and very low level of development together covers a bit less than a quarter of all villages respectively. Very low level of agricultural development is found in villages Bajori, Bhinsaura Lagga, Chachquarali, Jhipajhabar, Maitandola, Masore, Timilkhet, Jandriya Talla etc. These villages are located in very steep slope conditions. Low share of irrigated land and steep terrain are the main causes of low agricultural potentiality in these areas.

6.5 Infrastructural Parameters of Social Development

In the assessment of Social Development Services ten indicators like Distance of Post Office, bank, hospital, schools, ANME Centre, Drinking Water Source, and Veterinary Centre, Market Centre Link Road and blockhead quarter have been taken into considerations.

Those villages which are nearest to the motor road, school, hospital, market etc., have high level of development. Thus first rank is given to minimum distance value and last to the maximum distance which denote high to low level of infrastructural development respectively.

Figure 6.5 depicts the level of Social Development Services in the area. It reveals that most of the villages are under moderate level of Social Development Services followed by low level of development (about two thirds of villages are in these two classes). Only few villages are under high level of development. The main villages of high level of development are Jaspure, Paithana, Pipaura, Nagargaon and Bhakura Talla. Villages situated near the Block H.Q. and in the lower valley part have the high level of Social Development Services. Very low

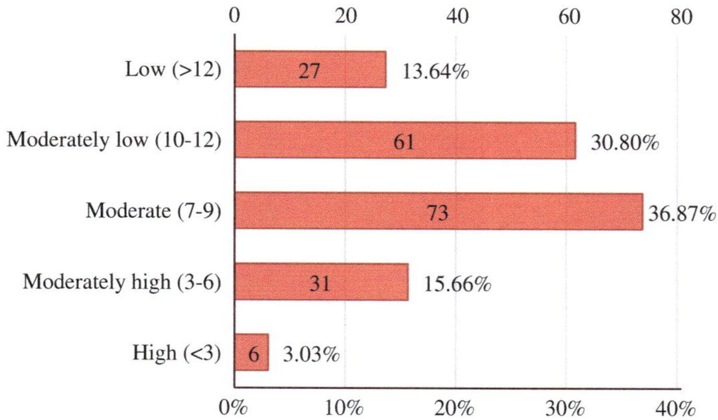

Fig. 6.5 Number of villages and their share in different social development services in Bino watershed, 2008–2009

level of Social Development is found in the villages which are located in the remote areas. Daira, Dharkot, Chantharkhani, Seramandai, Dumanikot, Kaliyalingur, Linguriya, Munani, Bagdiyalgaon, Chintoli, Chaunda etc. are the main villages in which low level of Social Development is observed. Upper part of the Bino watershed shows low and very low level of development. Middle part of the area shows high to moderate level of development just because of early development of transport and communication facilities. Pockets of moderate level also occur in between the rivers Bino and Ramganga. Very low and low level of development areas confirms the low level of overall development in the fringe areas of the north.

6.6 Parameters of Economic Development

In the assessment of economic development, total average income, per capita income and per household income have been taken for consideration. The average ranking of above three indicators is considered to evaluate overall economic in the study watershed.

The explanation of income pattern shows that 43.44% villages are under low level of income while 31.31% villages were under high level of income. Only 25.25% villages have moderate level of income in the study area. Average per capita income of the Bino watershed was estimated about Rs. 19,657. This average was about 1.91 times lower than the national per capita income of Rs. 37,490 in India. It shows that this region is economically very backward. The level of income is higher in those villages having the maximum number of service class people and pensioners. Most of the lower valley villages are under higher groups (>Rs. 25,000) of income while most of the upper valley villages are under the lower group (<Rs. 10,000) of income. It is also clear that the average per household income of the

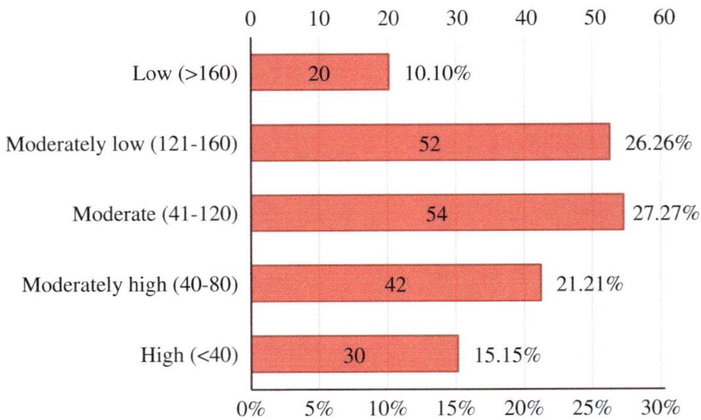

Fig. 6.6 Number of villages and their share in different economic development in Bino watershed, 2008–2009

Bino watershed is Rs. 100,469 per year. Almost similar type of distributional pattern is also found in the household income. The average ranking of three (total Income, Per-capita income and per-household income) income pattern is as follows.

Figure 6.6 shows the distribution of villages under different average income index. On an average 27.27% villages are under moderate level of income development while equal number (36.36%) of villages is under high and low level of development respectively. The distributional pattern of average income rank is considered here as one important parameter of the economic development. High and very high level of income development is found in those villages which are located near the rural service centres. Such villages are Pipora, Paithana, Gaddigaon, Basnalgaon, Jaspur, Jaurasi Kelani, Sarsaon, Bhakura Talla, etc. Beside this low level of development is found in the villages which are the hamlets of large size villages. Such villages are Kaniyal Bakhal, Chippa, Chantharkhani, Nagarkotiya, Bhaiswara, Kaphalgaon, Jandriya, Saogarh, Lambari, etc.

6.7 Modern Means of Economic Development

It was assumed that maximum utilization of modern means and technology reflects the level of high development in a backward region like Bino Watershed. In this assessment three parameters like percentage of electric connection, percentage of LPG connection and percentage of mobile sets/per household have been taken for consideration.

Higher number of electric, LPG gas and mobile connections in the village household are the best indicators of modernization and development of the

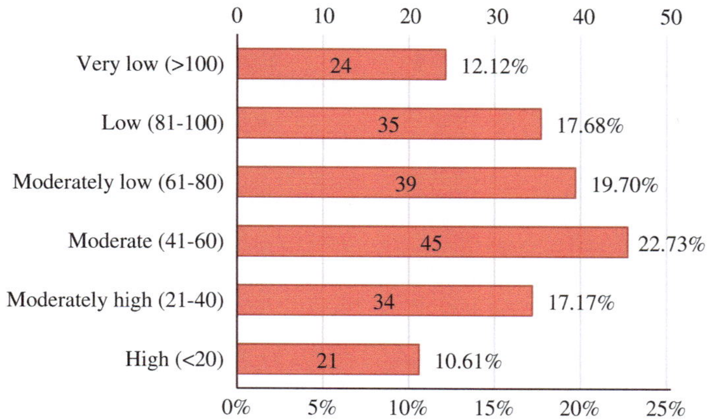

Fig. 6.7 Number of villages and their share under modern means of economic development in Bino watershed, 2008–2009

mountain region. Percentage of each indicator has been calculated in each village and arranged into descending order to calculate the average rank. Higher percentage of household in the villages is allotted first rank and so on. Thus all values of each parameter have been added and average value has been calculated. It shows that lowest score value which indicates a developed village while the highest value indicates a backward village. Finally, ascending order has been allotted to average values which show socio economic level of development of each village in the Bino watershed. Thus, socio economic level of development is categorized into 6 classes (Fig. 6.7).

Most of the high levels of development villages are situated near the Block headquarters in the lower Bino valley. The prominent villages are Basnalgaon, Newalgaon, Bharsoli, Palpur, Chamiyari, Jaspur, Patharkhola, Bungigaon Paithana, Kotsari, etc.

6.8 Composite Level of Development

The higher per capita income in a village does not always mean that its people are better off than those in villages with lower income, because there are many aspects of human well being that these indicators do not capture. In order to assess a better measure of composite economic development, experts use different methods of integrating data on average incomes with data on average health and education levels. These methods make it possible to assess an overall achievements in both economic and human development.

In the development process, geographers study spatial patterns of development. They seek to understand both the economic causes and consequences of varying

Table 6.2 Composite level of development in Bino watershed, 2008–2009

Index class (R)	Level of development	Served population	%
<20	High 16.67%	2959	6.51
20–40		9364	20.61
41–60	Moderate 51.51%	22,231	48.94
61–80	Low 31.82%	9946	21.89
>80		929	2.05
Total		45,429	100.00

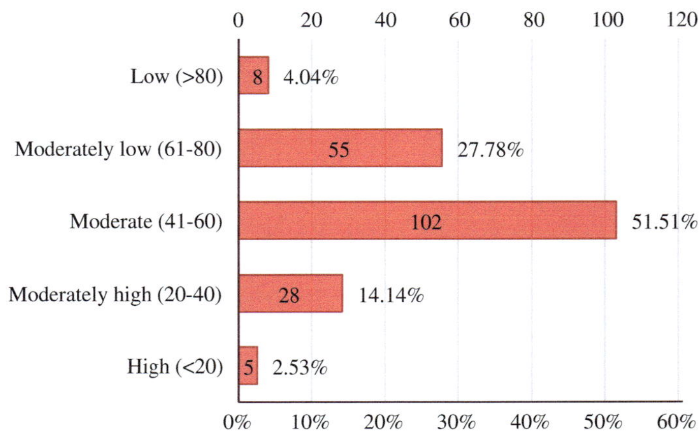

Fig. 6.8 Number of villages and their share under composite level of development in Bino watershed, 2008–2009

development. Based on the indicators already selected in preceding sections, a composite level of development has been worked out in the study area. All above indicator and their ranks were combined and average values have been found out for each village. After that ascending order is assigned to each average value. This composite level of development of villages has been grouped into five classes (Table 6.2, Fig. 6.8).

It is noted that higher level of development covers lower percentage of villages (16.67) but covers high percentage (27.11) of population. Beside this low level of development covers high percentage of villages (31.82) but it covers low percentage (23.94) of population. The basic reason is that most of the developed villages are located in the valleys which have higher population while low level of villages are located on the mid slope or higher ridges with less population.

It is concluded that more than two third villages are either moderate, moderately low or under low level of development while the remaining villages are under moderately high and high level of development in the Bino watershed which indicates that the majority of the area is low or moderately low developed.

6.9 Disparities in the Level of Development

The distribution pattern of high, moderate and low level of development in the study area varies from place to place. Topography, climate, altitude, slope, soil and water sources are the major responsible factors for the uneven distribution of development. These factors are previously explained in detail.

High level of development is found in the villages of Basanalgaon, Bharsoli, Bhakura Talla, Palpur, Paithana, etc. In these villages the value of all indicators is estimated to be very low. Most of the developed villages are located on the river terraces along Bino. Average income level is high in these villages because people are engaged in the tertiary and quaternary activities. The valley villages are well connected by roads and other essential amenities of life are also available here. The distances of roads, hospitals, schools, colleges and block headquarters are less so that they indicate high level of development. Basnalgaon, Paithana, Bhakura Talla, Saraikhet, Bungigaon, Jaspur, Patharkhola and Chintoli are the prominent villages of this category.

The maximum villages of moderate development are found in the central zone and along the mid slopes of the mountain ranges in the Bino watershed. Out of the total 198 villages about 102 are under moderate level of development. Most of these villages are located on the mid slopes or in between the sub watershed and located on the water divides. Most of the villages of Babliya Gad, Basola Gad, Kanchan Gad and Gurana Gadhera's are under moderate level of development. The low level development villages are mostly located in the upper valley zone of Bino watershed. Out of the 64 villages of upper valley (Garhwal region) about 36 (56.25%) villages are under low level of development. It shows that the Upper zone of the watershed which is the part of Pauri Garhwal is very backward by all means. As pointed out earlier that the upper valley has precipitous topography without level land. At the same time longitudinal and transverse ridges, tributaries valleys and coverage of forested area have been the main obstacle in the decentralized planned development. The prominent low level villages are Kaflekh, Jandariya, Sungargaon, Daira, Bhainswara, etc.

6.10 Comparison with State

While looking at economic indicators of Bino watershed, we observe it has very low per capita income compared to Uttarakhand state and national average (Table 6.3). It is more than three times lower than the state average (31.32%) and almost two and a half times lower than the national average (42.50%). The percentage of BPL families is also very high as compared to Uttarakhand state and India. This is true for all genders in the rural areas. According to Planning commission estimates nearly 40% population in the Uttarakhand state was living below poverty line during the year 2004–5, which is very high share as compared to the

Table 6.3 Comparisons of selected development indicators

Indicators	Bino watershed[a]	Uttarakhand state	India	Source
Per capita income (Rs.)	19,657	62,757	46,249	All India Central Statistical office N. Delhi (2009–10)
Share of below poverty line (BPL) (in %)	69.60	39.6	27.5	Planning Commission N. Delhi (2004–5)
Rural (BPL in %)	69.60	40.8	28.3	
Urban (BPL in %)	Nil	36.5	25.7	
Share of workers (in %)	51.77	30.5	32.1	NSS: Employment and Unemployment in India (2009–10)
Literacy (in %)	65.59	72.28	65.37	Census of India, 2001
Share of urban pop. (in %)	Nil	30.55	31.2	Census of India 2011
Unemployment rate (in %)	NA	4.6	9.5	Ministry of Labor and Employment (2009–10)
Sex ratio (per 1000 male)	1250	964	933	Census of India, 2001

[a]Data was collected from primary sources by the authors

national average, but still well below the share of Bino watershed where it is almost three percentage points above two thirds. As per capita income in Uttarakhand is well above the national average this shows severe social inequalities with big gap between rich and poor. The last predominantly live in areas such as Bino watershed.

About 51.77% of the people were workers in 2009–10 which is 21.27% points higher than the state and 19.67 points higher than national average. Out of the total main workers, 55.56% were female. It shows that the male workers have been migrating in search of employment. Out of the total population 13.15% has migrated. Similarly 18.27% families migrated from the watershed till the year 2009 (Chauniyal 2011). The tendency is continuously increasing every year. The high sex ratio (1250 per 1000) also supports the fact that the unemployed males migrated outside the region in search of job. Most of the youths joined the army or allied services.

Majority of workers were engaged in agricultural activities which does not fulfil their basic requirements. Low yielding agriculture is a predominant sector of employment, more so in hill districts of the state. In fact, lack of gainful employment opportunities is one of the major reasons for migration from the hills. It is understandable that poverty is widespread in the rural areas. However, the average size of holdings in hills is very low as compared to plain areas. Similarly, productivity of land is also less than half as compared to plain areas.

In fact natives of Bino watershed are largely facing a problem of working poor. This can be seen, in the relatively low level of development and higher occurence of poverty in Bino watershed. The state government has initiated a number of policy measures at block level to promote the overall development of villages including generation of productive employment opportunities.

6.11 Conclusion

Topography, climate, altitude, slope, soil and water sources are the major responsible factors for the uneven distribution of development from village to village even in the same watershed. These factors are reflected in different combinations and make their impact differently. It is therefore necessary to deal with these parameters while dealing with the developmental studies being conducted in a watershed basin in any mountain region. From this point of view the presented micro level study can be very helpful for implementing the watershed development plan in the mountains.

The following important conclusions are drawn from this study. The social development which is based on demographical parameters is poor in the study area while agricultural development is comparatively high. Infrastructural development is very poor particularly in the remote and high altitude villages. As far as the economic development is concerned, high, moderate and low levels of development are almost equally distributed all over the watershed. The area is far from modern means such as electricity, LPG connection, telephone and mobile services. The lower valley is well developed, mid slopes are moderately developed and high altitude ranges are less developed. Composite level of development indicates that a bit more than half of the villages were moderately developed in comparison with the others in the area, almost one third were low developed and only one sixth were high developed. Per capita income was very low in comparison with the state and national average. Unemployment, migration and disparities in development are serious problem in the study area.

References

Babita, (2006). *Regional disparities in social-economic development of upper Ganga-Yamuna Doab*. Unpublished D. Phil. thesis in Economics, C.C.S. University, Meerut.

Chauniyal, S. (2011). *Population, resource dynamics and socio economic development: A case study of Bino watershed*. Unpublished D. Phil Thesis in Economics, HNB Garhwal University, Srinagar (Garhwal) Uttarakhand.

Gosal, G. S., & Krishnan, G. (1979). Regional disparities in levels of socio-economic development in Punjab. Chandigarh: Department of Geography, Punjab University.

Gupta A. K. (1978). *Regional and dimensional imbalances in agricultural development in Punjab, Haryana and Bihar*. M. Phil Thesis, CSRD, JNU, New Delhi.

Haq, M. U. (1976). *The poverty curtain; choice for the third word*. Delhi: Oxford University Press.

Krishnan, G. (1981). *Issue of social development*. Paper presented to the seminar on Strategy for Rural Development in Western India. ICSSR, Punjab University Chandigarh.

Mamgain, R. P. (2007a). *Employment, migration and livelihoods in the hill economy of Uttaranchal*. Unpublished Ph.D. thesis in Economics. Centre for the Study of Regional Development School of Social Sciences, JNU. New Delhi.

Mamgain, R. P. (2007b). *Growth, poverty and employment*. Working paper no. 39, Institute for Human Development, New Delhi. P4.

Mishra, S. K., & Chopra, A. (1979). Dimension of inter-district disparities in level of development in M.P. *Indian Geographical Journal, 53*(1), 14–26.

Pal, M. N. (1966). Differential regional pattern of consumption of commodities in India. AICC Economi Review.

Papola, T. S. (2002). Poverty in mountain areas of the Hindu Kush-Himalayas: Some basic issues in measurement, diagnosis and ALleviation. Talking points 2/02, ICIMOD, Kathmandu.

Raza, M. (1975). Measurement of regional disparities in developing countries with special reference to India. In R. khan., et al (Eds.), India and Soviet Union. Allied Publishers, Bombay.

Sen, L. K., et al. (1972). *Reading on micro-level planning and rural growth center*. Hyderabad: NICD.

Sharma, H. R., Sharma, R. K., & Kumar, V. (2001). Diversification of rural economy: Effect on income, consumption and poverty. *Asian Economic Review, 43*(1), 107–114.

Singh, R. (1985). A study of integrated area development in Syalde and salt blocks, Almora District (U.P.).

Chapter 7
The Tales of Origin and Settling Process of Marginal Societies in Bhutan

Aditi Chand

7.1 Introduction

Tucked away from the mainstream population of Bhutan, the lesser known societies that exist in the fringes of the country in distant villages are the ones that are replete with tales and historical evidences regarding their origin. Almost every village has a story residing in the public memory. It is these stories of historical, mythical, religious and social value that enhance our understanding of marginalisation of these villages. Important in their own account as they may be, these settlements have lagged far behind in the context of modernisation and globalisation with a constant flux of villagers moving out of these places to larger towns and cities in search of employment and better lives. However, it is essential to keep in mind that each settlement at its beginning provided a fresh beginning and new opportunities to their initial settlers. It is also noteworthy that most of the villages tucked away in the Himalayan landscape of Bhutan are self sufficient and that people have been mobilised in the outward direction only after the effects of modernisation were apparent in the outside world. Globalisation has had a widespread influence in societies throughout the world, bringing the world closer in many respects. Modern technology and the modern ways of livelihood have changed the perspective of everyday life in most of our modern societies altering the basic requirements and needs of people. Socio-political expectations have been modified in keeping with the changing norms. In this context, it is of remarkable interest that we study the tales of origin and settling process of marginal societies in Bhutan. These tales are derived from both mythical and historical factors drawing heavily from local supernatural elements and imagination. At the same time they also provide us with an opportunity to understand the beginning of settlement of these places that lie in

A. Chand (✉)
Department of Literature, Faculty of Arts, Katholieke Universiteit, Blijde-Inkomststraat 21, 3000 Leuven, Belgium
e-mail: aditichand65@gmail.com

R. Chand et al. (eds.), *Societies, Social Inequalities and Marginalization*, Perspectives on Geographical Marginality,
DOI 10.1007/978-3-319-50998-3_7

the fringes of main stream society. It is only with such knowledge that our understanding of any marginal society can be made complete. It will help us to learn from these historical, mythical or fantastical relics about these marginal villages, their co-existence with the outside world treading the path of globalisation.

There is a rich tradition of storytelling in all Bhutanese villages with remarkable repository of tales and legends passed down from one generation to another. A good collection of tales is published by Choden (1997, 2006, 2008, 2009). These stories uncover both the simplicity and complexity of the Contemporary Bhutan. The Centre of Bhutan Studies (2004) also published many such stories related to the customs and traditional festivals celebrated in Bhutan. Chand (2009), Kinga (2004), Rigden and Pelgen (1999) and many others have also followed this folk tradition of Bhutan. While community studies are an expanding area of Studies, more literature is needed to describe how the relationships between various marginal communities evolved. This paper is based on the tales being narrated by the village elders about the origin of their villages.

Bhutan is a country of hamlets enfolded in the deep forests. Ecologically rich but with a difficult terrain many of these villages are disconnected from each other with very little communication between them. For the convenience of understanding different types of settling processes I have demarcated four different types of reasons and stories behind the settlement of these villages. The first is settlement in an area for agricultural and pastoral reasons. The second is villages of historical or religious significance. The third is villages that have strong historical background and the last is stories of migration form one village to the other for various reasons.

7.2 Part I: Villages that Were Born Due to Agricultural and Pastoral Reasons

A mountainous country like Bhutan has its limitation in its arable land. People went far and wide in search of cultivable lands or green pastures for their herds. Completely dependent on their livestock and yield from farms, it was of absolute significance to venture into high altitude pastures and cultivate lands covered by thick forests. Tokshingmang is one such village. It is located in Radhi gewog (block) in Trashigang Dzonglhag (district) in Eastern Bhutan. Its earliest settlers were people who migrated to this village because they discovered this place to have many fruit trees and found the land favourable for agriculture. The earliest settlers are believed to have come from Arunanchal Pradesh in India. The name "Trokshingmang" is also derived from "Trokshing" meaning "a fruit tree" and "mang" meaning "many". This village is also believed to have been the resting place of one of the local protective deity, *Aum Jomo*, of another Brokpa village by the name of Mirak Sakting. The village Radhi located in the same Radhi gewog (block) under Trashigang Dzongkhag (district) in Eastern Bhutan has a similar story where people started to build their houses due to the vast fertile lands that they discovered here. They started with small huts which they visited from their near-by villages, but slowly, this cluster of huts was developed into a full fledged village as more people came to settle down close to their agricultural lands. The village Tang which falls in the famous valley of Tang in Bumthang Dzongkhag (district) in Central Bhutan has a similar story. It was found that it was a very suitable place with ample forest cover for wood and enough plain area to cultivate without the need to build terraces. It was also favourable due to its proximity with other villages of Bumthang for bartering purposes. The tale of the village Kazhi in Wangdue District of western Bhutan is truly associated with agriculture as the Kazhi is a term used for an agricultural village in Dzongkha language. Village Tsamang located in north west of Mongor Dzong (fort) in Monggar Dzongkhag(district) in Eastern Bhutan. The village Tsamang was deemed most desirable for its wide green pastures that attracted many cattle herders to graze their animals. "Tsamang" means "ample grass". The summer pasture bamboo huts were then made into permanent houses as the need for more cattle arose and hence it took the form of a village. The predominantly mountainous environment and a characteristic cold climate afford very few opportunities for the cultivation of crops or exploitation of other resources for the economic betterment of the community. The abundance of high altitude alpine pastures for summer grazing, forest meadows and bamboo grass lands at lower altitudes for autumn and winter grazing gave rise to full scale pastoral activity (Chand 2002).

Thus, we can see that the tales of origin of the above mentioned villages suggest that these villages were identified as fertile lands that would provide the people with agricultural prosperity and so were self-sufficient in many respects until they found themselves left out of the process of modernisation and were left to exist on the margins of the developing country. The otherwise self-sufficient village economy that provided for its people through its internal economy is now only lacking in infrastructure and the plethora of modern facilities and technology. In the areas of difficult terrain, the extreme altitude and harsh climate and where it is hard to grow agriculture crops, cattle rearing seems to be still perceived and favoured as the best occupation. Similar stories are found for the formation of all villages of yak herdsmen living in desolate high valleys of Bhutan known as Bjops in western Bhutan and Brokpas in Eastern Bhutan who came from Tibet centuries ago (Chand 2004).

7.3 Part II: Villages that Grew Because of Historical, Religious or Mythical Significance

The religious history of Bhutan is steeped in myths and beliefs that are largely believed by the people. Certain settlements were thus formed out of this religious belief system and are of utmost religious value to the people. History and religion move hand in hand in the cultural history of Bhutan. Hence, strong historical evidence also suggests that the development of certain villages that became bigger settlements from a small cluster of huts took place because of certain historical figures who passed through these places or lived there for some duration of time. One such village is Nobgang which falls under Talo gewog (block) in Punakha Dzongkhag (district) in Western Bhutan. Village Nobgang was founded by the ninth Je Khenpo of Bhutan, Shacha Rinchen. He was a celebrated abbot. The myth has it that a valuable stone was found here that was luminous and shone like a precious gem. It was a gem or treasure (nob in dzongkha language). The name Nobgang means "treasure hill". Later a temple was also built where this precious stone was kept. Nobgang, also served as a home for great lamas and the Je khenpos (chief Abbot of Bhutan). Nobgang village is also the residence for Royal families. It is a home town for the Queen mothers. Indeed there are lineage connections with Zhabdrung Jigme Dorji (1905–1931), the sixth mind incarnation of Zhabdrung Ngawang Namgyal, who is the founder of Bhutan.

The village Drangla Gompa which is located at twenty kilometres from Palela Pass and about seven hours walk from Trongsa Dzong (Fort) in Central Bhutan was founded by a great monk Lama Oensey Tshering who dreamed of meditating in a cave. When he went out in search of the cave he reached Drangla Gompa. Hence this village came to existence as people began to settle down in this place. Similarly, the village Thinleygang was founded by a monk who came to meditate in this place. This is a place situated in Toepisa gewog (block) in Punakha dzongkhag (district) in western Bhutan. Soon people started to visit him to seek his blessings and so the place soon took the shape of a small village. Thinleygang village is associated with many historical saints like Lam Ngawang Chogyal, Zhabdrung Ngawang Namgyal and Dukpa Kuenley. In the past Zhabdrung Rinpoche always spent a night at Thinleygang Lhakang when he went to Thimphu in summer and winter in Punakha. The practice is carried out still today, the Central Monastic body of Bhutan also make a stay in Thinleygang village while it moves from between the summer residence in Thimphu and the winter residence in Punakha. The village Yurung too was founded by a monk from Tibet, who was travelling in Bhutan and came across a lake in the places near the present Yurung. He built a hut there and soon many other people came there and settled there too. Today Yurung is one of the eleven gewogs (blocks) in Pema Gatshel Dzongkhag (district) in eastern Bhutan. There is one settlement called Neykhang in Mongar gewog and dzongkhag which means "sacred place" it was established at a place where many butter lamps were found. The place assumed religious significance and people streamed into visit this landmark. This lead to many people staying back and living in Neykhang which soon became an important village of religious importance. Nanong is another small village in Pema Gatshel Dzongkhag (district). It was established by clearing a very dense forest, rousing the rage of the local deity residing in the forest. The story says when people destroyed the forest and made Shinglha (the god of tree) angry they all became ill. They asked advice of a monk who told them that they should not cut down any more trees and worship the tree god. This is why even today Nanong has a deep forest surrounding it and its inhabitants are very protective about the woods being of, and taking care not to cut down any trees.

The mythical stories behind the origin of the villages of Hathoe (Bjee block of Haa District, in the western most tip of Bhutan) and Tsangdroe (Metsho block of Lhuentse district in eastern Bhutan) speak of courage and bravery on the part of men who came here to "tame" the land. Agay Hapa, a very fearless man and the great ruler of Haa came to the place which is now the village of Hathoe. The myth revolves around Agay Hapa and how he fought many devils and evils who wanted to destroy the village. He was successful in all his endeavours and the village became a peaceful place to live in. The village Tsangdroe is believed to have had many treasures which were discovered by a monk. Many people followed him here and then stayed back to settle down in Tsangdroe. The people believe that there are still many hidden treasures in their village which shall be revealed only when the right time comes.

It is easy to see that myths play a very important role in the tales of the origin of these villages. The myths are at many times associated with a real monk or abbot who is believed to have supernatural powers. While these people do have historical evidence of their existence, the mythical aspect of the tales adds immense opportunity for analytical and observational purposes. I would like to highlight this point that historic evidence is not the only reason which makes these tales important from the point of view of oral traditions of story telling. It is of great cultural significance to observe that each tale has its characteristic association with an important monk or historic figure who is deemed to be of special powers. So these villages were not in the margins when they were conceived, rather they became important and were visited by many people. It is in the Twenty-first Century that these villages are long forgotten and their tales and traditions remembered only by the village elders.

7.4 Part III: Villages that Were the Pioneering Establishments in Bhutan

Many villages have a strong historical significance in their conception and settlement. They were settled because of political or social reasons keeping in mind their location and proximity to political and social centres. The village Samkhar of Trashigang dzongkhag was established by the sons of the Lhalung Pelgi Dorji, the minister of the King of Tibet from 629 to 649 AD. In their travels they reached Samkhar and settled down there. In time it became an important landmark in the history of Bhutan. Tshatsi, a village from Nanong gewog of Pema Gatshel dzongkhag has a security related reason behind its settlement, it was established so that people could live in small clusters in the Seventeenth Century. It was reckoned to be safer and more secure to live together in groups and so people belonging to the Khoche family came to Tshatsi to settle down in a new place which gave them more opportunity to make their fields and build their houses in a more planned way. The evolution of Trong village located very close to Zhemgang Dzong (fort) in

Zhemgang dzonkhag dates back before the arrival of the Lama (monk) Zhang Dorji Dragpa in the 12th century. The village only had a cluster of huts before his arrival but once he arrived it became a bigger settlement with permanent houses. The story of village Barchham of Trashigang zongkhag is derived from the relic of a Memey Chador Ku (as an important treasure) that was brought to this village from Pema Gatshel. Soon after this statue was brought to the village many people arrived here to pay their homage to the statue. This gave rise to many settlers arriving here to make their homes. The village Zordung was established 250–300 years ago. It has a very interesting tale behind its conception. A cowboy was killed in a nearby village. This gave rise to a lot of discontent amongst the villagers. When the situation could not be resolved a group of families moved out of this village and arrived at Zordung to settle down and cultivate its land. In time many more people joined them and the village became an independent society of its own. It is my observation that the tales of origin like the above offer to us enormous prospects to explore the marginal societies of these villages that were once the centre of trade and political activities. The process of modernisation in these places could rehabilitate them to assume their prior importance in the socio-political scenario of the country. This could be done by re-establishing the traditional structure of agriculture, trade and commerce in these settlements. In doing so, new avenues of livelihood can be created by promoting agriculture and trade among the villagers. Such neglected settlements, with rich and fertile land, could also be rehabilitated by modernising the infrastructure. In time, this could help many marginalised villages to re-emerge as prosperous settlements providing good living to its people and their cultural and historical position could also be restored.

7.5 Part IV: Villages that Emerged Due to Migration

Many villages were not founded by any important monks or royal descendents but simply evolved as people arrived here seeking new pastures or farm land. Thus such villages are very prosperous and fertile and very well planned in terms of agriculture and habitation. Yebsa in Chhubu gewog close to Punakha dzong (fort) in western Bhutan is one such village. "yeb" means "wealth" and "sa" means "place". After the abolition of serfdom by the third king of Bhutan, Yebsa evolved as the village where kheps, the agricultural labourers who were earlier suppressed were settled on farming land and in the village. The village Manthung in Kanglung gewog of Trashigang dzongkhag of eastern Bhutan became a settlement when people found out that its forests were very rich in medicinal plants and trees. They migrated to its forests and open areas and settled down to cultivate the land and explore its medicinal greenery. The village of Mongar situated in Dechheling gewog in Pema Gatshel dzongkhag was discovered by a blacksmith named Memi Jogtang who realised that the gentle slopes of the valley were more suitable for agricultural purposes and migrated here. Many people followed him when they realised the land was more fertile and the village became inhabitable. It is believed that the footprints of this blacksmith are still preserved in a slab of stone. The village Ramchongma of Trashigang dzongkhag was similarly only used for its pastures to graze animals by the nearby villagers but soon the people felt the need to stay closer to the pasture and migrated to the gentle slopes of Ramchongma. The village became well known for its dairy products and the beautiful green pastures. The village Cheaya located under Uzarong gewog of Trashigang dzongkhag has a somewhat different tale of migration from the others. An old couple, Memi Tenzin and Abe Lungten migrated to Cheaya with their relatives in search of new cultivable lands and when they reached Cheaya they found fertile land with water resources close by and settled down here. The people practiced shifting cultivation and so they were always in search of new fertile lands. Many more villages have been evolved in search of new farmlands.

These villages became permanent homes to people who migrated from far away places and assumed important role in the social network of the earlier times. Since they were largely initiated because of their fertile lands and rich ecological resources they were also prosperous and looked up to by neighbouring villagers.

7.6 Conclusion

In these modern times, when we speak of global village with the comfort of people at par with the latest developments in the world in various fields of education, technology, political power, etcetera, it is vital for us to look at the margins of our society and read the responses to globalisation by these local populations. It is also important for us to understand the crux of their societies in terms of social and cultural past and future development. It is also important to see that the margins are created and re-created after every major reconstruction of the society at large and that what lies in the margins today need not necessarily be a part of the margins in the past. That globalisation has shrunk the places and made them very close to one another is true, but it is also true that it has made distant the places that have not been within the reach of globalisation. That globalisation has redefined margins in many societies in keeping to the development of the people vis-à-vis the characteristics of globalisation and modernisation. The margins that have been created now in the society are not the margins that must remain "backward" and under-developed forever but that the margins need to be stretched to include these villages that have now become far flung due to exclusion from the processes of modernisation, like infrastructure and education facilities. Existence in the margins of the people in these villages grows in its intensity due to continual indifference from the rest of mainland society. But in the past a significant amount of development has taken place in these villages of Bhutan and the government is trying to pull them of the oblivion of the margins by introducing them to numerous modern facilities and bringing them closer to global awareness than they were before.

The oral histories of most villages suggest the rise in political power or emergence as a socio-cultural and religious centre for many villages. Through their stories of creation and construction of societies and traditions it is clear that oral history is an important medium to preserve knowledge of these villages, the stories of their emergence and the records of their trade and commerce. This knowledge helps us to understand how societies are created and how, despite being in their prime in some point of time, every successful settlement runs the risk of subsiding to the margins if they do not develop with the changing times and are not able to keep up with the pressing competition that is brought upon it by the ever-changing technological, commercial, political and social restructuring and growth. They also indicate the relevance of development for any society and settlement. Many tales of origin indicate migration for the purpose of either agriculture or trade. Some aspects of globalisation can be seen in its micro form when we read the tales that tell us of migration to increase agricultural produce by farming more land or tales about settlements that are established as centres of trade or centres for religious and cultural activities. These settlements too aim at integrating different societies by serving the purpose of being a centre for any traditional or cultural practice. Like the concept of globalisation, some of these societies too, once thrived on the ideals of integration, trade and communication-social or commercial. Thus, oral historical records about the emergence of now marginalised villages are good indicators of how development plays an important role in the sustenance of a successful settlement. Oral histories help us to understand globalisation and marginalisation through their tales about new and emerging settlements that describe both growth and development. Hence, marginalisation can be understood both in the context of globalisation and modernisation. While modernisation of infrastructure, educational institutes and equal opportunities are yet to reach the people who are still in the fringes of the modern world, the call of globalisation to become one small village too bypasses these societies that are yet to be integrated and readapted into the current tide of the modern world. However, the marginal societies are not to be mistaken as necessarily backward in their knowledge. Oral histories help us to realise the importance of traditional knowledge and that it is critical to preserve and remember this rich tradition of preserving this storehouse of knowledge through the tradition of story-telling. Thus, we learn from these oral histories the significance of evolution for any society as a way to combat marginalisation and be augmented in the progressive drive towards globalisation while preserving the traditional and cultural knowledge that oral histories showcase.

References

Chand, A. (2009). Yeti and the man–tiger: The other worlds of the Bhutanese, Nepali, and Indian Folktales. *Sherub Doennme, 9*(1&2), 31–38.

Chand, R. (2002). Pastoral economy & pasture ecology of Brokpas of Eastern Bhutan. In T. Gyeltshen & Sadruddin (Eds.), *Environment and life support system of Bhutan Himalaya* (Vol. 1, pp. 233–249). Kanglung, Bhutan: Sherubtse College.

Chand, R. (2004). *Brokpas: The hidden highlanders of Bhutan, PAHAR* (164p). Nainital.

Choden, K. (1997). *Bhutanese tales of the Yeti*. Thimphu: White Lotus Press.

Choden, K. (2006). *Folktales of Bhutan* (197p). Thimphu: White Lotus Press.
Choden, K. (2008). *Dawa: The story of a Stray Dog in Bhutan* (Oling Revised Edition Thimphu, 131p).
Choden, K. (2009). *Tales in colour and other stories*. New Delhi: Zubhan. 145p.
Kinga, S. (2004). A *brief history of Chendebji Village and Lhabon Celebration, Wayo, Wayo-Voices from the Past, monograph no. 11*. Thimphu: The Centre of Bhutan Studies.
The Centre of Bhutan Studies. (2004). *Wayo, Wayo-Voices from the Past, monograph no. 11*. Thimphu.
Rigden, T., & Pelgen, U. (1999). *Khengrig Namsum: A historical profile of Zhemgang Dzongkhag* (105p, Integrated Sustainable Programme). Zhemgang, Bhutan.

Part II
Indigenous Communities, Identity, Livelihood Practices and Biodiversity

Chapter 8
Consequences of the Ottoman Land Law: Agrarian and Privatization Processes in Palestine, 1858–1918

Ruth Kark

8.1 Introduction: The Ottoman Land Laws

This paper examines the objectives of the Ottoman Land Code, the Tabu Law of 1858 and the 1867 Law that permitted foreign citizens to acquire urban and rural land, and it will assess their significance, impact, and success or failure in Palestine in the light of its objectives. It will touch only briefly upon the failure of the abolition of the *musha'a* (communal holding of village state land), and issues such as land surveys, systematic mapping, land registration and land settlement (Gavish and Kark 1993; Kark 2004).

The main purpose of the Land Code of 1858 was to define landholdings and categories precisely, abolish the system of tax farming, and consolidate and retrieve the state's rights to its *miri* lands in order to increase agricultural production and therefore tax revenues. *Miri* were the lands belonging to the State, which constituted the bulk of rural land, usually assigned or leased to the local population for the purpose of individual or collective use and cultivation (*musha'a*) with usufruct rights (*tasaruf*) given to the landholder, who had to pay a fee for the title deed for the holding rights or usufruct of the property, the so-called *tapu*. The state maintained ultimate 'ownership' of the land (*raqaba*), while the farmers had 'possession'.

The Code intended to extend and confirm the rights of use, of possession, and of ownership. However, it is doubtful that it aimed to create a body of peasant title-holders. The code entrusted the tax collection to the *tapu* officials and aimed to abolish the *musha'a* system, in which all the inhabitants of the village held the land of the village collectively in shares that were periodically redistributed among them every few years. In that same year, the Ottomans also published the Tabu Law, which set up the system of registering land and issuing title deeds of *miri* lands, and

R. Kark (✉)
Department of Geography, The Hebrew University of Jerusalem, Jerusalem, Israel
e-mail: ruthkark@huji.ac.il

R. Chand et al. (eds.), *Societies, Social Inequalities and Marginalization*, Perspectives on Geographical Marginality,
DOI 10.1007/978-3-319-50998-3_8

a few years later, also *mulk* (private) and *waqf* (religious endowment) lands (Kark 1997a; Kark and Grossman 2003; Solomonvic and Kark 2015). Solomonovich and Kark (2015) argued that privatization began as informal violations of the law, proceeded with the struggle of landholders against authorities who tried to reverse the process, and ended in victory for the landholders after the state acceded to their demands, inter alia, as a result of pressure from foreign nations and their consuls. Thus did de facto land privatization become de jure privatization. They suggested that: "expansion of the usufruct rights of *miri* landholders should be viewed only as partial privatization, first, because the Ottoman state compensated itself for the de facto use of the land as *mülk* by increasing the tax rate and, second, the formal ownership of the land remained in the hands of the state. One must remember that a building constructed on *miri* land was considered *mülk* even if it was built without a permit but if it was later destroyed, the property would revert to its *miri* status, and would require a new building permit."

The Land Law and other processes in nineteenth century Palestine brought about the beginning of land survey and land settlement, land registration and systematic mapping, as well as the creation of new estates, new settlements, and new cities. After the publication of the Ottoman Land Law in 1858, we find in Palestine a trend towards concentration of land for purposes of development or speculation in the hands of families of urban *effendis* (absentee landlords), who lived in the cities of Lebanon, Syria, Egypt and Palestine (Kark 1997a).

The chapter will discuss, albeit briefly, to the main issues related to land privatization following the 1858|Ottoman Land Code, which include: legal aspects, processes and scope of privatization including the private lands of the Ottoman sultan Abdülhamid II, architectural aspects, the establishment of new villages and cities, and the implementation of land surveys, mapping, and land registration.

8.2 Changing Patterns of Landownership in Palestine: Legal Aspects

Increasing the area of cultivated land was indeed one of the main objectives of the 1858 Ottoman Land Code by bringing as much *mahlul* (land uncultivated for over 3 years), and *mewat* (dead) State land under cultivation as possible (Bunton 2000).[1]

[1]Most uninhabited and uncultivated land was defined as 'mewat' (dead) land. According to Article 6 of the Ottoman Land Code, '*mewat*' land was land that was located "at such a distance from a village or town from which a loud human voice cannot make itself heard at the nearest point where there are inhabited places, that is a mile and a half, or about half an hour's distance from such." Likewise, Article 103 defined *mewat* as "dead land … [meaning] vacant (*khali*) land, such as mountains, rocky places, stony fields, *pernallik* and grazing ground which is not in the possession of anyone by title-deed or assigned ab antique to the use of inhabitants of a town or village, and lies at such a distance from towns and villages from which a human voice cannot be heard at the nearest inhabited place". Taken from Kedar (2001): "The legal transformation of ethnic

In spite of the authorization given in the Hatt-i-Humayun Edict (1856) for foreigners to buy land, and the publication of the Ottoman Land Law in 1858, final legal permission was withheld for another decade. On 10 June 1867, a law was published permitting foreign citizens to acquire urban and rural land in all areas in the empire, with the exception of the Hijaz, which was excluded due to fact that Islam's Holy City of Mecca is located there. Permission was granted on condition that foreigners accepted equality with Ottoman citizens, and were subject to the jurisdiction of Ottoman institutions and courts on all questions relating to property. Most of the interested governments signed separate agreements in the years following 1869 to extend this privilege to their citizens. While the law was issued under the pressure of the European powers, it attempted to remove responsibility for land matters from the sphere of influence of the consuls. The purchasers included the churches and missions, the Temple Society (German Templers), private initiators, and Jews (Kark 1984, 1997a).

Although the British Mandate's formal policy was not to introduce large changes in the existing law of the land, such changes occurred, and not only as a result of the interpretation of Ottoman laws by British colonial officials in Palestine (Bunton 1999; Levin et al. 2008, 2010). In fact, the British administration used the existing laws in order to free as large an area as possible to be vested in the state, defining indigenous land rights as narrowly as possible (Bunton 2000; Levin et al. 2008).

8.3 The Process of Land Privatization (Post 1858)

After 1858, we find a dominant and important process of privatization of land ownership and the resulting phenomenon of large agricultural estates and estate buildings in Palestine. This had a long-term impact on the land and landscape. The main catalyst of this process and even of the establishment of new villages and new cities in the Ottoman Empire in general and in Palestine in particular, were the Ottoman Land Laws (Kark and Grossman 2003).

The Land Laws led to a change in the ownership of village lands, particularly in uninhabited regions. Thus, large tracts of State lands that were available for sale changed hands (the "fluid inventory" of land). These lands were purchased by certain groups of people who purchased large plots, and concentrated the lands in their hands. The main purchasers were families of urban *effendis* (absentee landlords), who lived in the cities of Lebanon, Syria, Egypt and Palestine. At first, they did not buy the land for purposes of speculation or prospecting but for developing modern and profitable agriculture and cash crop farms. Eventually they considered

(Footnote 1 continued)

geography: Israeli law and the Palestinian landholder 1948–1967", *New York University Journal of International Law and Politics Quarterly,* 33, pp. 923–1000. This translation of the Ottoman Land Code is based on Goadby and Doukhan (1935): *The Land Law of Palestine*, Tel Aviv: Shoshani Printing Co. (reprinted Holmes Beach: 1998 by Gaunt).

developing the land. This often involved the introduction of new technologies and had a large-scale impact on the landscape. But after a while they sold some of it to other groups who came to settle in the country, including Jews and Christian Churches, agencies, missions and settlers. Whether the increased production also led to increased taxation and state income requires additional investigation (Kark 1984, 1997a, b, 2004; Gavish and Kark 1993; Kark and Gerber 1984; Rubin 1999).

The private investors in real estate included Arabs (both Muslim and Christian) mainly from Syria and Lebanon, Egypt, North Africa and Anatolia, the Ottoman Sultan Abdülhamid II who privately owned land in many parts of the Ottoman Empire and approximately 3% of Palestine, European Christian Churches, and individual Christians and Jews from Europe and America. Urban entrepreneurs (absentee landlords) engaged in a process of concentration of huge tracts of land in their hands (Fischel and Kark 2008; Kark and Frantzman 2010; Amit-Cohen 2012; Galilee and Kark 2006; Galilee 2007).

The investors built large estate building complexes in the form of small khans, with functions such as residence, farm and storage rooms. In some they kept agricultural tenants and developed modern agricultural infrastructure (wells, reservoirs, canals, fences, roads and plans for a railway). One example of such an estate is the "Castle", which still exists today, at the former Arab village of Igzim in the Carmel region. It was built by a local Arab landlord from Igzim as a castle-like stone building. The trend of the privatization of state land increased and influenced the implementation of improved legislation to ensure systematic mapped surveying, registration, and assessment of real estate at the beginning of the 20th century (Kark 1984, 1997a).

8.4 Scope of the Privatization Process

Ownership by effendis facilitated the purchase of part of this land by Jewish immigrants from Europe for settlement in nucleated villages (Grossman 1992). Thus over 52 estate buildings became the initial core of the new Jewish agricultural

Table 8.1 Area and number of estates in Palestine, 1948

Data categories	Palestine
Total area	27,000 km^2
Arable (fertile) land	4000 km^2
Rural estates, according to written sources	200–300 (an attempt will be made to locate all)
Rural estates, location identified by Kark	150
Mansions, existing; total number	50–80
Mansions, historically important (including ruins)	Est. 30–50
Mansions to be examined (including ruins)	~100
Mansions, in-depth-study	7

Table 8.2 Detail from a table reconstructing estates in Palestine

N	Region	Coordinate	Name of the site	Location	Owners—Efendis	Structure	Bib.	Notes
1	Western Galilee	1596/2658	Khan in El Mazraá	North of the Arab village 'El Mazraá	The structure was used as a farm constructed by Abbdala for one of his wives. Victor Gue'rin wrote that the owner of the place was Gasar Basha. Irit Haiman in her work, wrote that the owner was Said Basha	Square structure (37 × 37 m′)	Haiman, pp. 13–14	In the maps of the PEF S.W. P. from 1878, we can find another structure north east from this structure—Kusr Muhammed Bek. We do not have any information about it

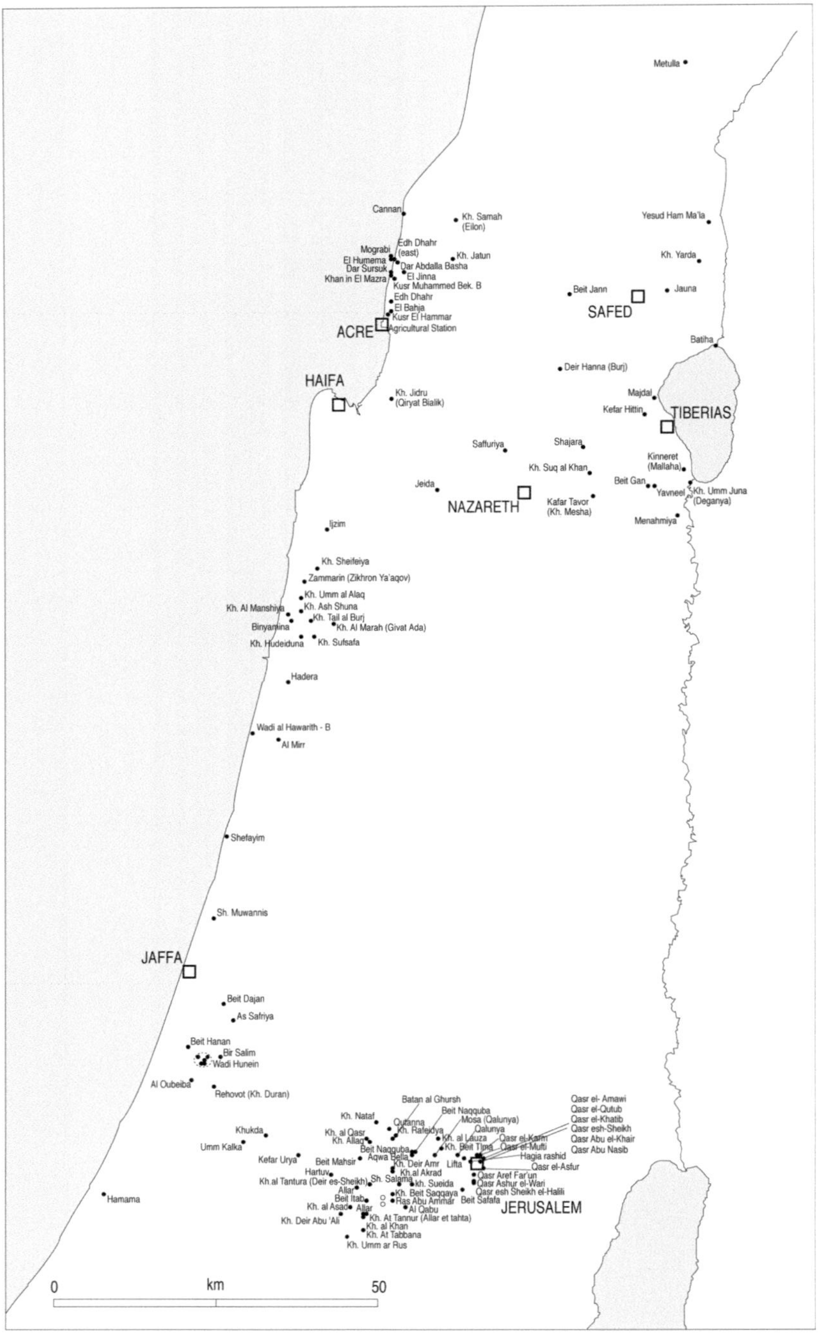

Fig. 8.1 Spatial distribution of estates in Palestine, 1948

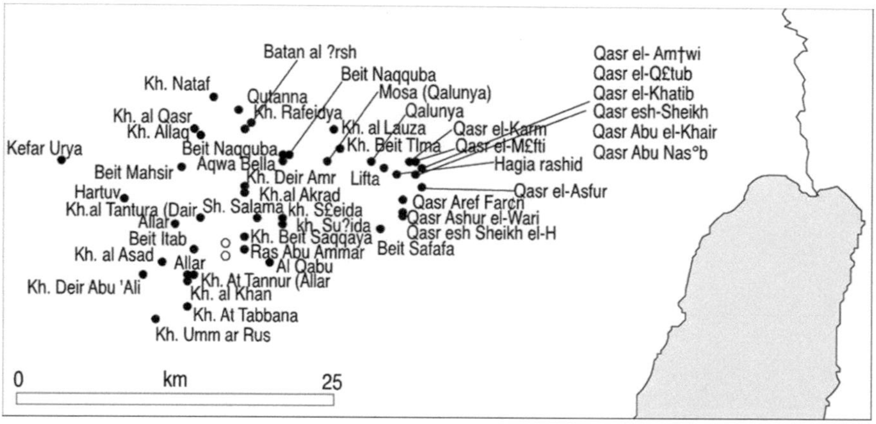

Fig. 8.2 Distribution of estates around Jerusalem, 1948 (Ibid.)

settlement in Palestine (Ben-Artzi et al. 1988; Kark 2001). At least 15 became the initial core of German Christian settlement or religious activity. The trend to privatization of state land increased and influenced the implementation of improved legislation to ensure systematic mapped surveying, registration, and assessment of real estate at the beginning of the 20th century (Kark 1984, 1997a, 2001). By the end of Ottoman rule in Palestine private estates covered over one million dunams out of a total land area of 27 million metric dunams of Western Palestine (1 metric dunam = 1000 m^2), comprising over a quarter of the arable land (4 million metric dunams) (Stein 1984; Kark 2005). This estimate does not include the 0.8 million dunams owned by the Sultan Abdülhamid II, who by 1908 privately owned approximately 3% of the land area of Palestine (Fischel 2006; Fischel and Kark 2008). Land ownership and settlement is shown in Tables 8.1 and 8.2 and Figs. 8.1 and 8.2.

8.5 Policy and Distribution of Sultan Abdülhamid II's Private Lands in Palestine

At the end of the nineteenth Century, the Ottoman Empire faced severe problems which threatened its very existence. Abdülhamid II (ruled—1876–1909), the last potent Sultan of the Ottoman Empire, took various measures aimed at dealing with the Empire's problems and holding back the ever-growing threats introduced by Western powers. Employing both traditional and modern methods, Abdülhamid II aspired to modernize and centralize the State (Fischel 2006).

These initiatives of the Sultan required a substantial fortune which the treasury could not provide. Abdülhamid II, therefore, employed a unique instrument in order to increase his funds. During his reign, he acquired large tracts of lands and became

Fig. 8.3 Abdülhamid II's private lands: distribution according to region (Fischel 2006)

one of the largest landowners in the Empire. He privately purchased extensive tracts of land in different parts of the Empire (northern Syria and Iraq and other regions) (see Figs. 8.3 and 8.4). In Palestine the Sultan purchased more than 800,000 metric dunams, comprising approximately 3% of the country. Those included Palestine's

Jordan Valley around Beisan (close to 400,000 metric dunams, or about 100,000 acres bought in 1879–1882) and Jericho. Those tracts were registered in the Sultan's name in the *tapu* (land registry office) and were considered as his private property. Various measures were taken in those estates aimed at increasing their profitability. The revenues were used to fulfill the goals of the Sultan, for instance to enable his charitable initiatives and the construction of the Hijaz railroad (Fischel 2006).

In 1900 an urban administrative center was established in Beisan, in order to control and settle the Bedouin, and to develop the Sultan's fertile private lands in the region. The maps presented in Kark (2004), were drawn in that context, and transferred to the German expert Dr. Krueger, the Director of Agriculture Affairs, in 1906, in order to enable him to engage in agricultural planning of the area, on behalf of the Sultan. The new railway branch from Dera'a to Haifa, with stations in Sammakh and Beisan, which opened in 1905, was intended to facilitate commercial agriculture. The Young Turks, after the 1908 Revolution confiscated the Sultan's private lands, which then reverted to state ownership (Kark 2004).

However, economic interests were not the only motive which led Abdülhamid II to purchase these estates. During his reign, the Sultan used his private lands to solve some of the crucial problems which threatened Ottoman sovereignty throughout the Empire, and particularly in Palestine. Private lands were used for the settlement of

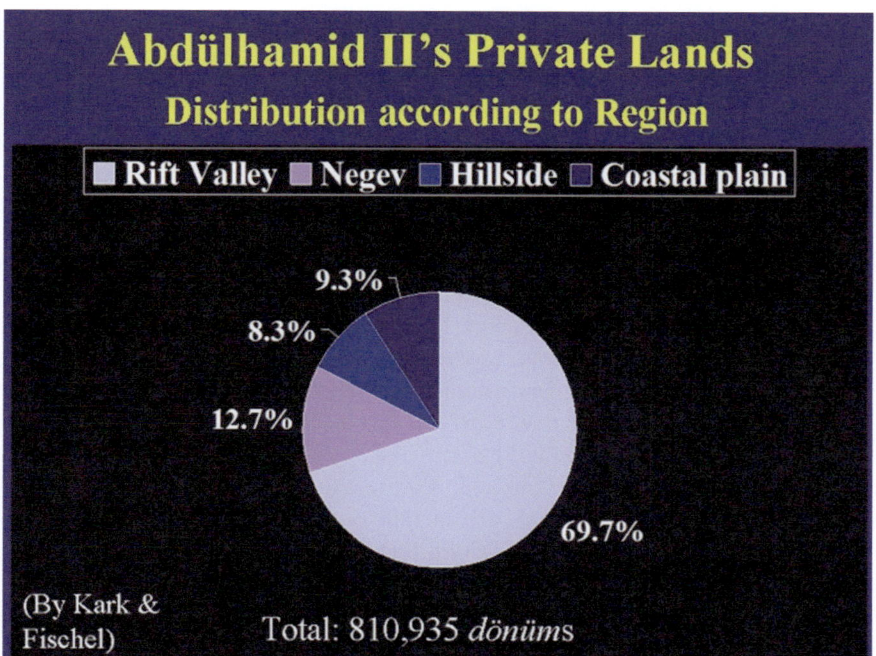

Fig. 8.4 Abdulhamid II's Private Lands: Distribution according to Region (Fischel and Kark 2008)

Bedouins, who were a constant factor of disorder in the country. Muslim refugees from the lost territories of the Empire, especially the Caucasus and the Balkans, were occasionally resettled on the Sultan's estates. In addition, it is evident that the Sultan acquired strategically important tracts, and thus their transfer to foreign control was prevented. Moreover, some of those problems were solved simultaneously, and the settlement of refugees on the Sultan's lands, for instance, was used to gain control on lands and Bedouin groups at the same time. The private lands, therefore, were used by Abdülhamid II as an instrument in his efforts to maintain Ottoman authority and actual control in the Provinces of the Empire (Fischel and Kark 2008).

8.6 The Architectural Aspect

The process of privatization in the agrarian system brought about a new pattern of settlement—estates and estate buildings and facilities which were built around Palestine. The analysis of the architecture provides information about the agricultural forms, functions and technological innovations of the estates (stables, dairies, barns, mills, olive-oil presses, plantations etc.), and the daily life of the inhabitants. It also allows for determining if the estates were run by the owners themselves or by tenants, and for assessing the manpower requirements.

In the framework of this paper only two examples are given:

8.6.1 I The Kfar Uria (Kafruria) Estate Between Jaffa and Jerusalem

The Kafruria estate was built in the second half of the nineteenth century (1874 according to a date in Arabic that appears today over the 2nd floor frame head). This estate was bought and developed according to a few nineteenth century sources by two Christian Arab brothers who were absentee landlords from Jerusalem. The structure, as seen from a map from the beginning of the twentieth century, and from field work done in the building by Kark in 2005–2008, includes a main 2 storey stone structure built in an L-shape by those entrepreneurs as the center of an agricultural estate (around 4000 metric dunams). Today only part of structure is preserved. It is located in the middle of an agricultural settlement, and is not in use (Brill 1978; Kark 2005–2008). Recently, new conservation plans were approved for the site, and restoration work is currently underway (see Fig. 8.5).

Fig. 8.5 Kfar Uria estate building. *Photo* Ruth Kark, July 2008

8.6.2 II The Dubrovin Farm in the Galilee

The estate was bought and built by a Sobotnik Christian family from Astrakhan on the Volga, who settled in Palestine in 1909, adjacent to a Jewish agricultural settlement of Yesud Ha-Maala, near the Hula swamps in the Galilee. The structure comprises a main one storey stone house around a square courtyard with a well and semi-modern irrigation system (around 700 metric dunams) (see Fig. 8.6). At present the structure and agricultural and irrigation installations are preserved. The current use is as a historical museum and elite restaurant.

8.7 Establishment of New Villages

One of the outcomes of the of agrarian change and the privatization process in the period after the Ottoman Land Law was the establishment of new villages by the land lords including the Sultan. Two good examples are:

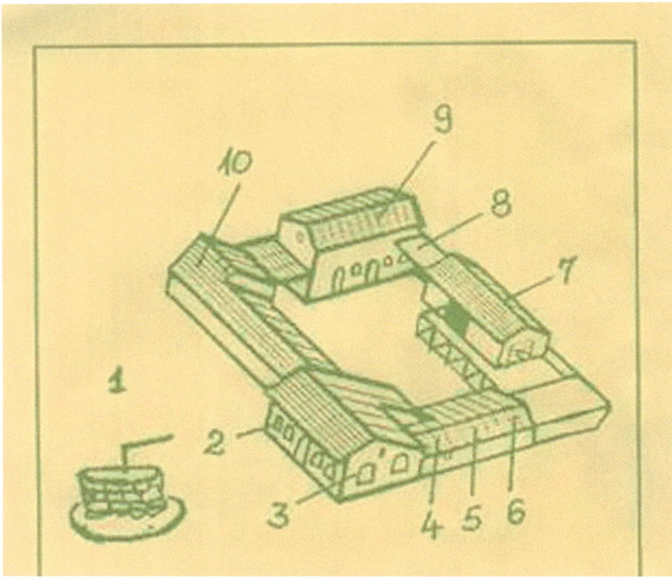

Fig. 8.6 Plan of the Dubrovin family farm house (2–10), central yard and water well (1) plan and legend of the Dubrovin farm (from the museum's brochure)

8.7.1 I The Establishment of Planned Villages on Sursuq's Private Lands

The establishment of villages on the estates of the Sursuq family lands in the Valley of Yizrael (Marj Ibn Amar), where they accumulated in the last quarter of the nineteenth century over half a million metric dunams (Kark and Galilee 2006). The Sursuqs were absentee landlords who owned extensive plots of land (around half a million metric dunams) in Palestine, by the end of the nineteenth century. The Sursuqs built several new tenant villages (including Jalud and Tel el-Fir) in the Valley of Esderalon (Yizrael) for their tenants on their lands. We also have a few semi-modern cadastral maps of several villages in the valley, which testify to the intention of developing these locations. Among these are the Ottoman Plan of Djindjar (1910) and Ruin of Tel el-Fir (Galilee 2007).

An example of the Sursuqs activity is found on two Ottoman period maps which are discussed in detail by Kark and Gerber (1984), and a photograph of a third one (Djindjar), also in the Valley of Yizre'el, which was later found. The first two are general scaled, beautifully coloured maps of the boundaries, area, and cultivation categories of the villages of Semmoune and Ummu-Kbey (1:10,000), owned by the Beirut Sursuq brothers and probably intended for sale. The mapping of the two villages was officially completed by an Ottoman Committee on November 30, 1901, and the maps were approved on November 25, 1902. The plan includes the boundary of the village, and area by land fertility categories.

8.7.2 II The Establishment of Planned Villages on Abdülhamid's Private Lands

The Sultan Abdülhamid II promoted three new planned villages in the Negev on his private lands: Koufaha, El-Muharraka, and El-Jaladia. From analyzing the 1894 map of El-Muharraka, we notice the modern grid plan of the village, which was a new and untypical concept of village planning in the area.[2]

8.8 Soil Betterment and Irrigation on Abdülhamid II Lands

A good example of the intention to improve the soil, dry swamps, and irrigate the land, are the development plans in the Palestine section of the Rift Valley, from the Hula (1894) and Beisan Valleys in the north to the *jiftlik* area (in the Palestinian context, the term *jiftlik* was generally identified with the private lands of the Sultan. Here it relates to a specific area located in the Rift Valley, east of Nablus and in the surroundings of Jericho). In the latter the Sultan built an Aqueduct in Wādī Qelt, Jericho (Fig. 8.7).[3]

Examples elaborating this category—maps of *jiftlik* land, and reclamation and engineering projects of state lands in the Hula and Beisan Valleys were discussed in another study. They do offer a type of cadastral inventory. I have elaborated elsewhere on the Ottoman nature of cadastral surveying which dealt with village units with defined geographical boundaries (Kark 2004).

8.9 New Settlements: Urban Centers

From the mid-nineteenth century onwards we can find evidence to the establishment of new towns, which were to serve as administrative and regional centers. The first mention of such a plan initiated by the governor of the Jerusalem District in 1850 was that of Caesarea.[4] In 1888 the German Templer architect from Haifa, Gottleib Schumacher, prepared a Plan of a Modern New City for the Bosnians in Caesarea (City plan, 1888 by Schumacher). Caesarea was settled by Muslim Bosnians refugees in the 1870s and 1880s.

[2]Ottoman map, 1894, in: Ruth Kark Archive, Jerusalem.

[3]Banāt Ya'qūb/Daughters of Jacob Bridge and a Bridge near Bāysān photographed by Kark (1982). Aqueduct and dedication inscription photographed by Kark (1982).

[4]James Finn doc. In Kark, "Settlement Plans".

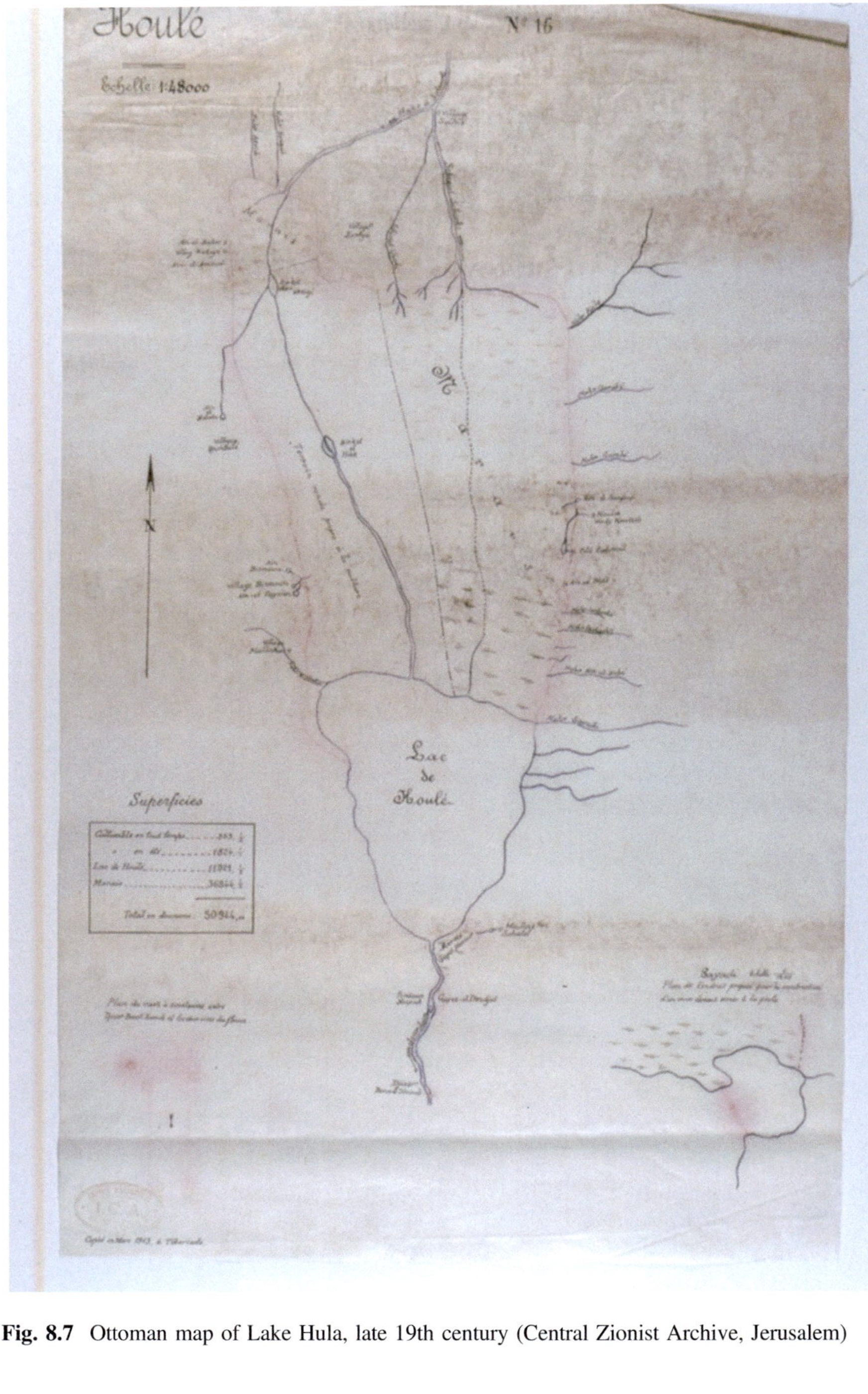

Fig. 8.7 Ottoman map of Lake Hula, late 19th century (Central Zionist Archive, Jerusalem)

As part of a policy to block or settle the Bedouins in southern and eastern Palestine, Transjordan, and the Syrian desert, an urban modern administrative center was established in 1900 in southern Palestine, in the townlet of Beersheba. A court was opened there to settle tribal disputes. In the same year, another new city with a government house (*saray*), mosque, and other government facilities were established in Beisan in north-eastern Palestine. A few years later another new town was planned in Auja el-Hafir, in southern Palestine near the new (1906) border with Egypt. The Ottomans also rebuilt new government centers in the ruins of Jarash and Amman in Transjordan (Kark and Gerber 1984; Kark 1990; Kushner 1995; Avci 2009).[5]

8.9.1 *Abdülhamıd II's Possible Motives and Objectives in Palestine*

- Investment for economic profit; Entrepreneurship
- Development of new urban and rural centers
- Settlement of Bedouins
- Development of rational agriculture.

8.9.2 *The Abolition of the* Musha'a

The main catalyst of the privatization of land and the development of big estates in Palestine was the Ottoman Land Law (1858), which also aimed to end the common land system (*musha'a*). The advantages and disadvantages of the *musha'a* phenomenon and the objectives of the Land Laws in regards to abolishing the *musha'a* and their failure have been elaborated in two additional papers published by Kark and Grossman (2003), Grossman and Kark (2003) elsewhere.

8.9.3 *Land Surveys, Systematic Mapping, Land Registration and Settlement*

At first glance, it might appear as if the Ottomans did not bother to map Palestine, leaving this to zealous French, British, German and American teams of surveyors and explorers. However, Dr. Arthur Ruppin who was head of the Palestine Office of

[5]Beer Sheba ruin, Conder 1874; Beer Sheba new town (established: 1900), Aerial photo and city plan, 1917–18.

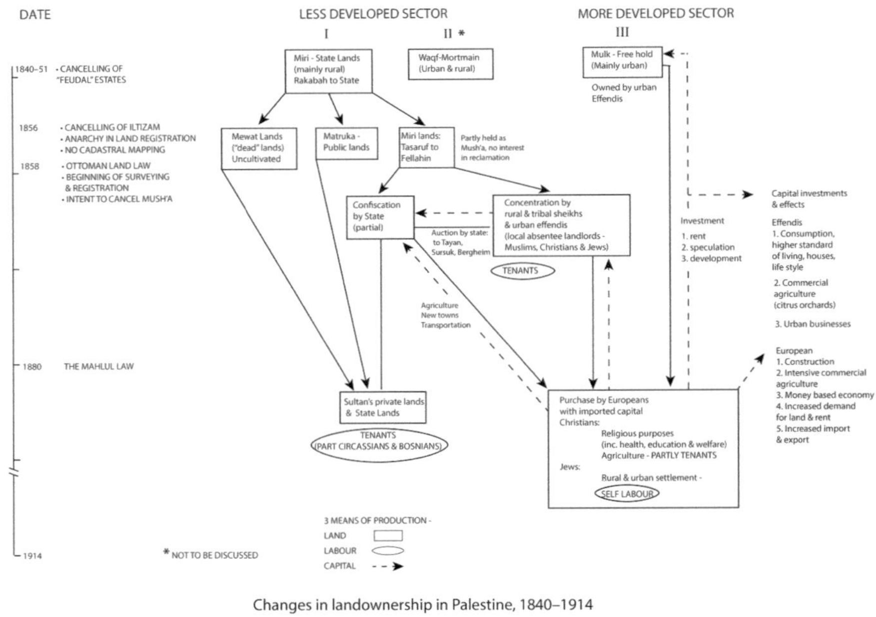

Fig. 8.8 Changes in landownership in Palestine, 1840–1914 (in Kark 1984)

the Zionist Movement wrote at the beginning of the twentieth century that the Ottomans undertook with partial success a general measuring of all the state lands in Palestine. Avitzur, a geographer and Shavit, a historian, in a general description of villages and agriculture in Palestine published in 1983 also wrote (without giving a reference): "The Ottoman regime conducted an extensive activity of land registration in the land registers. This was based on cadastral maps of the different regions. It initiated a systematic mapping of *jiftlik* [Ottoman land held by the Sultan] and *mahlul* lands [lit. vacant—state land that reverts to the state for various reasons—such as not being cultivated by its holder, or the holder having no heirs]."[6]

Kark, Gavish and Gerber uncovered in several Israeli archives a few original maps and copies of large-scale maps drawn by Ottoman officials. Attention in these Ottoman maps was focused on three major categories: first—*jiftlik* lands; second—major blocks of state land, or land owned by absentee landlords, earmarked for sale and purchase; and third—reclamation, improvement or engineering projects. The common denominator of these maps is their individual nature, detached from other or any standard reference system such as a triangulation network, and none were part of a cadastral system (Kark 2004) (Fig. 8.8).

[6]Rupin A., *Der Bodenerwerb in Palästina,* (a brochure published by Rupin around 1908 according to Rupin (1968, I, pp. 96–97); Avitzur and Shavit (1983).

8.10 Conclusion

This study has highlighted the importance of manorial estates and structures (which developed as an outcome of the Ottoman Land Laws) for the historical development of Middle Eastern settlement patterns and land use.[7] Ownership by effendis facilitated the purchase of part of this land by Jewish immigrants from Europe for settlement in nucleated villages (Grossman 1992). Thus over 52 estate buildings became the initial core of the new Jewish agricultural settlement in Palestine (Ben-Artzi et al. 1988). At least 15 became the initial core of German Christian settlement or religious activity. The trend of privatization of state land, including land bought by the Ottoman Sultan, influenced the implementation of improved legislation to ensure systematic mapped surveying, registration, and assessment of real estate at the beginning of the 20th century (Kark 1984, 1997a, 2001).

In summary, the Land Laws led to privatization and estate formation in Palestine which had an impact on spatial change (geographical traits, settlement patterns, building and architecture, landscape), economic development (entrepreneurship, rational economy, new technologies), population and immigration, social change (classes, elites, stratification), and political implications, and conflict, including wide scale Arab and Jewish settlement.

References

Aaronsohn, R. (1990). *Baron Rothschild and the Colonies*. Jerusalem: Yad Izhak Ben-Zvi. (in Hebrew).

Amit-Cohen, I. (2012). *Zionism and free enterprise*. Jerusalem and Berlin: Magnes and De Gruyter.

Atran, S. (1986). "Hamula Organisation and Masha'a Tenure in Palestine", Man. *The Journal of the Royal Anthropological Institute, 21*(2), 271–295.

Avci, Y. (2009). The application of the *tanzimat* in the Desert: The Bedouins and the creation of a New Town in Southern Palestine, 1860–1914. *Middle Eastern Studies, 45*(6), 969–983.

Avitzur, S., & Shavit, Y. (1983). The land and its settlement: Population, settlement and economy. In Y. Ben-Arieh & I. Bartal (Eds.), *The history of Eretz Israel: The last phase of Ottoman Rule (1799–1817)* (p. 61, In Hebrew).

Avneri, A. (1984). *The claim of dispossession: Land settlement and the Arabs, 1878–1948*. New Brunswick, NJ: Transaction Books.

Baier, G. (1990). *Die Gutsanlagen des 17. bis 19. Jahrhunderts in Mecklenburg und Vorpommern als Zeugnisse territorialer, gesellschaftlicher und architektonischer Entwicklungstendenzen.* Zeitschrift Icomos.

[7]Aaronsohn (1990), Amit-Cohen (2012), Atran (1986), Avneri (1984), Baier (1990), El-Eini (2003), Gerber (1987), Grossman (1992), Issawi (1982), Kark (1995), Katz (1994), Karpat (1974), Kayyali (1978), Khalidi (1983), Kimmerling (1983), Owen (1981), Phillip (1988), Poliak (1940), Shafir (1989), Stein (1984), Sasson (2002), Schölch (1993), Warriner, (1948).

Ben-Artzi, Y., Kark, R., & Aaaronsohn, R. (1988). Kahns or estate buildings? The question of the original function of Jewish settlement sites in the Land of Israel Landscape (1882–1914). *Zionism, 13,* 263–283. (in Hebrew).

Brill, Y. (1978). *Yesud Ha-Maala*. Jerusalem: Yad Izhak Ben-Zvi (in Hebrew) [facsimile of 1883], 107.

Bunton, M. (1999). Inventing the Status Quo: Ottoman land-law during the Palestine Mandate, 1917–1936. *The International History Review, 21,* 28–56.

Bunton, M. (2000). Demarcating the British Colonial State: Land settlement in the Palestinian Jiftlik Villages of Sajad and Qazaza. In R. Owen (Ed.), *New perspectives on property and land in the Middle East* (pp. 121–158). Cambridge MA and London: Harvard University Press.

Conder, C. R. (1874). *Drawing of Beersheba desert & ruin, 1874*. Israel National Library, Jerusalem. MS VAR 1209D.

El-Eini, R. (2003). *Mandated landscape, British imperial rule in Palestine 1929-1948*. New York and London: Routledge.

Fischel, R. (2006). Abdülhamid II and Palestine: Imperial policy and private lands (M.A. Thesis). Jerusalem: The Hebrew University of Jerusalem.

Fischel, R., & Kark, R. (2008). Sultan Abdülhamid II and Palestine: Private lands and imperial policy. *New Perspectives on Turkey, 39,* 129–166.

Galilee E. (2007). *Nomadism, land and settlement in the Valley of Yizrael, 1858–1910* (M.A. Thesis). Jerusalem: The Hebrew University of Jerusalem.

Galilee, E., & Kark, R (2006, December). Planned New Arab Villages in the Valley of Izrael. *Paper presented at the Meeting of the Israel Geographical Association*. Tel Aviv, Israel

Gavish, D., & Kark, R. (1993). The cadastral mapping of Palestine, 1858–1928. *The Geographic Journal, 159,* 70–80.

Gerber, H. (1987). *The social origins of the Modern Middle East*. Boulder: Lynne Reinner.

Goadby, F. M., & Doukhan, M. J. (1935). *The land law of Palestine*. Tel Aviv: Shoshani Printing Co. (reprinted Holmes Beach: 1998 by Gaunt).

Grossman, D. (1992). *Rural process–pattern relationship* (pp. 175–176). New York: Praeger.

Grossman, D., & Kark, R. (2003). Communal holding and the economic impact of land privatization. In W. Leimgruber, R. Majoral, C.-W. Lee (Eds.), *Policies and strategies in marginal regions* (pp. 20–34). London: Ashgate.

Issawi, C. (1982). *An economic history of the modern Middle East and North Africa*. London: Methnen & Co.

Kark, R. (1910). *Arab village exercise file; map of Kafruria estate, ca. 1910*. In R. Kark, Archive, Jerusalem.

Kark, R. (1984). Changing patterns of landownership in 19th-century Palestine: The European influence. *Journal of Historical Geography, 10,* 357–384.

Kark, R. (1990). *Jaffa: A City in Evolution, 1799-1917* (pp. 38–39). Jerusalem: Yad Izhak Ben-Zvi.

Kark, R. (1995). *Land and settlement in Eretz Israel 1830–1990. Selected Papers by Prof. Ruth Kark,* Jerusalem: The Land-Use Research Institute, (in Hebrew and in English).

Kark, R. (1997a). Mamluk and Ottoman Cadastral surveys and early mapping of landed properties in Palestine. *Agricultural History, 71,* 46–70.

Kark, R. (1997b). Land purchase and mapping in a mid-nineteenth century Palestinian village. *The Palestine Exploration Quarterly, 130,* 150–161.

Kark, R. (2001). Khans or estate buildings. In N. Rudin (Ed.), *The Khan Museum Catalogue* (pp. 19–23). Hadera: The “Khan” Museum.

Kark, R. (2004). The lands of the Sultan—Newly discovered Ottoman cadastral maps in Palestine. In D. Loupis & G. Tolias (Eds.), *Mediterranean cartographies* (pp. 197–222). Athens: Institute for Neohellenic Research INR/NHRF.

Kark, R. (2005). Napoleon to Allenby: Processes of change in Palestine, 1800–1918. In P. Scham, W. Salem, & B. Pogrond (Eds.), *Shared histories: A Palestinian Israeli dialogue* (pp. 13–61). Jerusalem: Left Coast Press, Truman Institute, Panorama Center, and Yakar Center.

Kark, R. (2005–2008). *Field work*.

Kark, R., & Frantzman, S. (2010). One of the most spectacular lawsuits ever launched": Abdülhamid's heirs, his lands and the land case in Palestine, 1908–1950. *New Perspectives on Turkey*, 145–175.
Kark, R., & Galilee, E. (2006). Privatization of land in Palestine during the end of the Ottoman, and the mandate periods: Land ownership and cultivation of Northern Samaria Landlords in the Valley of Yizrael. In Y. Eshel (Ed.), *Judea and Samaria Research Studies* (vol. 15, pp. 203–214). Ariel: Ariel College (Hebrew with English Abstract).
Kark, R., & Gerber, H. (1984). Land registry maps in Palestine during the Ottoman period. *The Cartographic Journal, 21,* 30–32.
Kark, R., & Grossman, D. (2003). The Communal (musha') Village of the Middle East and North Africa. In W. Leimgruber, R. Majoral, & C.-W. Lee (Eds.), *Policies and Strategies in Marginal Regions* (pp. 223–236). London: Ashgate.
Karpat, K. (1974). Ottoman immigration policies and settlement in Palestine. In I. Abu-Lughod & B. Abu-Laban (Eds.), *Settlers regimes in Africa and the Arab World* (pp. 58–72). Wilmett: Medina University Press.
Katz, Y. (1994). *The "business" of settlement*. Jerusalem: Magnes Press.
Kayyali, A.-W. S. (1978). *Palestine, a Modern history*. London: Croomhelm.
Kedar, A. (2001). The legal transformation of ethnic geography: Israeli law and the Palestinian landholder 1948–1967. *New York University Journal of International Law and Politics Quarterly, 33,* 923–1000.
Khalidi, T. (Ed.) (1983). *Land tenure in the Middle East.* Beirut: American University of Beirut.
Kimmerling, B. (1983). *Zionism and territory: The socio-territorial dimensions of Zionist politics*. Berkeley: Institute of International Studies, University of California.
Kushner, D. (1995). *A governor in Jerusalem: The City and Province in the Eyes of Ali Ekrem Bey, 1906–1908* (pp. 101–116). Jerusalem: Yad Izhak Ben-Zvi, (in Hebrew).
Levin, N., Kark, R., & Galilee, E. (2010). Maps and the settlement of southern Palestine, 1799–1948: An historical/GIS analysis, *Journal of Historical Geography*, *36*, 1–18, and on-line with color maps: http://www.sciencedirect.com/science/article/pii/S0305748809000486
Levin, N., Kark, R., & Galilee, E. (submitted for publication 2008). Application of GIS to verify accuracy of historical mapping: The case of Southern Palestine, 1799–1948.
Owen, R. (1981). *The Middle East in the world economy, 1800–1914*. London: Methuen.
Phillip, T. (1988). The social origins of the Middle East. *Middle Eastern Studies, 24,* 505–507.
Poliak, A. N. (1940). *History of land relations in Egypt, Syria and Palestine: Late medieval period and modern times.* Jerusalem (in Hebrew).
Rubin, R. (1999). *Image and reality: Jerusalem in maps and views* (Vol. 11–16, pp. 149–172). Jerusalem: Magnes Press.
Ruppin, A. (1968). *Chapters of my life, diaries and letters* (Vol. I). Tel Aviv, Am Oved (in Hebrew).
Sasson, A. (2002). From estate house to well house. *A paper presented in the Annual meeting of the Israeli Geographical Association*. Beer Sheva, Israel, December 2002.
Schölch, A. (1993). *Palestine in transformation, 1856–1882: Studies in social and economic development*. Institute of Palestine Studies: Washington, DC.
Shafir, G. (1989). *Land, labor and the origins of the Israeli Palestinian Conflict, 1882–1914*. Cambridge: Cambridge University Press.
Solomonovic, N., & Kark, R. (2015). Land privatization in nineteenth-century Ottoman Palestine. *Islamic Law and Society,* 22(3), 221–252.
Stein, K. W. (1984). *The land question in Palestine, 1917–1939*. Chapel Hill: University of North Carolina Press.
Warriner, D. (1948). *Land and poverty in the Middle East*. London: Royal Institute of International Affairs.

Chapter 9
Negev Bedouin and Indigenous People: A Comparative Review

Havazelet Yahel, Ruth Kark and Seth Frantzman

9.1 Introduction

In the last decade, the use of the term *indigenous* in relation to various population groups became widespread. Most of the studies relating to indigenous people throughout the world stress their marginalization in both the past and the present. We can find examples of these marginal and subjugated societies across continents (i.e. Australia, North America, Europe, Asia and Africa) (Wiessner 1999, 60; Aidi 2016; Al Jazeera 2015; National Post 2016). The concept of indigenous people and white settler states often relates to states where the Europeans became the majority and the "natives" became a tiny group, shunted off to reservations (Frantzman 2014a, b). One may deduce from current situations that "protected" and "special rights" serve to keep these minority groups, such as Aboriginals or Native-Americans, at barely subsistence levels of poverty in primarily wealthy white states. Statistics bear this out. The "first peoples" have become permanently dependent on the government, remain deprived, in poor health and in need of far more health care, and with extensive social problems.

An abbreviated version of this paper was published in the *Middle East Quarterly*, Summer 2012, pp. 3–14. The authors thank *MEQ* for permission to publish the full-length paper (*Ben-Gurion University of the Negev, ***The Jerusalem Post,*** ***The Hebrew University of Jerusalem).

H. Yahel
The Ben-Gurion Research Institute for the Study of israel & Zionism,
Ben-Gurion University of the Negev, Sde-Boker, Israel
e-mail: havatzelety@gmail.com

R. Kark (✉)
Department of Geography, The Hebrew University of Jerusalem, Jerusalem, Israel
e-mail: ruthkark@huji.ac.il

S. Frantzman
The Jerusalem Post, Jerusalem, Israel
e-mail: sfrantzman@hotmail.com

R. Chand et al. (eds.), *Societies, Social Inequalities and Marginalization*,
Perspectives on Geographical Marginality,
DOI 10.1007/978-3-319-50998-3_9

However, the meaning of this term and its applications are less clear. In Israel, the concept of indigenous peoples has been raised recently, mainly in the context of the Negev Bedouin. They are a formerly nomadic, Arabic-speaking group that exists throughout the region, as well as in the southern marginal arid part of the country, the Negev.

In order to examine the compatibility of the term to this portion of a regional group, and due to the absence of a straightforward definition, there is a need to address the general question of indigeneity and its developing meaning and interpretation. This study focuses on the current framework of the indigenous regime. It does not intend to question the justifications of the concept of *indigenous people*, these issues were the focus of other studies (Waldron 2003).

In the pre-modern and modern periods, the term "indigenous peoples" derives from the interaction of different cultures—the meeting between the original inhabitants, known by various names, such as "first nations," "natives," "indigenous," or "aborigines" in a specific region and new, foreign "settlers" or "colonizers." The latter arrived in waves of immigration, bringing with them a normative, cultural, legal, and administrative system foreign to the area. In the end of the fifteenth century the first confrontations between these groups began, and received impetus from the sixteenth century onward as western nations attempted to discover new territories in America, Asia, Africa, and Oceania. The newcomers imposed their alien values and legal systems and way of life on the indigenous population (Anaya 2004, 3). The foreign legal system did not recognize the indigenous groups as having a normative or sovereign system of their own. The outcome of this meeting was the pushing aside of unique, thousands-years-old local cultures and their relegation to limited territories.

Relations between the new settlers and the colonies they established, on the one hand, and the ethnic groups that lived there prior to their arrival, on the other, were structured by diverse means. As far as the European colonizers were concerned the legal status of the rights of the European discoverers, both internally vis-à-vis the population that lived in these territories prior to their arrival, and—of greater importance—externally, between the colonizing nations themselves, was based on the doctrine of discovery. This maintained that sovereignty over and full ownership of a territory belongs to the nation whose nationals discovered it (Miller 2010, 2). This doctrine was upheld multiple times by the United States Supreme Court in the nineteenth century, and courts of additional nations followed suit (Johnson v. M'Intosh 1823; Worcester v. Georgia 1832; St. Catharines Milling and Lumber Company v. The Queen 1887; Mabo and Others v. Queensland 1992) In Australia, for instance, the British Crown used the argument of "*terra nullius*" (an empty land, i.e., an unoccupied territory with no sovereignty or recognized system of rights), so the territory would be considered Crown land (Daes 2001, 11). From the eighteenth century it was argued in courts of various states that the population that lived in a territory *before* the advent of the Europeans held rights. The legal argument focused on the question whether prior to the arrival of the colonizers there had already existed in a specific territory a system of land rights that must be taken into account, and if so, in what manner (Western Sahara, Advisory Opinion 1975).

9.2 The International Arena and Indigenous Rights Regime

The Law of Nations, or, in its more modern terminology, international law, developed in Europe from the seventeenth century onwards, parallel to the emergence of sovereign nation states, with the objective of regulating relations between states (Ben Naftali and Shani 2006, 12, 202–203; McNeil 2008). Traditionally, international law made no mention of group rights, which were considered a domestic concern of the state (Lerner 2003, 112; Ben Naftali and Shani 2006, 195, 205; Sabel and Adler 2010, 241). The 1945 United Nations Charter, dealt with relations between states and did not touch upon domestic issues, including relations between indigenous groups and the state (Lerner 2003, 112; Ben Naftali and Shani 2006, 195). Until the establishment of the UN, defending human rights was considered mainly as a domestic issue, but whereas individual human rights were dealt with in the Universal Declaration of Human Rights 1948 and then in two international covenants of 1966, no parallel change occurred regarding recognition of group rights (UN General Assembly 1948, 1966a, 171, b, 3). Defence of human rights included several issues important to indigenous peoples, such as the right to cultural integrity, and non-discrimination on the basis of language or religion (Article 27 of the International Covenant on Civil and Political Rights) and even the right of self-determination (Article 1 of both covenants). However, these articles did not defend group rights, only those of individuals, and were interpreted narrowly (Sabel and Adler 2010, 241; Wiessner 2009).

International law was reluctant to further group rights for several reasons (Lerner 2003, 111; Dirlik 2003; Khan and Rahman 2009), among them concern for the integrity of the state and fear of separatism that would undermine its stability (Gabizon and Balfur 2005). Furthermore, group rights were considered contradictory to the concept of a modern state based on a direct social contract between the citizen and the sovereign, with no mediation (Hobbes 1651, Part 2, Chap. XIII). Additional reservations were voiced by human rights organizations who argued that one group's rights might come at the expense of those of another group and that group conflicts at times were intended to preserve traditions that sometimes discriminated between individuals on the basis of gender, blood kinship, status, and more (Reuter 2009, 203; Kuper 2003).

As we shall see, over time the idea of group rights for indigenous groups began to emerge. The indigenous groups succeeded in convincing international bodies that due to their unique history they should be afforded special protection under international law (Thornberry 1991, 335; Lerner 2003, 115; Wiessner 1999, 99).

Indigenous societies claimed that their position was unique in view of the great damage to the independent political frameworks that they had maintained from time immemorial, their subjugation to a regime and lifestyle that was foreign to their culture, and the limitation of the physical area in which they were forced to live. This process resulted in difficult social, economic, and mental conditions and their members being relegated to the side-lines. The indigenous people demands, therefore, centered on revoking this injustice (Wiessner 1999, 98), by recognition of the existence of a unique spiritual relationship to the common territory of the group (Daes 2001, 9; Williams 1990, 682). They included a demand to preserve sacred sites, traditional crafts, and customs, as well as to honor pre-existing treaties, to the extent that such had been signed. These societies also insisted on their right to self-determination, whether in the choice of group members or in the wider sense of sovereignty. The rights demanded were for and on behalf of the indigenous group and of a common and collective character.

9.2.1 *Advancing the Cause of Indigeneity in International Forums*

Early attempts by indigenous peoples to bring their case before international forums began in the 1920s. Two examples are Cayuga Chief Deskaheh's journey to protest in Geneva and that of W.T. Ratana (UN Permanent Forum). Their first successes, however, came in the late 1960s and early 1970s when activity shifted from domestic arenas. For instance, the establishment of the National Congress of American Indians in 1944, (National Congress of American Indians website) to the establishment of regional, and later, international organizations. On the international level, the issue of indigeneity was advanced in three major frameworks. The first comprised two covenants adopted by the International Labor Organization, an affiliate of the UN: Convention No. 107—Indigenous and Tribal Populations 1957, and Convention No. 169—Convention Concerning Indigenous and Tribal Peoples 1989 (Lerner 2003, 112; ILO Convention No. 107 1957; ILO Convention No. 169 1989). The conventions did not differentiate between natives and tribal populations, nor were they implemented with much success. The second framework consisted of the efforts of organizations, such as the World Bank to make development loans conditional upon ensuring the rights of natives. This policy started in the 1990s when the Bank began to list indigenous rights as an issue of concern in its dealing with countries, especially in the Third World (World Bank 1991). The third framework *was* via *less formal* actions within various forums of the UN dealing with human rights. This included initiating conferences and promoting study of the topic such as the Conference on Discrimination against Indigenous Peoples of the Americas in Geneva (1977), the World Council of Indigenous Peoples Kiruna, in Sweden, (1977) and the Secretariat of the United Nations Permanent Forum on Indigenous Issues (Center for World Indigenous Studies; United Nations Permanent Forum).

In 1971 the UN Economic and Social Council (ECOSOC) decided to conduct an extensive study of the issue of native populations (UN Session I 1971) and appointed José Martinez Cobo as Special Rapporteur. The results of his study, carried out over a period of about ten years, were published in a series of reports submitted between 1981 and 1986. In 1982, after the submission of one of his reports, the UN Working Group on Indigenous Populations (UNWGIP) was established, charged with furthering protection of indigenous populations and the development of international standards relating to their rights (Williams 1990, 676). Erica-Irene Daes was appointed in 1985 to head the working group to prepare a draft Declaration on the Rights of Indigenous Peoples (Lerner 2003, 115). Work on the draft continued for about a decade, but without states representatives. The Commission on Human Rights, to which the draft was submitted, decided in 1995 to establish a new working group to include representatives of states. The UN General Assembly (GA) declared the years 1995–2004 as the decade of the indigenous peoples and established a permanent forum on this issue within the framework of the Sub-commission on Prevention of Discrimination and Protection of Minorities (UN General Assembly 1993; U.N. GA Resolution 1994). During this period, there was a dramatic increase in the number of groups which attended the forum (Hodgson 2002). The assembly declared a second decade on December 20, 2004 (U.N. General Assembly 2004). During the 1990s and 2000s work on the draft declaration continued. Most differences of opinion centered on "sovereignty" (Anaya 2004, 97) and the definition of "indigenous." (Pritchard 1998, 43; Wiessner 2009; Lerner 2003, 112) Since no consensus was reached on this crucial definition, the problem was circumvented by deleting it from the draft (Mills 2002, 57). In 2006, the draft was finally submitted to the General Assembly and approved in 2007 after receiving the support of more than 140 nations. Four nations that voted against it (United States, Canada, Australia, and New Zealand) in time withdrew their opposition. Israel did not participate in the voting (Ha'aretz 2007).

Numerous countries, mainly from Asia and Africa, made qualifying statements regarding their support for the declaration.

During the session at which the draft was adopted, some delegations made some qualifying statements. The United Kingdom, for example, declared:

"[T]he Declaration was non-legally binding and did not propose to have any retroactive application on historical episodes. National minority groups and other ethnic groups within the territory of the United Kingdom and its overseas territories did not fall within the scope of the indigenous peoples to which the Declaration applied" (Pierce 2007)

Various Asian states, that over the years had expressed reservations about the Declaration, voted in favor but delivered clarifying statements. One of their arguments was that they were entitled to decide which groups living within their borders are indigenous, this in view of their specific histories. China, for example, had claimed in the past that its history was completely unique, typified by ongoing immigration over the years that resulted in a mosaic of cultures and peoples. It maintained that there were no indigenous peoples in China, and that a call to grant special status and land rights to certain groups would create new confrontations, undermine stability, and disrupt the delicate balance achieved by arrangements reached over the years (compare with Kingsbury 1998, 434). After the voting, the representative of China declared that it had voted in favour despite the Declaration's faults because these were to be corrected in the future. Japan made it clear that "regarding property rights, the content of the rights of ownership and others relating to land and territory is firmly stipulated in the civil law and other laws of each State. The Declaration would not counteract "civil law and other laws of each State," while Thailand noted that it "does not create any new rights." The Government of Suriname noted that "it has a responsibility to all its constituents … as well as the responsibility to ensure a just balance between the different ethnic groups. Granting special rights to one part of our population may run contrary to the concept of equal treatment." Bangladesh abstains and said that "… indigenous peoples have not been defined or identified in clear terms." (UN General Assembly 2007) And Indonesia with its hodgepodge of ethnicities and languages, argued that "the rights in the Declaration accorded exclusively to indigenous people and did not apply in the context of Indonesia." (UN General Assembly Media Release 2007)

A few Middle Eastern countries that voted in favor of the Declaration also made some statements for the record: "Turkey did not have any people in its territory that could be interpreted as indigenous peoples in the Declaration"; "The protection of the rights of indigenous people around the world was a matter of principle for Iran, although Iran did not have any indigenous peoples, as such"; while the representatives of Egypt and Jordan emphasized that the Declaration did not affect the sovereignty or territorial integrity of states (UN General Assembly 2007). The traditional attitude of African states was that—with but a few exceptions—all their citizens were equally "indigenous." The representative of Nigeria, which had abstained in the voting, emphasized that the indigenous rights "affected the rights of all Nigerians with its more than 300 ethnic groups speaking more than 300 languages."

The position stated by some African and Asian states in their statements during the vote, was consistent with their view that the indigenous concept does not apply to their domestic situation. This view was supported by the report submitted already in 1999 by Miguel Alfonso Martinez, then Special Rapporteur of the UN working group, who expressed his reservations about applying the concept of indigenous peoples to these continents except in a few cases, such as the San in Botswana and the Masai in Kenya. "[T]he term 'indigenous'—exclusive by definition—is particularly inappropriate in the context of the Afro-Asian problematique and within the framework of United Nations activities in this field." (Martínez 1999, para. 78 and 91) He believed that groups that were inappropriate for consideration as indigenous should be referred to UN bodies dealing with minorities, a stand for which he was heavily criticized (Lerner 2003, 112). His view is consistent with a concept that indigeneity is relevant only to countries where there is a "two-stage model" of first inhabitants and colonizers, and is less relevant or completely irrelevant in an environment of multi-stage historical development.

It should be noted that the exclusion of a definition of indigenous populations in the Declaration left the way open for diverse groups to turn to the UN claiming indigeneity and special rights to land and natural resources, even if, prime facie, many of their claims lacked justification for inclusion or caused other difficulties. Examples are the Bahai, the Rehoboth Baster in Namibia, and others (Corntassel and Primeau 2006, 55; Lehmann 2006, 517–8). Due to the great heterogeneity between such groups, friction developed among them, preventing the UN bodies from adequately focusing on the issue of indigeneity. This contributed to low level of of de facto implementation of the Declaration among UN member states.

A detailed analysis of the Declaration on the Rights of Indigenous Peoples (DRIP) is beyond the scope of the present article and shall be dealt with in a future study. We shall only indicate that the Declaration considers an indigenous nation to be a separate political entity with unique characteristics within the framework of the state. These nations have the sovereign right to determine the structure of their institutions, their identity, and their membership (DRIP 2007). Furthermore, the Declaration differentiates between rights accruing to individuals and to the collective body; and of particular relevance to the land claims of the Negev Bedouin, the articles dealing with land rights refer only to the rights of indigenous peoples as a collective body, not as individuals.

9.2.2 "An Indigenous PEOPLE"

The final version of the Declaration, adopted in 2007, did not include a definition of an "indigenous people," mainly because the relevant UN bodies were unable to agree on the matter. However, generally speaking, a definition or at least a clear term of reference is required when relating to those who are eligible for special treatment under law. According to John Packer, these are necessary in three contexts: mutual understanding, sustainable policy formulation, and effective law-making (Packer 1999, 223). Siegfried Wiessner, who argued in favour of including a definition in the Declaration, wrote that the need for an indigenous definition makes "intuitive sense", despite pointing to the difficulties that could arise from one that was overly inclusive or too narrow. An overly inclusive definition, such as leaving it to "self-identification", is problematic. Too narrow a definition, on the other hand, would leave certain worthy groups without the protection of the Declaration and might be exploited by governments (Wiessner 1999, 113). Karin Lehmann also supported a definition of the term and pointed to the problematical situation created by its absence. She argued that lack of a precise definition, together with lumping together of tribal and indigenous groups, undermines the ability of those groups about which there is consensus as to their indigeneity of acquiring "whatever rights inhere in indigenous peoples." (Lehmann 2006, 517) She brings examples of groups claiming indegeneity while disregarding the issue of precedence in time and the circumstances of their arrival in the territory, such as the Afrikaners who came to South Africa about three centuries ago and in the past even had a special status and enjoyed quasi-autonomy. Today they are demanding the right to preserve their unique lifestyle and autonomy in that territory, which is now part of Namibia (Lehmann 2006, 517).

In contrast, there are others, including organizations representing groups seeking recognition as indigenous peoples—such as the African Commission on Human and Peoples' Rights (ACHPR) and the International Work Group for Indigenous Affairs (IWGIA)—who would like to see a broad definition that does not merely focus on aboriginality and would enable many more groups to take part in the indigenous discourse. Erica-Irene Daes was involved in the non-inclusion of a binding and strict definition in the Declaration, maintaining that any definition would be either too inclusive or too narrow. But even she, like all those opposed to an exact definition, admitted that there must be guiding parameters, albeit with some flexibility.

Numerous definitions of indigenous peoples have been provided in literature. Below, we present three examples which include commonly used terms.

One of the most widely quoted definitions was included by José Martinez Cobo in his report:

Indigenous communities, peoples and nations are those which, having a historical continuity with *pre-invasion and pre-colonial societies* that developed on their territories, consider themselves *distinct* from other sectors of the societies now prevailing in those territories, or parts of them. They form at present *non-dominant sectors* of society and are determined to preserve, develop and transmit to future generations their ancestral territories, and their ethnic identity, as the basis of their *continued existence as peoples*, in accordance with their *own cultural patterns, social institutions and legal systems* [authors' emphasis] (Cobo 1987, 29).

Siegfried Wiessner, suggested the following definition:

> … groups traditionally regarded, and self-defined, as *descendants of the original inhabitants of lands* with which they share a strong, often spiritual bond.

These people are, and desire to be culturally, socially and/or economically *distinct* from the dominant groups in society, at whose hands they have suffered, in past or present, a pervasive pattern of subjugation, marginalization, dispossession, exclusion and discrimination [authors' emphasis] (Wiessner 1999, 60).

The final definition of indigenous people is by Ronald Niezen, an anthropologist, and includes the following attributes:

> … descent from *original inhabitants of a region* prior to the arrival of settlers who have since become the dominant population, maintenance of cultural differences, distinct from a dominant population, and political marginality resulting in poverty, limited access to services, and absence of protections against unwanted "development" [authors' emphasis] (Niezen 2003, 19).

From the literature and judicial decisions dealing with this issue as well as DRIP articles, we are able to draw up a list of recurring parameters that are typical of indigenous peoples:

Original Inhabitants: These are descendants of the *first known inhabitants* of a territory, people who "*were there first*" (Maybury-Lewis 2002, 6; UN Conference on Environment and Development 1992).

Time Duration: Indigenous people have lived on the land from *time immemorial*—thousands and even tens of thousands of years. The Australian aborigines, for example, have lived in their territory for anywhere between 40,000 and 60,000 years, while American Indians could claim a history of thousands of years.

Another time-related attribute is that they were there before imperialist or colonialist newcomers arrived (Anaya 2004, 5).

Pre-colonial sovereignty

> ... *Experience of oppression by a foreign culture and legal regime.* While many groups sense their being oppressed, here oppression refers to "colonialism or something like colonialism." (Anaya 2004, 5) This also describes as a clash of cultures. Waldron reviewed with some criticism, the relative aspects of the term which is sometimes used "as doubly relative, so that a people is called indigenous, first in relation to a certain land or territory, and secondly, in relation to some other people, who arrived in the land at a time subsequent to the people now called indigenous... (As one commentator has noted, this second definition more or less treats "indigenous" as synonymous with "colonized.")" (Waldron 2003).

Group Attachment to Land: Indigenous peoples are groups that maintain a unique common relationship, of a spiritual nature, with the land on which they lived or have lived. (Daes 2001, 11) Communal attachment of a spiritual nature to land is reflected in the belief that land is a gift from God (Erueti 2006, 544; Kark 1992).

Distinct, Non-dominant (Marginalized) Populations.

Separate Customary, Cultural, Economic, Social, and Political Institutions.

Self-identification and Recognition by Others as Indigenous.

The differentiation between indigenous peoples and minorities is related primarily to those parameters that relate to the historical dimension (Lerner 2003, 113; Thornberry 1991, 331). This prevents a group of settlers who arrived in the territory a century or even two or three centuries ago and gained control of it—with or without the concurrence of the sovereign—to claim preferential rights vis-à-vis other groups by dint of being "indigenous."

Such approaches have recently been challenged primarily by groups in Africa for whom proving the historical connection is problematic (Hodgson 2009, 7) These groups maintain that the differentiation between "indigenous people" and "minorities" depends on the *subjective aspirations* of the specific group and should be identified as either collective or individual rights: "The crucial difference between minority rights and indigenous rights is that minority rights are formulated as individual rights whereas indigenous rights are collective rights." (Indigenous Peoples in Africa 2006, 13). In the view of Dorothy L. Hodgson, these claims are "constructivist," "structural," or "relational" definitions of "indigenous," as opposed to what she calls "'essential', 'substantial', or 'positivist' definitions that depend on evidence of territorial precedence." (Hodgson 2009, 7).

This section of the article has presented the general framework relating to indigenous peoples and their demands. The next section will focus on a specific case: the development of ruling regimes in the Middle East in general, and Palestine in particular, and the specific case of the Negev Bedouin.

9.3 Middle Eastern Regimes and the Negev Bedouin in the State of Israel

9.3.1 General Background

During the past few years, the Negev Bedouin in Israel have begun claiming the status of an indigenous people. In effect, they are the only group in Israel continuously engaged over several years in furthering their interests through UN indigenous forums (Frantzman et al. 2012. For Palestinian Arabs as indigenous, see: Bishara and Rosenberg 2004. See also, with some reservation: Or et al. 2003). Their claim is based on the un-based argument that in Israel there was a colonialist regime. Thus, in Israel, as in other areas that fell to imperialist powers, domination of the territory by foreigners was accompanied by denial of the rights of the Bedouin and refusal to admit their lengthy presence in the Negev (Forman and Kedar 2003; Shamir 2000. For the Bedouin as an indigenous people, see: Abu Rabia 2002, 202–11; Yiftachel 2009; Amara et al. 2012; Frantzman 2014a, b). This line of argument is consistent with the position of the Arab leadership, voiced as early as the early 1920s, that disparaged the Jewish national revival as an alien, colonial intrusion into the pan-Arab patrimony.

These arguments are both erroneous and misleading (Troen 2008; Aaronsohn 2000, 30–35, 296–99). They choose to ignores other parts of the Middle East which were ruled by foreign powers (i.e. Egypt from 1882), and the Mandates of the post—WWI period. The Bedouin are by no means the only people who can lay claim to the notion of being a "first people" in Palestine. In contrast for instance to British common law in its former colonies, no foreign legal system, was instituted in Palestine/Israel, which was not considered *terra nullius* and to which the doctrine of discovery did not apply. As we shall see, there is a great difference—especially in parameters relevant to the concept of indigeneity—between the history and the development of the legal system in the colonies such as in North America or Australia and in the Middle East, including Palestine (Karsh 2008; Friling 2003).

9.3.2 Governance of Palestine

Many studies have dealt with the history of Palestine and its populations. This ancient territory, in which the State of Israel was established in the mid-twentieth century, has seen many migrations and conquests since antiquity.

One of the peoples which lived there for lengthy periods throughout its history, and even ruled it part of the time, is the Jewish People (Hertz 2011). About three millennia ago the Kingdoms of David and Solomon (also known as the First Kingdom of Israel) was established in territories extending from the Negev in the south to the Golan Heights in the north. At one stage it was split into two kingdoms: Israel and Judah. The northern kingdom of Israel was conquered by Assyria in the eighth century BCE and a portion of its population was exiled. The southern kingdom of Judah which exercised sovereignty over the Negev, continued to exist until it fell to the Babylonians in the sixth century BCE, who destroyed the First Temple and exiled most of the people. The Babylonians were replaced by the Persians who allowed the exiled Jews to return to their homeland in 538 BCE. In addition to them, the land was peopled by Edomites, the remnant of Philistines, and Samaritans (a mixture of Israelites and Assyrian colonists), and some Arab groups, likely the ancestors of those who would come to be called the Nabateans. The Persians were defeated by Alexander of Macedon who brought Hellenistic civilization to the country. At a later date, from 141-63 B.C.E., the sovereign Jewish kingdom of the Hasmonean dynasty was established. With its fall Palestine came under the rule of Roman Empire, with some minor hiatus, for the next seven centuries. When the latter collapsed in the late 5th century, Byzantium was its successor, until the conquest of the country by the Muslims in the seventh century CE. With the Muslim conquest there began an increased movement of Arab tribes into the area. In 1099 the Crusader conquest of the Holy Land began, only to be defeated and chased out by the Ayyubids, whose short-lived domination (1187–1260) was replaced by the Mamluks and finally by the Ottoman Empire in 1517, whose four hundred years rule in Palestine shall be discussed in slightly more detail (Gil 1997, 23–40).

Since its advent in the seventh century, Islam constituted the organizing principle of the socio-political order underpinning the long string of great Muslim empires (Karsh 2006). The Ottoman Empire developed out of nomadic Turcoman tribes who converted to Islam in the eleventh century (Vainshtein 1989, 65; Lapidus 1991). Islam was a unifying element during the shaping of the empire and aided in making the Arab tribes partner to this process. Islamic principles were the framework that brought the tribes together, served as a unifying force for social organization, and invested the empire with political legitimacy, with the sultan-caliph recognized as the religious and temporal head of (most of) the world Muslim community (Lapidus 1991, 42, 44). The tribal lifestyle and customs from their nomadic period also became an integral part of the systems of government and law (Sochowolski 2001, 74). Courts were established throughout the empire that passed judgment according to Muslim Shari'a law. Civil aspects of this body of law were codified in 1876, known as the *Mejelle*, which served as a civil code of law. The earlier Ottoman Land Law of 1858 was also in force (Albek and Fleisher 2005, 7; Friedmann 1975, 196; Banyan 2002, 174; Krämer 2008, 81–87).

In the First World War Britain took control of the territory and in 1922 was appointed as the Mandate administrator for Palestine by the League of Nations, (For the Mandate system, see: Sabel and Adler 2010, 96) with the specific goal of facilitating the establishment of a Jewish national home in Palestine as envisaged by the 1917 Balfour declaration. The British Mandate in Palestine continued utilizing most of the existing Ottoman legal system, including laws related to land. Paragraph 46 of the Palestine Order in Council, July 1922, declared that all Ottoman laws in force at the outbreak of World War I would remain valid (Bernard 1932, 39–46). With the establishment of Israel, the Provisional Government (the temporary parliament antecedent to the Knesset) enacted the Law and Administration Ordinance of 1948 that maintained the existing legal system with its roots in Ottoman law (Law and Administration Ordinance 1948).

9.3.3 The Bedouin of the Negev

Until the twentieth century the Bedouin of the Middle East, including the Negev Bedouin, were livestock-raising nomads whose movements were dictated by constant search for pasture and water (Marx 1977, 345; Stewart 1986, 3). As a rule, what characterizes the Bedouin is their relationship to the tribe, rather than to a specific place or territory (Bailey 1969). Among the Bedouin tribes living in the Negev today, according to Aref al-Aref and others, most viewed themselves as descendants of nomadic tribes from the Arabian Peninsula (al-Aref 2000; Aharoni 2007, 30–1). According to Touvia Ashkenazi, "A great desert is spread to the east of Palestine which is known as Jazeerat el-Arab, 'the Arabian Peninsula' or al-Jazeera, 'the peninsula' which is the homeland and the metropolis of the Bedouin." (Ashkenazi 2000, 30) Joseph Ben-David, too, maintains that the origin of the Negev Bedouin was the Arabian Peninsula and that it is the wish "of every tribe to be reckoned as belonging to the tribes of the Arabian Peninsula, or in other words: to be a pure Arab" (Ben-David 2004, 36). In the opinion of Helmut V. Muhsam, the only tribe that considered itself as originating in the Negev was the *Hanajra* tribe, which no longer exists there (Muhsam 1966, 24, 26).

Ottoman tax registers demonstrate that the tribes which lived in the Negev in 1596–97 are not those residing there today (Wolf-Dieter Hütteroth and Abdulfattah 1977, 3). According to Hütteroth and Abdulfattah, the tax registers that reflect material collected in those years show names of forty-three Bedouin tribes living in what became Mandatory Palestine, including six in the Negev. There is not much information on what became of those tribes (Wolf-Dieter Hütteroth and Abdulfattah 1977, 51–3). However, the names of the tribes currently living in the Negev do not appear on the tax registers from 1596 (Wolf-Dieter Hütteroth and Abdulfattah 1977, 51–3). The Ottoman government did not maintain reliable records for this area after 1596, so these registers are the best indicators of which tribes existed in the early Ottoman period. Clinton Bailey, a scholar of Bedouin culture, also found no evidence from the thirteenth and fourteenth centuries of the continuity or existence of the eighteenth and nineteenth-century Bedouin tribes in the Negev (Bailey 1980, 42, 45, 1985, 21–4). Bedouin tribes whose descendants in the Negev claim the status of an indigenous people arrived there during the late eighteenth and the nineteenth centuries, from the deserts of Arabia, Transjordan, Sinai and Egypt (Sharon 1964, 21–24; Galilee et al. 2013). Part of this migration occurred during the upheaval of Napoleon's invasion of Egypt and Palestine in 1799 and the subsequent ten-year Egyptian rule there under Muhammad Ali and his son Ibrahim (1831–1841). During this period, Egyptian forces moved through Sinai and into the Negev using the coastal road that runs through Rafah, accompanied by numerous camp followers, peasants, and Bedouin. Some of the Egyptian peasants who followed in the footsteps of the army established new settlements and new urban neighborhoods in Palestine, others joined Bedouin tribes in the Negev (Kressel and Aharoni 2004; Kressel 2012).

The tribes' consolidation of their foothold in the Negev was achieved, inter alia, though armed struggles between Bedouin tribes for control of pasture lands, and raids on established Arab settlements, and even their total destruction (Sharon 1964, 49; Ben-David 1998, 64; Marx 1967, 7). Anthropologist Anatoly Khazanov notes that although the nomads were dependent upon the sedentary population for survival, they looked down upon them (Khazanov 1994, 199; al Aref 1974, 202). British surveyor and archaeologist Claude R. Conder, wrote in the 1880s "The Bedawi (sic) speaks with greatest contempt of the Fellah…" (Conder 1895, Vol. II, 271(According to Ben-David, Hütteroth and Abdulfattah the nomadic Bedouin were perceived by settled Arabs as waiting for opportunities of vulnerability in which they would attack and destroy agricultural settlements, or as heartless robbers (Ben-David 2004, 17; Wolf-Dieter Hütteroth and Abdulfattah 1977, 11).

Numerous authors documented the role of Bedouin in conquering the Negev, expelling settled Arabs in other parts of Palestine and plundering them (Granovski 1949, 32; Amiran 1953, 69; Wolf-Dieter Hütteroth and Abdulfattah 1977, 62; Sawaed 1992, 147–49; Epstein 1939). Conder described a situation of unending war between the Bedouin tribes and the settled villagers (Conder 1895, Vol. II, 271).

Nomadism continued in Palestine until the beginning of the twentieth century, when the transition to semi-nomadic life and settlement began (Ashkenazi 2000, 23; Marx 1977, 348). Concurrently, there was a gradual shift in the manner in which the Bedouin related to land: from common exploitation as grazing land by all members of the tribe to private use (Meir 1994, 76), without official recognition of ownership.

Simultaneously, there was a gradual transition from animal husbandry to agriculture (Kressel et al. 1991, 29). By 2000, animal husbandry was practiced by only about 10% of the Bedouin, and many of the younger generation have expressed reservations about maintaining the lifestyle of their parents (Degen et al. 2000, 130, 142).

Prior to the establishment of Israel there were about 65,000 Negev Bedouin (Marx 1967, 12). During the 1948 war and its immediate aftermath most of them left for neighboring states, so the Negev Bedouin population was reduced to about 11,000 (Muhsam 1959, 542). Since then, however, it has dramatically increased to 220,000 persons in 2016. During the years there was also significant improvement in education and in health indexes. However, when compared with other groups in Israeli society, including urban and rural Arabs, they remain at the lowest socio-economic level (Goldberg et al. 2008, 39).

In the 1970s about 3000 claims were filed by Bedouin demanding that Israel recognize their full private ownership, including the right to sell, of hundreds of thousands of dunams of land in the Negev (1 dunam = 1000 m^2). Today, they are no claims before Israeli courts for collective land rights and there is no expressed interest in land for collective grazing or for the maintenance of nomadic traditions.

Israeli courts, basing their decision on Ottoman and British law, have consistently refused to sanction the Bedouin claims. The courts have decreed that the lands claimed were never allocated for private use, and that they are of the category of *mewat* lands (literally "dead"—defined by the Ottoman Land Law as the area of waste land which lies beyond the carry of the human voice when uttered from the nearest habitation). It cannot be assigned as privately owned (el-Uqbi et al. 2015; Hawashla and Others 1984; Tute 1927, 179; Yahel 2006, 2).

9.4 Comparison of Indigenous Peoples and the Negev Bedouin

Until the last few years, the research literature has not referred to the Negev Bedouin as an indigenous community. Indeed, no other Bedouin tribe in the entire Middle East has raised a claim to indigeneity (Frantzman et al. 2012).

Despite the increasing use of the term "indigenous" in relation to the Bedouin during recent years, no serious study has been conducted to ascertain whether they fit that connotation. Those that did touch upon the issue did no more than provide general references (Boteach 2005, 2; Abu Saad 1997; Human Rights Watch 2008, 78–80). When referring to the question "Are the Bedouin Indigenous?" jurist Alexandre Kedar lists the components of the definition supplied by Martinez Cobo and four of the principles declared by UNWGIP in 1995 and resolutely maintains: "there is no doubt that today the Bedouin are a minority group, being part of the Palestinian minority group in Israel, which is considered by some to be an indigenous people in its own right. And if we apply the criteria noted above, it seems that we should include the Bedouin in the definition of indigenous groups." (Kedar 2004, 1–5). James Anaya in a recent report also failed to provide a comprehensive analysis (Anaya 2011). It is worthwhile to examine how the Bedouin fit the generally accepted parameters for what constitutes an indigenous people.

- *Original inhabitants*. As demonstrated, many groups preceded the Negev Bedouin in Palestine. Many groups preceded the Bedouin in Palestine in general and in the Negev in particular, including the Jewish people, which has maintained uninterrupted presence in the land since biblical times. Hence, the Bedouin can hardly claim to be the country's original inhabitants.
- *Time dimension*. This requires a lengthy presence in a territory—the so-called "time immemorial" parameter. But the Negev Bedouin have been there for only two centuries. Nor can they claim presence in the land before the arrival of the foreign power as the imperial Ottoman presence there predated that of the Bedouin by centuries. By contrast, the Jewish presence in Palestine fully corresponds to the "from time immemorial" parameter.
- *Sovereignty*. In the case of the Negev Bedouin, they were never sovereign in the area. When they arrived, the Negev was already under Ottoman rule, before coming under British Mandate, and then Israeli sovereign authority.

- *Oppression by a foreign culture and legal regime—colonialism or something of the kind.* During their immigration, it was the Bedouin who imposed themselves on established settlers in the Negev, displacing them and destroying their villages. The Ottoman Muslim regime, which they met upon arrival, was not foreign to them and was similar to that they had experienced in the other parts of the empire from which they migrated to Palestine. John Sheehan, disregarding Ottoman history, maintained that the British were the relevant colonialists (Sheehan 2010). But Britain, although a foreign power, had never attempted to colonize Palestine. Its presence there was transitory from the start in line with the League of Nations mandate. The British were charged with preparing the territory's residents for independent rule, they did not introduce significant changes into the laws. As for the claim that the Jewish democratic State of Israel is a colonialist power, this claim can hardly holds water. The Jews are far from being colonial intruders, as they were descendants of the country's ancient inhabitants, authorized by the international community—as represented by the League of Nations—to re-establish their independence in the ancestral homeland. In addition, Israel, like its predecessor, the British Mandate, at first preserved Ottoman and Muslim laws relating to land—and also in relation to other claims which have been extensively discussed in the literature (see: Salzman and Divine 2008).
- *Unique spiritual relationship to the territory.* While nomadic life, by definition, precludes permanent attachment to specific territory, pastoral lands do become an important element in Bedouin life given their vitality for tribes' subsistence. Furthermore, even today, control of an area is a matter of honor among the Bedouin, and any challenge to this control, or any infringement of their property, even if the property had been acquired recently on the free market, would be considered an insult to honor (al-Aref 2000, 273; for extended discussion on honor see Stewart 2008). We did not find any other arguments in the research literature that could justify a specific spiritual claim to Negev lands. The literature also does not yield any traces of lengthy Bedouin traditions relating to the Negev, a logical situation considering their fairly short presence there and nomadic lifestyle. We can also point to the fact that the Bedouin consider the Arabian Peninsula, not the Negev, as their historical homeland, and the fact that, at least until the end of the nineteenth century, Bedouin evinced a strong attachment to the tribe, not to a specific place.

Consideration should also be taken of the lengthy media and legal campaign waged by the Negev Bedouin to receive land rights in the Negev. These claims are not for collective rights but are all formulated as individual claims (overwhelmingly by males, with almost total exclusion of women), demanding full private rights which would enable them to sell the land and transfer it to a third party (Al-Rifaaya 2008). These private demands are not congruent with the spiritual dimension parameter and even contradict it.

In addition, since these private land claims are the center of the Negev Bedouin campaign for recognition as indigenous it leads to the conclusion that their main *subjective aspirations* are private and not collective.

Another issue is whether nomadic tribes can be considered an indigenous people. Nomadic populations are mentioned in the ILO Conventions No. 107 and 169 and also in the judgment of the Inter-American Court of Human Rights that referred to indigenous rights stemming from the Inter-American Convention (ILO Convention No. 169 1989; Mayagna (Sumo) Awas Tingni Community v. Nicaragua Community v. Nicaragua 2001). However it should be stressed that both conventions related to the rights of 'tribe' in general and not only to the rights of indigenous peoples.

We believe that the answer to that question is linked to the theoretical indigeneity perception, this paper does not address this broader question in detail.

On one hand it can be claimed that nomads do not have the necessary attachment to lands, since indigeneity is linked to a specific territory to which there is a strong relationship, and mobile groups cannot establish that kind of relationship. On the other hand, it may be claimed that they are an indigenous people provided they have wandered for long spans of time throughout a specific geographically defined area. The purpose of such recognition of indigeneity is the preservation of past traditions, and the right granted the group is the ability to continue to wander through the area with which it is traditionally identified.

- *A minority with an identity different from that of the general population. The* Bedouin are, without doubt, a numerical minority in Israel. In the past, anthropologists pointed out that for years the various rulers (the Ottoman Empire, the British Mandate, and the State of Israel) considered the Bedouin to be a separate group. For many years, in effect until the middle of the Mandate period and the beginning of the national confrontation between Arabs and the Jews in Palestine, and continuing into the Israeli period, the Bedouin were considered as enemies of the Palestinian Arab fellahs (Conder 1895, Vol. II, 271). Recently there have been signs of abandonment of an independent Bedouin identity and the gradual adoption of a Palestinian Arab identity accompanied by increasing involvement in Muslim fundamentalism (Ben-David 2004, 21, 29). In 2003 Musa el-Hujeirat maintains, on the basis of a study he conducted, that the Bedouin should no longer be considered a "society unto themselves." Rather, their identity today is Palestinian Arab, lacking in any common tribal element, and is in the process of being shaped anew. Furthermore, he claims there was an ulterior motive behind the categorization of a separate Bedouin identity: to negate the national Palestinian Arab identity (El-Hujeirat 2003, 6).
- *A group with separate economic, social, cultural, and political institutions*. In the past, Bedouin tribes behaved as separate units with an accepted leadership in the person of the tribe's sheikh. Tribes had a system of customs that governed all aspects of life. Each tribe was an independent economic and social group; several tribes joined together politically to form a confederation. Today, the

situation has changed dramatically. Studies attest to a significant weakening of the framework that handled the tribes' affairs and of tribes' ability to come to decisions acceptable to all its individuals. Institutions that formerly made decisions within the tribe or in inter-tribal relations no longer exist today (Ben-David 2004, 21).

Customary law and values necessary in the past when the Bedouin were nomads, such as mutual responsibility, are no longer relevant (Ben-David 2004, 335–36, 352). It would seem that today one cannot speak of Bedouin tribes in the Negev, alone or in confederation, as an operational administrative framework. The end of nomadism and the transition to permanent settlements during the past century have done away with identification of the tribe as a separate economic entity. Today, every household has its own occupation as part of the general economy and there is no universally acceptable authoritative leadership. Nor are there consequential political frameworks whose decisions are accepted by all, even in areas that are of primary importance to indigenous peoples, such as lands. Decisions relating to land are taken only by individuals; any declaration "in the name of the tribe" or "in the name of the Bedouin" is therefore not legitimate. In all claims by the Bedouin there is no demand by individuals to subordinate themselves once again to an internal, independent tribal framework. The opposite is the case: the tendency today is to increase individual rights. Authority that formerly rested with the sheikh vis-à-vis his tribe, including matters relating to lands, was abrogated after members of the tribe claimed that such authority was superfluous and that the sheikhs exploited it to further their own interests at the expense of the ordinary tribesmen. Despite the disappearance of an authority to manage and operate tribal matters, and the absence of tribal political frameworks, specific customs and traditions continue to exist as part of Bedouin customary law, but mainly in certain spheres of personal life such as marriage and inheritance rights (Abu Rabia 2011, 13–4).

- *The group identifies itself, and is considered by others, as an indigenous people in the territory*. As has been demonstrated in a parallel study, the Bedouin claim to indigeneity is very new, having been raised for the first time only a few years ago (Frantzman et al. 2012). Earlier studies did not report that the Negev Bedouin consider themselves as such, nor did the researchers make the claim that they are an indigenous people. Since Bedouin tribes in other Middle Eastern countries have never claimed indigenousness, the validity and authenticity of this new claim by the Negev Bedouin is doubtful. Are the Bedouin somehow indigenous only in relation to the Negev land in Israel but not in their homeland —Arabia—or in other Middle Eastern countries that border on Israel? Even parts of the same tribes as in the Negev that live elsewhere (e.g. the Sinai) do not claim indigenousness in their countries of residence.

9.5 Summary and Conclusions

In this article the global concept of indigeneity and the extent to which it can be applied to the Negev Bedouin, a marginal group in southern Israel, has been examined. In the first part we reviewed the legal foundations of the concept and described the international forums dealing with it. Since there is no definite and accepted definition included in the DRIP, it was necessary to find and analyze what parameters are typical of indigenous peoples, setting them apart from other ethnic minorities.

In the second part of the research, we presented the regional and local levels and included an outline history of changing rule in Palestine. We then described the Bedouin living today in the Negev, noting their late arrival in the area.

The third part contains implementation of the major parameters characteristic of indigenous people vis-à-vis the Negev Bedouin. The outcome showed a lack of concordance in the most important definitional areas. The major elements that differentiate between indigenous peoples and ethnic minorities do not apply to the Negev Bedouin. As presented, the Bedouin are not the original inhabitants of a territory settled from time immemorial, for the Bedouin living today in the Negev arrived there only two centuries ago, many years after other peoples.

Rather than being akin to indigenous peoples, the Bedouin have more in common with the European settlers who migrated to other lands, coming into contact with existing populations with often unfortunate results for the latter. Moreover, the Bedouin migrated mainly from one part of the Ottoman Empire to another, governed by the same system of administration and legislation with which they were well familiar and which the British and the Israelis have subsequently largely maintained.

As clearly demonstrated the Negev Bedouin do not presently prefer to be a separate and independent entity. Their aspirations are of an individual nature. They are not interested in maintaining nomadic traditions of collective ownership of lands for the maintenance of a collective community but rather in an exclusively male proprietorship that would enable Bedouin men to sell the land to others at their own discretion. No studies have shown the existence today of functioning, independent institutions in various spheres of daily life that could point to the Bedouin being an indigenous people.

That no other Bedouin tribe in the entire Middle East has raised a claim to indigenousness raises questions regarding the motivations and authenticity of such an argument. Since the Bedouin in the Negev in some cases are from the same tribe as those found in neighboring countries, it is not logical that they can only be indigenous when they are on the Israeli side of the border.

The ultimate conclusion arising from our study is that the Negev Bedouin cannot be afforded recognition as an indigenous people in the sense that this term has been defined and commonly accepted in international law. The attempt to apply this recognition to the Negev Bedouin is inconsistent with its accepted meaning and content inherent in the concept of indigeneity (and may be intended to advance other objectives).

Within the scope of this study it is important to emphasize additional aspects. **First**, the broader questions of whether the indigenous concept is relevant in and applicable to Israel, with its history layers and extremely complex social components. Similar questions were raised by African and Asian states, and even by some in Europe, and were given negative answers. Some of the countries not only reject the relevancy but also pointed out the problematic applications of such recognition on their overall situation. The fear is that instead of providing remedies and established order, it will create new disputes. Israel, alike, can justly hold the position that the concept of indigeneity is less relevant to it due to its past and complex present.

Israel's history, which is marked by constant waves of immigration and invasion by various peoples, brings us to the **second** question of whether there may be other groups that have a potential parallel demand for indigenity? Clearly, a decision to the effect that the concept is suitable for application to the Bedouin in Israel will provide a legal basis for claims of indigeneity by other groups who arrived in the area before the Bedouin. If the parameters and preconditions are made more flexible, why should the Jews for example not claim to be an indigenous people in Israel—including extensive regions they once controlled—to which they came thousands of years before the Bedouin? The Jews always saw the Land of Israel as their national homeland, once lived in it as a sovereign, maintained at least a toehold there despite persecution, and returned to it time and again after being exiled. This spiritual relationship is also expressed in both Jewish daily prayers and Israel's Declaration of Independence. And what of Jews living today in Judea and Samaria? Should a Palestinian state be established in these territories, once the claim of Bedouin indigeneity is acknowledged in the region, would this not provide a basis for Jewish settlers to claim indigenous rights as a minority in their ancestral homeland (Bender 2010). It may also be expected that other ethnic groups, such as Druze, Christian Arabs, and Samaritans, would then claim indigenous status. No doubt, this would add to confrontations already existing over control of lands and holy places, such as those between Christians and Muslims in Nazareth, between Jews and Muslims over the Temple Mount, and more.

The concept of indigeneity was intended to help remedy the injustices of the past by giving indigenous peoples the means by which they could preserve their separate identity, common lifestyle, and the customs of their past. We found it inapplicable to Israel in general, and to the Negev Bedouin in particular. Despite the many hardships the Negev Bedouin experience at present as a less educated and low income minority and marginal group located in the southern part of the country, they do not meet the criteria of indigenous peoples. Their present-day aspirations are not in line with those of other generally recognized indigenous peoples.

References

Aaronsohn, R. (2000). *Rothschild and Early Jewish Colonization in Palestine*. London: Rowman & Littlefield and Jerusalem: Magnes Press.

Abu Rabia, A. (2002). Displacement, Forced Settlement and Conservation. In D. Chatty & M. Colchester (Eds.), *Conservation and Mobile Indigenous Peoples: Displacement, Forced Settlement, and Conservation* (pp. 202–211). London: Berghahn.

Abu Rabia, K. (2011). *Three Circles in the Desert: The Conflict between Bedouin Customs, the Sharia Law, and the Law of the State of Israel*, Beersheba: Ben-Gurion University of the Negev (Hebrew).

Abu, Saad I. (1997). The education of Israel's Negev Bedouin: Background and prospects. *Israel Studies, 2*(2), 21–39.

Aharoni, R. (2007). *The Pasha's Bedouin*. London and New York: Routledge.

Aidi, H. (2016). Australia's 'history wars' heat up again. *Opinion Human Rights*, March 29, 2016.

Al-Rifaaya, H. (2008). Testimony to the Goldberg Committee, Protocol February 7, 2008 (Hebrew).

Albeck, P., & Fleisher, R. (2005). *Land Law in Israel*. Jerusalem: the authors (Hebrew).

Al Jazeera. (2015). Indigenous Australians may soon Lose Ancestral Land, September 10, 2015.

Amara, A. et al. (Eds.). (2012). *Indigenous (in)justice*. Cambridge: Harvard University Press.

Amiran, D. H. K. (1953). The Pattern of Settlement in Palestine. *Israel Exploration Journal, 3,* 65–78.

Anaya, J. (2011). Report by the Special Rapporteur on the rights of indigenous peoples, A/HRC/18/35/Add.1, August 22, 2011, Annex VI, 24.

Anaya, S. J. (2004). *Indigenous Peoples in International Law* (2d ed.). Oxford and London: Oxford University Press.

al-Aref, A. (1974). *Bedouin Love Law and Legend: Dealing Exclusively with the Badu of Beersheba*, Jerusalem: Cosmos, 1944; repr. 1974.

al-Aref, A. (2000). *Bedouin Love Law and Legend* , [al-Aref A., *Toldot Be'er Sheva Ve-Shvateha: Shivtey Ha-Beduim Bemahoz Be'er Sheva* , photocopy of first edition, Tel-Aviv :Bustenay, 1937], Jerusalem: Ariel (Hebrew edition).

Ashkenazi, T. (2000). *The Bedouin of Palestine*. Ariel : Jerusalem (Hebrew).

Bailey, C. (1969). *The Bedouin*. Sede-Boqer: Midreshet Sede-Boqer, Tuviyahu Archives of the Negev (Hebrew).

Bailey, C. (1980). The Negev in the 19th Century. *Asian and African Studies, 14,* 35–80.

Bailey, C. (1985). Dating the arrival of the Bedouin Tribes in Sinai and the Negev. *Journal of the Economic and Social History of the Orient, 28*(1), 20–49.

Banyan, M. (2002). *Land Law: Principles and Rulings*. Tel Aviv: Institute for Research of Law and Economics (Hebrew).

Ben Naftali, O., & Shani, Y. (2006). *International Law between War and Peace*. Tel Aviv: Ramot (Hebrew).

Ben-David, J. (1998). More on the conflict over land between the Negev Bedouin and the State. *Karka, 44*, 63–92 (Hebrew).

Ben-David, J. (2004). *The Bedouin in Israel: land conflicts and social issues*. Jerusalem: Jerusalem Institute of Israel Studies (Hebrew).

Bender, A. (2010). *Initiative to Recognize the Settlers in Judea and Samaria as Indigenous*. Ma'ariv, October 8, 2010 (Hebrew). (http://www.nrg.co.il/online/1/ART2/164/557.html) (Accessed July 10, 2011).

Bernard, J. (1932). 'Palestine Legislation under the British', *Annals of the American Academy of Political and Social Science*, 164 (1), 39–46.

Bishara, S., & Rosenberg, R. (2004). 'Land rights and the Indigenous Palestinian Arab Citizens of Israel: Recent cases in law, land and planning. *Adalah* http://www.adalah.org/eng/intl04/unIndigpop.pdf (Accessed July 10, 2011).

Boteach, E. (2005). *The Bedouins in the Negev as an Indigenous Population: A Report Submitted to the UN Working Group on Indigenous Populations*. Beersheba: The Negev Coexistence Forum Newsletter.
Center for World Indigenous Studies. http://cwis.org/fwdp/International/wcipinfo.txt (Accessed October 26, 2010).
Cobo, J. R. M. (1987). Study of the problem of discrimination against indigenous people. 29 (UN E/CN.4/Sub.2/1986/7/Add.4).
Conder, C. R. (1895). *Tent Work in Palestine*. London: A. P Watt.
Corntassel, J. J., & Primeau, T. H. (2006). Indigenous 'Sovereignty' and International Law: Revised Strategies for Pursuing 'Self-determination'. *Hawaiian Journal of Law and Politics, 2,* 52–72.
Daes, E.-I. (2001). Indigenous peoples and their relationship to land. UN E/CN.4/Sub.2/2001/21.
Degen, A. A., et al. (2000). Bedouin households and sheep production in the Negev Desert, Israel. *Nomadic People, 4*(1), 125–147.
Dirlik, A. (2003). *Globalization, indigenism, and the politics of place*. New Bulgarian University, anthropology dept., http://www.nbu.bg/PUBLIC/IMAGES/File/departments/anthropology/publications/32.pdf (Accessed February 23, 2011).
DRIP. (2007). DRIP, Articles 33 and 35.
El-Hujeirat, M. (2003). The collective identity of the Bedouin in the State of Israel. *Notes on the Bedouin, 35,* 6–13 (Hebrew).
el-Uqbi (2015). el-Uqbiet al v. *The State of Israel*. Court of Appeal 4220/12 (2015 unpublished).
Epstein, E. (1939). Bedouin of the Negeb. *Palestine Exploration Fund Quarterly, 71,* 59–78.
Erueti, A. (2006). The demarcation of indigenous peoples' traditional lands: Comparing domestic principles of demarcation with emerging principles of international law. *Arizona Journal of International and Comparative Law, 23,* 543–612.
Forman, G., & Kedar, A. (2003). Colonialism, colonization and land law in Mandate Palestine: The Zor al-Zarqa and Barrat Qisarya land disputed in historical perspective. *Theoretical Inquiries in Law, 4,* 496–534.
Frantzman, S. (2014a). The politization of history and the negev bedouin land claims: A review essay on indigenous (in)justice. *Israel Studies, 19*(1), 48–74.
Frantzman, S. J., et al. (2012). Contested indigeneity: The development of an indigenous discourse on the bedouin of the Negev, Israel. *Israel Studies, 17*(1), 78–104.
Frantzman, S. J. (2014b). Palestinians: The new "Noble Savage". *Mida*, May 2014.
Friedmann, D. (1975). The effect of the foreign law on the law of Israel: Remnants of the ottoman period. *Israel Law Review, 10,* 192–206.
Friling, T. (2003). *An answer to a Post-Zionist Colleague*. Tel Aviv: Lamiscal (Hebrew).
Gabizon, R., & Balfur, T. (2005). Minority group rights, working paper, submitted to the constitutional committee (Hebrew). http://www.huka.gov.il/wiki/material/data/H13-09-2005_8-54-06_miutim.doc (Accessed January 20, 2011).
Galilee, E., Kark, R., & Kressel, G. (2013). 'Like Footprints in the dessert: Reflection of tribal boundaries and immigration processes in Negev bedouin cemeteries 1800–1966. *Hamizrah Hahadash, 52,* 116–137 (Hebrew).
Gil, M. (1997). *A History of Palestine, 634–1099*. Cambridge: Cambridge University Press.
Goldberg, E. et al. (2008). Report of the Committee [Appointed] to Propose Regularization of Settlement of the Bedouin in the Negev, Jerusalem, (Hebrew).
Granovski, A. (1949). *Israel Land Regime*. Tel Aviv: Dvir (Hebrew).
Ha'aretz. (2007). An interview with Advocate Ady Schonmann, Legal Adviser Israel's Permanent Mission to the UN, *Ha'aretz*, October 2, 2007 (Hebrew).
Hawashla. (1984). Hawashla and Others v. *The state of Israel and others*, Court of Appeal 218/74, 38(3) P.D. 141.
Hertz, A. Z. (2011). Aboriginal rights of the Jewish People. *American Thinker*, October 30, 2011.
Hobbes, T. (1651). *Leviathan, or the matter, forme and power of a common-wealth ecclesiastical and civil*. London: Andrew Crooke.
Hodgson, D. L. (2002). Introduction: Comparative perspectives on the indigenous rights movement in Africa and the Americas. *American Anthropologist, 104*(4), 1037–1049.

Hodgson, D. L. (2009). Becoming indigenous in Africa. *African Studies Review, 52*(3), 1–32.
Human Rights Watch. (2008). *Off the map*. March 30, 2008.
ILO Convention No. 107. (1957). International Labour Organization Convention (No. 107) Concerning the protection and integration of indigenous and other tribal and semi-tribal population in independent countries. June 26, 1957.
ILO Convention No. 169. (1989). International Labour Organization Convention (No. 169) Concerning indigenous and tribal people in Independent Countries. June 27, 1989.
Indigenous Peoples in Africa. (2006). The forgotten peoples?: The African commission's work on indigenous peoples in Africa (Banjul, Gambia: African Commission on Human and Peoples' Rights (ACHPR) and Copenhagen: International Work Group for Indigenous Affairs.
Johnson v. M'Intosh. (1823). 21 U.S. 543, 595 (U.S. 1823).
Kark, R. (1992). 'Land-God-Man: Concepts of land ownership in traditional cultures and in Eretz Israel. In A. R. H. Baker & G. Biger (Eds.), *Ideology and Landscape in Historical Perspective* (pp. 63–82). Cambridge: Cambridge University Press.
Karsh, E. (2006). *Islamic Imperialism: A History*. New Haven: Yale University Press.
Karsh, E. (2008). The missing piece: Islamic imperialism. In P. C. Salzman & D. R. Divine (Eds.), *Postcolonial Theory and the Arab-Israel Conflict* (pp. 120–141). London: Routledge.
Kedar, A. (2004). Land settlement in the Negev in international law perspective. *Adalah Newsletter, 8,* 1–5.
Khan, B. U., & Rahman, M. M. (2009). *Protection of minorities: A South Asian discourse*. The European Academy of Bozen/Bolzano, Italy, 2009 http://www.eurac.edu/en/research/institutes/imr/Documents/EURASIANet_Del_17_South_Asian_Discourse.pdf?AspxAutoDetectCookieSupport=1 (Accessed December 21, 2010).
Khazanov, A. M. (1994). *Nomads and the Outside World* (2d ed.). Madison: University of Wisconsin Press.
Kingsbury, B. (1998). 'Indigenous peoples' in international law: A constructivist approach to the Asian Controversy. *The American Journal of International Law, 92*(3), 414–457.
Krämer, G. (2008). *A History of Palestine: From the Ottoman Conquest to the Founding of the State of Israel*. Princeton: Princeton University Press.
Kressel, G. (2012). *Telephone interview with Gideon Kressel*, March 8, 2012.
Kressel, G. M., & Aharoni, R. (2004). Egyptian Immigrants in the Bilad Esh-Sham. *Jama'a, 12,* 206–245 (Hebrew).
Kressel, G. M., et al. (1991). Changes in the land usage by the Negev Bedouin since the Mid-19th Century: The Intra-Tribal perspective. *Nomadic People, 28,* 28–55.
Kuper, A. (2003). The return of the native. *Current Anthropology, 44*(3), 389–398.
Lapidus, I. M. (1991). Tribes and state formation in islamic history. In P. S. Khoury & J. Kostiner (Eds.), *Tribes and State Formation in the Middle East* (pp. 25–47). London and New York: I. B. Tauris.
Law and Administration Ordinance, 5708–1948, Provisional Government of the State of Israel, Tel Aviv, May 19, 1948.
Lehmann, K. (2006). Symposium: Lands, liberties, and legacies: Indigenous peoples and international law: Regional issues in the international indigenous rights movement: To define or not to define—The definitional debate revisited. *American Indian Law Review, 31,* 509–530.
Lerner, N. (2003). *Group Rights and Discrimination in International Law* (2d ed.). The Hague: Martinus Nijhoff Publishers and Kluwer Law International.
Mabo. (1992). Mabo and Others v. Queensland (No. 2) (AU 1992) 175 CLR1.
Martínez, M. A. (1999). Human right of indigenous people: Study on treaties, agreements and other constructive arrangements between states and indigenous populations, para. 78 and 91, E/CN.4/Sub.2/1999/20.
Marx, E. (1967). *Bedouin of the Negev*. Manchester: Manchester University Press.
Marx, E. (1977). The Tribe as subsistence unit: Nomadic pastoralism in the middle east. *American Anthropologist, 79*(2), 343–363.
Mayagna (Sumo) Awas Tingni Community v. Nicaragua. (2001). Inter-American Court of Human Rights (Ser. C) No. 79.

Maybury-Lewis, D. (2002). *Indigenous Peoples, Ethnic Groups, and the State* (2d ed.). Boston: Allayn and Bacon.
McNeil, K. (2008). Judicial treatment of indigenous land rights in the common law world. *Comparative Research in Law & Political Economy, 4*(5), 1–37.
Meir, A. (1994). The evolution of territoriality among the negev bedouin in transition from nomadism to sedentarization. *Studies in the Geography of Israel, 14*, 71–92 (Hebrew).
Miller, R. J. (2010). The Doctrine of Discovery. In R. J. Miller, J. Ruru, L. Behrendt, & T. Lindberg (Eds.), *Discovering Indigenous Lands: The Doctrine of Discovery in the English Colonies*. Oxford: Oxford University Press.
Mills, J. A. (2002). Legal constructions of cultural identity in Latin America: An argument against defining 'indigenous peoples'. *Texas Hispanic Journal of Law & Policy, 8,* 49–78.
Muhsam, H. V. (1959). Sedentarization of the Bedouin in Israel. *International Social Science Journal, 11*(4), 539–549.
Muhsam, H. V. (1966). Bedouin of the Negev. In H. V. Muhsam (Ed.), *Eight Demographic Studies*. Jerusalem: Jerusalem Academic Press.
National Congress of American Indians. http://www.ncai.org/About.8.0.html (Accessed March 19, 2012).
National Post. (2016). Attawapiskat First Nation declares state of emergency after 11 suicide attempts in one day. April 10, 2016.
Niezen, R. (2003). *The Origins of Indigenism: Human Rights and the Politics of Identity*. Los Angeles: University of California Press.
Or, T. et al. (2003). Commission of Inquiry into the Clashes between the Security Forces and Israeli Citizens in October 2000, Jerusalem (Hebrew).
Packer, J. (1999). Problems in defining minorities. In D. Fottrell & B. Bowring (Eds.), *Minority and Group Rights in the New Millennium* (pp. 223–231). The Hague: Martinus Nijhoff.
Pierce, K. (2007). Representative of United Kingdom, in the UN 61st GA, New-York, September 13, 2007, in A/61/PV.107 also accessible in: http://daccess-dds-ny.un.org/doc/UNDOC/GEN/N07/504/29/PDF/N0750429.pdf?OpenElement
Pritchard, S. (1998). Working group on indigenous population: Mandate, standard-setting activities and future perspectives. In S. Pritchard (Ed.), *Indigenous Peoples, the United Nations and Human Rights* (pp. 40–64). London: Zed Books and Leichhardt Federation Press.
Reuter, T. K. (2009). Dealing with claims of ethnic minorities in international law. *Connecticut Journal of International Law, 24,* 201–237.
Sabel, R., & Adler, H. (Eds.). (2010). *International law*. Sacher Institute: Jerusalem.
Salzman, P. C. and Divine, D. R. (2008). *Postcolonial Theory and the Arab-Israel Conflict* London & New York: Routledge.
Sawaed, M. Y. (1992). *The Bedouin in Palestine between 1804 and 1908*, M.A. thesis, Bar-Ilan University, Ramat Gan (Hebrew).
Shamir, R. (2000). *The Colonies of Law: Colonialism, Zionism and Law in Early Mandate Palestine*. Cambridge: Cambridge University Press.
Sharon, M. (1964). *The Bedouin in Palestine during the Eighteenth and Nineteenth Centuries*. M. A. thesis, The Hebrew University of Jerusalem (Hebrew).
Sheehan, J. (2010). *Lecture in a symposium at the Technion*, Haifa, November 28, 2010.
Sochowolski, A. (2001). *Man and Land, Law and Language*. Tel-Aviv: Hagigim (Hebrew).
St. Catharines (1887). St. Catharines Milling and Lumber Company v. *The Queen* (Canada 1887), 13 S.C.R. 577.
Stewart, F. (2008). What is honor. *Acta Histriae, 8,* 13–28.
Stewart, F. (1986). *Bedouin Boundaries in Central Sinai and the Southern Negev: a Document from the Aḥaywāt tribe*. Wiesbaden: O. Harrassowitz.
Thornberry, P. (1991). *International Law and the Rights of Minorities*. Oxford: Clarendon Press.
Troen, S. I. (2008). De-Judaising the homeland: Academic politics in re-writing the history of Palestine. In P. C. Salzman & D. R. Divine (Eds.), *Postcolonial Theory and the Arab-Israel Conflict* (pp. 196–207). London: Routledge.

Tute, R. C. (1927). The law of state lands in Palestine. *Journal of Comparative Legislation and International Law*, *9*(4), 165–182 (Third Series).
UN Conference on Environment and Development. (1992). *UN Conference on Environment and Development*, Rio de Janeiro, June 13, 1992, A/CONF.151q26 (Vol. 3) at 16 annex 2 (Agenda 21), Chapter 26.
UN General Assembly. (1994). U.N. GA resolution 49/214 of December 23, 1994, A/RES/49/214.
UN General Assembly. (1948). *Universal Declaration of Human Rights*, December 10, 1948, 217 A (III), available at: http://www.ohchr.org/EN/UDHR/Pages/Language.aspx?LangID=eng (Accessed March 10, 2013).
UN General Assembly. (1966a). International covenant on civil and political rights, December 16, 1966, United Nations, Treaty Series, Vol. 999, p. 171. Available at: http://www2.ohchr.org/English/law/ccpr.htm (Accessed March 10, 2013).
UN General Assembly. (1966b). International Covenant on economic, social and cultural rights, December 16, 1966, United Nations, Treaty Series, Vol. 993, p. 3. http://www2.ohchr.org/English/law/cescr.htm (Accessed March 10, 2013).
UN General Assembly. (1993). UN GA resolution 48/163, December 21, 1993, A/RES/48/163. http://www.un.org/documents/ga/res/48/a48r163.htm (Accessed March 10, 2013).
UN General Assembly. (2004). U.N. GA resolution 59/174, New York, December 20, 2004.
UN General Assembly Media Release. (2007). UN GA/10612, September 13, 2007 http://www.un.org/News/Press/docs/2007/ga10612.doc.htm (Accessed March 10, 2013).
UN General Assembly. (2007). UN GA, September 13, 2007, A/61/PV.107. http://daccess-dds-ny.un.org/doc/UNDOC/GEN/N07/504/29/PDF/N0750429.pdf?OpenElement (Accessed March 10, 2013).
UN Permanent Forum. The Secretariat of the United Nations Permanent Forum on Indigenous Issues, State of the World's Indigenous Peoples, http://www.un.org/esa/socdev/unpfii/documents/SOWIP_introduction.pdf (Accessed July 1, 2011).
UN Session I. (1971). The Problem of Indigenous Population, ECOSOC Res. 1589 (L), 50th Session Supplement No.1, at 16, U.N. Doc. E/5044, May 21, 1971. http://daccess-dds-ny.un.org/doc/UNDOC/GEN/NR0/762/94/IMG/NR076294.pdf?OpenElement (Accessed October 30, 2010).
Vainshtein, S. I. (1989). The Turkic peoples, sixth to twelfth centuries. In V. N. Basilov (Ed.), *Nomads of Eurasia* (pp. 55–65). Seattle and London: University of Washington Press.
Waldron, J. (2003). 'Indigeneity? First peoples and last occupancy. *New Zealand Journal of Public Law, 1,* 55–88.
Western Sahara, Advisory Opinion. (1975). International Court of Justice, October 16, 1975, 12 in the website of the Court: http://www.icj-cij.org/docket/files/61/6195.pdf
Wiessner, S. (1999). Rights and status of indigenous peoples: A global comparative and international legal analysis. *Harvard Human Rights Journal, 12,* 57–128.
Wiessner, S. (2009). The rights and status of indigenous peoples. UN Audiovisual Library of International Law, United Nations, New York, July 2, 2009, UN Archives.
Williams, R. A., Jr. (1990). Frontier of legal thoughts III: Encounters on the frontiers of international human rights law: Redefining the terms of indigenous peoples' survival in the world. *Duke Law Journal, 39*(4) 660–704.
Wolf-Dieter Hütteroth, W. D., & Abdulfattah, K. (1977). *Historical Geography of Palestine, Transjordan and Southern Syria in the Late 16th century*. Erlangen: Palm and Enke.
Worcester v. Georgia. (1832). 31 U.S. 515, 542 (U.S. 1832).
World Bank. (1991). *The world bank operational manual, operational directive 4.20.* http://www.ifc.org/ifcext/enviro.nsf/AttachmentsByTitle/pol_IndigPeoples/$FILE/OD420_IndigenousPeoples.pdf (Accessed September 10, 2011).
Yahel, H. (2006). Land disputes between the Negev Bedouin and Israel. *Israel Studies, 11*(2) 1–22.
Yiftachel, O. (2009). Towards recognizing indigenous rights? Beer Sheva metropolitan planning after the Goldberg Commission. *Tichnun, 11*, 56–71 (Hebrew).

Chapter 10
Changing Identity, Livelihood and Biodiversity of Indigenous Communities in the Eastern Himalaya with Special Reference to Aka Tribe

Gibji Nimasow and R.C. Joshi

10.1 Introduction

Arunachal Pradesh lies in the Eastern Himalayas that spreads over a geographical area of 83,743 km^2 which shares international border with Bhutan, Tibet, China and Burma and state boundaries with Assam and Nagaland. It is a mountainous state and is situated at the northeast tip of India in between the latitudes of 26°28′ N and 29°20′ N and longitudes of 90°30′ E and 97°30″ E. The territory of the state lies in the catchments area of the Brahmaputra and its tributaries—the Kameng, Subansiri, Siang, Lohit and Tirap rivers. The mountains raise steeply northwards, some peaks even rising up to 7000 m above mean sea level. The Eastern Himalaya is divided into four physiographic divisions i.e. the Himalaya, Mishmi Hills, Purvanchal Hills and Foot hill plains (Joshi 2006). Geologically this area is covered by the Siwaliks (Outer Himalaya) and Lesser Himalaya. The main rock types found in this area are sandstone, conglomerate, shale, quartzite, dolomite, gneiss and schist. This area is crossed by Himalayan Frontal Fault (HFF), Tipi Thrust (TT) and Main Boundary Thrust (MBT) from east to west (Geology and Mineral Resources of Arunachal Pradesh 2010). The average slope calculated for the area ranges from level to 42°. The climate varies from hot and humid in the foothills to very cold in the north. As one moves northwards to higher altitudes it becomes progressively colder. In the interior parts of the state, rain practically falls all the year round specially from mid-April to mid-August when sun can hardly seen with clouds

G. Nimasow (✉)
Department of Geography, Rajiv Gandhi University, Rono Hills, Itanagar 791112, Arunachal Pradesh, India
e-mail: gibji26@yahoo.co.in

R.C. Joshi
Department of Geography, Kumaun University, DSB Campus, Nainital 263002, Uttarakhand, India
e-mail: joshircj@yahoo.co.in

R. Chand et al. (eds.), *Societies, Social Inequalities and Marginalization*, Perspectives on Geographical Marginality,
DOI 10.1007/978-3-319-50998-3_10

making it overcast for most part of the day. The north-west region faces heavy snowfall from December to February. The varying climate in north and south are responsible for differences in dress and culture of the people of the respective regions. The state is characterized by diverse species of flora and fauna. The variety of species of flora and fauna provides ample resource to these tribal people. In other words, they are totally dependent on the nature for their day-to-day requirements and their way of living is determined by the availability of the nearby forest resources. All kinds of primary as well as secondary needs and requirements from those of an individual to a group of people come from the nearby dense forests. Besides, the state is one of the more important areas in the world in terms of orchid production. It is estimated that more than 5000 different species of orchids flourished in the state. At present the state has 15 districts, 38 sub-divisions, 149 circles and 61 C.D. Blocks (March 2002), which was created from time to time for administrative convenience and developmental purposes. As per the census of 2011 the total population of the state was 13,83,727 with a density of population of 17 persons km^2. The decadal growth rate from 2001 to 2011 was 26.03%. The sex ratio was 938 female per 1000 male (Statistical Abstract of Arunachal Pradesh 2011).

The term ‘tribe’ refers to a group of families or communities, linked by social, economic, religious, or blood ties, and usually having a common culture and dialect and a recognized leader. Akas are the tribe who inhabits the south eastern part of West Kameng and western part of East Kameng district (Fig. 10.1). They are divided into two sub-tribes namely the *Hrusso* and *Koro*. They have no script of their own and speak the Aka dialect, which is grouped under the Tibeto-Burman

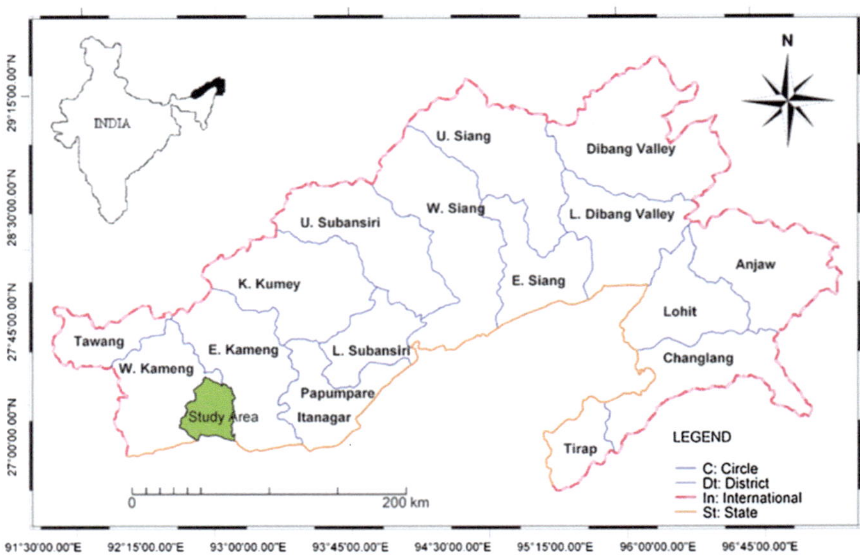

Fig. 10.1 Location of study area in Arunachal Pradesh, India. *Map Source* Digitized by Dr. Gibji Nimasow using GIS software ILWIS 3.4 version

family of language. Akas believe in *Nyezino* (*Hrusso*)/*Mene Alan* (*Koro*), (meaning the sky and the earth). The major festival is *Nyetriidow* (*Hrusso*)/*Sarok* (*Koro*), which means to clean the village from the evil influences, sufferings and natural calamities. Traditional *jhum* cultivation is practiced on a large scale basis because it is the only activity for their subsistence. Besides, they undertake some subsidiary means of sustenance, such as fishing, hunting and food-gathering to supplement the shortage of their food supply. The total population of the tribe prior to the post independence period was very less about 1000 approximately but, according to 1991 census the population of the Akas was 3531, which increased to 5027 according to the recent household survey conducted for the area in 2008. Out of the total population, 3905 inhibit the three circles of West Kameng namely Thrizino (2598), Bhalukpong (290) and Jamiri (1017). The remaining 1122 person inhibits the Rechukrang circle of East Kameng district. There is a considerable increase in the literacy rate i.e. 62.32% as per the recent survey. The socio-economic life of Akas is largely determined by the environmental conditions of their habitat. The family in the Aka society comprises of the husband, wife(s) and their children. They are socially organized on the basis of a compact social group, through the village community and through the clan. Marriage in the society is a socially sanctioned institution. Marriage can take shape in two ways, either by negotiation or by the romantic way of capture (Nimachow 2011). Plenty of works has already been done on the changing society, economy and culture of the Indian tribal people by Soumen (1988), Parihar (1989), Balan (1992), Ghosh (1992), Gohain (1994), Choudhary (1996) and Riba (1997). The aim of the study is to assess the changing identity, livelihood pattern and biodiversity of the Aka tribes of Arunachal Pradesh due to the impact of globalization. The changing demography, society, infrastructure, culture, livelihood pattern and biodiversity have been discussed in detail below.

10.2 Aka Habitat

The Aka inhabited area is delineated on Survey of India topographical maps as narrated by the elderly people during the field study. The Aka area is located in between 27°0′ N to 27°30′ N latitudes and 92°30′ E to 92°55′ E longitudes covering an area of about 1262.21 km^2. It comprises 38 (thirty eight) villages (Fig. 10.2). The altitude ranges in between 200 to 2936 m above mean sea level. The Aka study area is drained by Kameng River and its tributaries. The main tributaries are Tenga and Bichom rivers. The temperature of the study area ranges in between 1 and 30 °C. However, higher reaches experience snowfall during winter. The annual rainfall received by the study area is about 235 cm (S&I Unit-1, BLD-II/NEEPCO. Ltd. Tippi, Arunachal Pradesh, 2006). The area is located in a mountainous tract with deep gorges, valleys, hills, etc. So it is obvious that the soils are of mountainous type. The soil is not very fertile, except the small patches of

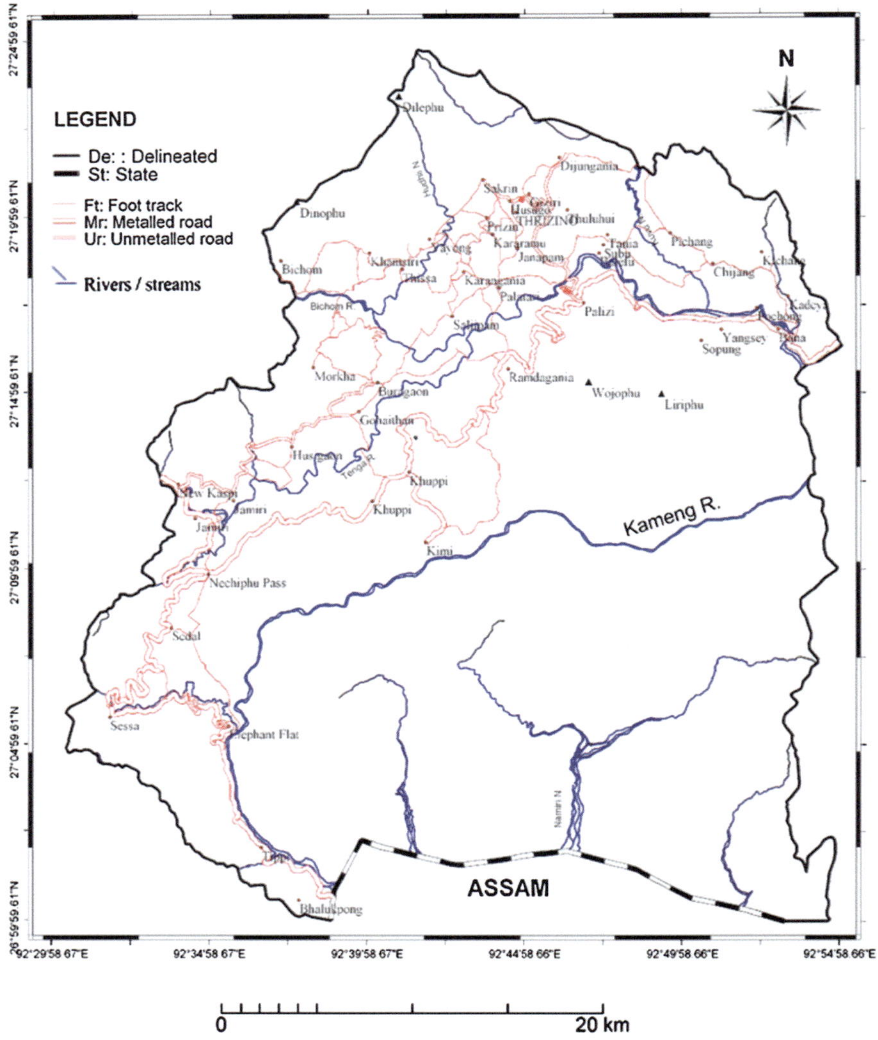

Fig. 10.2 Base map of the study area. *Map Source* Digitized with GIS software from topographical sheets No. 83A/11 to A/16

valleys and river terraces. The study area is rich in its flora and fauna. In general three types of forests i.e. tropical rain forest and semi-evergreen forest, subtropical forest and pine forest are found in the area. Earlier workers (Sinha 1962; Verrier 1968; Kholey 1997) studied Akas from historical and anthropological point of view by selecting a few villages. The present work is based on a field survey carried during 2006 using questionnaires at household level. A comparative study is made by using census data of 1971, 1981, 1991 and 2001. Different data collected from the field were analyzed using SPSS 9.05 package for generating appropriate quantitative and qualitative results.

10.3 Demography

The total Aka population was 5027 (Survey 2006). There has been a rapid increase in the population in the recent decades. The crude birth rate was 49.73 and the crude death rate was 9.55 in 2006. By putting the value in the exponential population growth equation (excluding migration) it is found that the Aka population will double in approximately 17 years i.e. 10,000 in 2024. The Aka population has a share of 44.87% of the total population and 78.46% of the total Scheduled Tribe (as per constitutional provision made in a particular schedule) population of the West Kameng district, Arunachal Pradesh. The Infant Mortality Rate was estimated to be 92 per 1000 live births (Nimachow 2008). The demographic statistics reveals that the population has entered the stage 2 of demographic transition characterized by high birth rate and low death rates. The total population in the age group of below 14 years is very high due to the high crude birth rate. Hence, the age sex pyramid of the population (see Fig. 10.3) shows the model of an area passing through the second stage of demographic transition. The sex ratio of the total population was 877 females per 1000 males as per 2001 census.

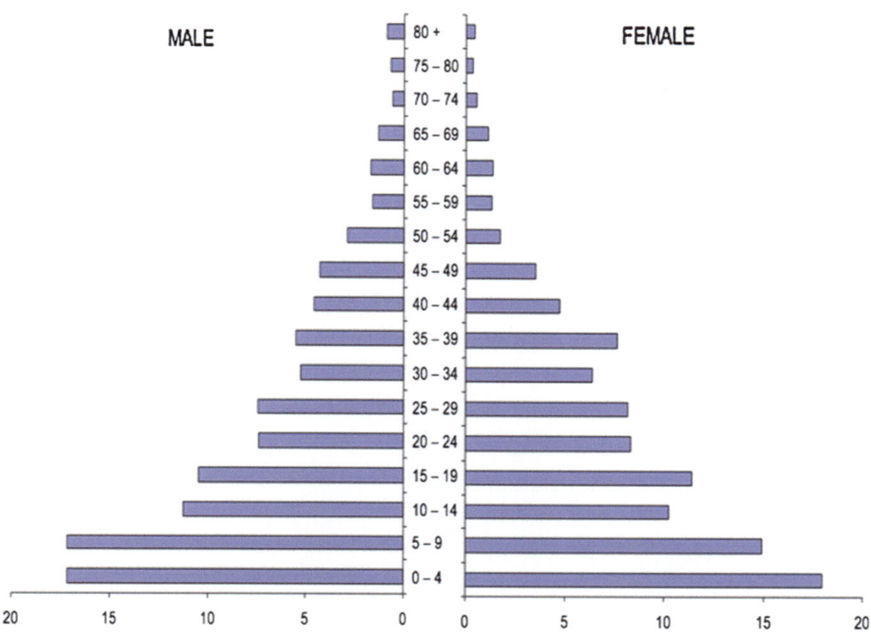

Fig. 10.3 Age sex pyramid of Aka population, 2006. *Source* Household survey 2006

Table 10.1 Age sex distribution of Aka population

Male	Percentage	Age group	Female	Percentage
430	17.16	0–4	453	17.97
430	17.16	5–9	377	14.95
281	11.21	10–14	258	10.23
262	10.45	15–19	287	11.38
185	7.38	20–24	209	8.29
186	7.42	25–29	206	8.17
132	5.27	30–34	161	6.39
137	5.47	35–39	192	7.61
115	4.59	40–44	119	4.72
107	4.27	45–49	89	3.53
72	2.87	50–54	44	1.74
40	1.6	55–59	33	1.31
42	1.67	60–64	34	1.35
33	1.32	65–69	27	1.1
15	0.6	70–74	13	0.51
17	0.68	75–80	8	0.32
22	0.88	80+	11	0.43
Total = 2506	100		Total = 2521	100

Source Field survey 2006

The sex ratio of the scheduled tribe population was 1017 females per 1000 males (2001 Census) and the sex ratio of the Aka population as per the survey of 2006 was estimated as 1006 females per 1000 males (Table 10.1). The work participation rate was 40.82% of the total Aka population, which shows a high dependency ratio.

10.4 Changing Identity of Aka Tribe

10.4.1 Society

A change in a society is inevitable as society adopts new strategies to cope up with the newly emerging necessities with the passage of time. The Akas are also one of the tribal groups of the state, who have been exposed to changes in their social, economic and cultural life. Societal changes seem to be more prominent among the Akas. Traditionally, there was a joint family system in Aka society where all the members lived together with separate fire places. The study reveals that the family in the society at present is now more individual than in the joint system. The separation of sons from their parents is much more frequent than the past in recent years and has led to the emergence of more and more individual families. The composition of the family is also changing due to the change in the family type.

In the past, a family consisted of grandparents, parents, children and grand-children. Nowadays families consist mainly of parents and children only, which shows the emergence of a more individual type of family. The parents preferred to live with the youngest son in the past due to tradition and emotional attachment. In some rare cases they used to live with the eldest son if he was capable of taking care of them. The parents living with the intermediate son or the daughter is almost negligible in the society. In recent days, parents select any son or the daughter to stay with, keeping in view the ability, love and care provided during their old age. The institution of marriage is one of the aspects, where changes are taking place rapidly. The concept of marriage among the younger generation is becoming different than the meaning of the traditional marriage. The traditional marriage practices are losing their effectiveness in the present context. The time duration of marriage ceremony has been shortened in recent years due to less time availability as people are involved in government jobs, business and education. It may be the result of the influence of marriage systems practiced by other surrounding tribes in which less time is involved. The number of mithuns (*Bos frontalis*)—a semi-domesticated gaur to be gifted as bride price is also reduced to five or less. Another impact on the traditional marriage institution is caused by the conversion of people into other religions in the recent times. Changes in religion lead to changes in the societal aspects of a community. Nowadays marriages are taking place in churches, temples, etc. Such changes are influencing the valued traditional marriage system of the tribe (Nimachow 2011).

The traditional village council is also losing its importance in the society due to the introduction of the Panchayati raj system (PRIs). The constituents of the council are mostly filled up with the elected/selected members of PRIs instead of the appointed members of the village council. The induction of the younger generation as a member of Zila Parishad, Anchal Samiti, Gram Panchayat (three tier system of Panchayat Raj of India) has also reduced the importance and active role of the village elders in the decision making process of the village (Nimachow 2011). The function of village council has also taken a different form. The functions which were necessary during the past like war, defense, etc. are being replaced by other developmental works. The changes in the constituents and functions of traditional village council have profound impacts on the social organization of the village as a single entity. The villagers are becoming more and more individualistic than a group. The village as a community has significant role in the house construction, hunting, fishing, festivals, rituals, etc. but, the modern political system has forced the villagers to remain aloof from each other even within the village.

The literacy rate was very low in the past, but now it has increased to 63.57% with 71.26% male and 55.74% female respectively as per the survey of 2006. The majority of the literates have primary and secondary level of education. The number of graduates and post graduates are significantly less. The status of women in the Aka society is better than many communities, because they are allowed to participate in all activities ranging from social, economic, cultural to political. The form of marriage by capture, child marriage, polygamy and is now very rare baring few examples. More and more women are becoming educated and employed in different jobs, which has improved the status of women. The division of labour has changed tremendously in the society. The introduction of education, medical and other infrastructural facilities has encouraged them to leave those arduous tasks. Nowadays education has been given more emphasis and the children are being sent to schools irrespective of sex.

10.4.2 Infrastructure

The absence of good infrastructural facilities in the area was one of the major factors in the underdevelopment of the area. Educational institutions were introduced very late to the area. There were few public schools in the 32 villages in 1971 with little infrastructures. Consequently, the literacy rate of the area in that particular year was only 12.11%. However, the number of educational institutions has increased rapidly since then. In a recent survey there were nine (9) community schools, fifteen (15) primary schools, four (4) middle schools, two (2) government secondary schools and eleven (11) adult literacy class centres attached to some of the schools in 39 villages. Some more private schools started by the missionaries and individuals have also been established at Palizi, Thrizino, Buragaon, Bhalukpong and Yangsey villages. Christian missionaries established St. Xavier secondary school at Palizi village. Their branches have also spread to Thrizino and Buragaon, though these schools are at their initial stages of development. Apart from this there is a Navajyoti school and a primary missionary school at Bhalukpong. The impact of the increasing number of educational institution is the abrupt rise in the literacy rate to 64%. At the same time, some of the schools in the area are not functional and exist in name only such as the government primary schools of Yayong and Thissa. The area lags behind in medical amenities. As per the survey conducted in 2006, there has been no increase in the number of health sub-centre since 1991. It remained three and out of these the PHS of Thrizino has been taken over by an NGO the Voluntary Health Association of India (VHAI). The hospitals are also not well equipped to cater the needs of the people. Every year, there is an outbreak of epidemics of one form or the other. The villagers are facing acute shortage of medicines during epidemics. The villagers of Thissa, Yayong, Karangania, Kararamu, Palatari, Morkha, Pichang, Chijang, etc. have to cover 10–20 kms on foot to avail the medicines and doctor. However, the area enjoys good connectivity of pipelines of drinking water and the villages which do not avail the pipelines are located near to water sources. Therefore, the area has good drinking water connectivity but, there is no treatment of water which causes water borne diseases. Water treatment system and awareness programme could play a significant role in reducing the occurrences of water borne diseases. Banking facilities are provided by the only bank i.e. the State Bank of India (SBI), Bhalukpong. However, the SBIs situated at Chindit top, Bomdila and Seppa are also catering for the needs of the people. A branch of SBI at Thrizino, the headquarters of the sub-divisional office as a central place could be helpful for the villagers. Post and

Telegraph facilities are available at Thrizino, Bhalukpong, Bana and NEEPCO at Kimi that cover the all villages. The other villagers have to travel more than 10 kms to avail such facilities. Poor transport and communication is one of the hurdles to the proper development of the area. The majority of the villages are connected by footpaths only. There are few villages located along the metaled roads. The two important concentrations of the population i.e. Buragaon and Thrizino are connected by un-metaled roads, which are not drivable in the summer season. There is an overall lack of transport infrastructure in the area. The number of buses and bus stands are also very low in comparison to other parts of the district and state. Though all the villages are electrified, power is only available during one season of the year only for domestic use. In Thrizino and Bhalukpong power supply is available throughout the year because of the arrangement of generators in case of a main power supply failure. Bhalukpong, located at the border with Assam, is a main market centre. Therefore, the villagers are dependent on the market of Bhalukpong only. Some minor day to day requirements are acquired from the market at the district headquarters. Major commodities are availed from Bhalukpong, Balipara and Tezpur (Assam).

10.4.3 Changing Culture

Change is inevitable, but change or transformation has two aspects; some changes sound healthy and some changes sound unhealthy for the society. The change in marriage system (especially marriage by capture) sounds good to prevent the exploitation of the women. However, the abandonment of the practice of rituals in marriage is regarded as unhealthy, because of associated loss of culture. They are very much attached to the priests and the rituals and even in this age of science and technology; they are still practicing a number of rituals in marriage ceremonies. Though, the rituals performed in the marriages through capture acts as a restriction forbidding the girl from running away, it has acquired some social importance during the olden days. In recent years there is a rapid decline in the number of priests in the Aka area. In some villages it has fallen so fast that even a single priest is not available throughout in some villages. Lack of interest among the younger generation is one of the important reasons for the rapid decline in the number of priests. Scientific knowledge, medicinal developments and other related developmental activities has reduced the belief and faith of the people in the priests. However, the conversions of the people towards different organized religions as well as the forces of modernity are also responsible for the decrease in the number of priests. The decrease in the number of priests as well as the changing attitude of the society has led to a decline in the practice of rituals on all fronts, such as marriage, offerings, construction of house, hunting, fishing, etc. Any kind of realization among the people regarding the importance of rituals is still essential at present. In the Aka society the oldest religion, which prevails from the earliest times is only the *Nyezino* or *Nyezi aou* (sky the father), *No ain* (earth the mother). Though, there is not any form of statue or idol for worship as in other religions, the word *Nyezino* has such a place in the minds and souls of the Akas that they use to take the name of *Nyezino* at every moment of grief, problems or sufferings. The penetration of the scientific knowledge and technology into the land of this forest people had caused deep impacts on their lives. Due to the assimilation of knowledge and the early adaptation by the local people, there are lots of changes in the sphere of religion as well as festivities. People are converting to Christianity, Hinduism, etc., which may be the outcome of either the local people's inclination towards new faiths or for getting better facilities from various organizations of the other recognized religions. During the field survey, out of a total Aka population (5027), 3905 are still continuing their own cultural practices i.e. *Nyezino* and *Mene Alan*, whereas 1117 persons have abandoned their own cultural practices and adopted Christianity. However, only 5 persons declared themselves as Hindu. They consider that the existing Aka cultural practices were among the Hindu believers during Vedic period.

10.4.4 Changing Livelihood Pattern

Forests and forest products have a profound impact on the economy of the Akas. Their traditional economy revolves around forest which includes various occupations such as agriculture (*jhum*/shifting cultivation) and subsidiary means like hunting, fishing and food gathering. The products of *jhum* field such as vegetables, cucumbers, fruits and maize are sold for the requirement of money. Farmers personally do not sell their product in the market. Usually, they handover their products to the owner of a shop to sell. The shopkeeper gets some money in the form of a reward for selling those products. Likewise wild vegetables, mushrooms, wild fruits, bamboo, etc. are collected from the forest for self-consumption. Only the excess amount is given to the shopkeeper for selling. The meat of wild animals and fish are also sold in exchange for money. Whenever they kill a big wild animal, they sell the half portion in exchange of money and keep the remaining portion for the family members. In case of an excess fish catch, some portion is sold and rest retained for the family members. The *jhum* cycle, which was 5–10 years during the post-independence period, has declined to less than 3 years at present due to increase in population. Proper suggestions and alternatives to *jhuming* could help in the mitigation of adverse linkages emerging out of man—nature interactions in the area. The inhabitants along the main road or the settlements having good transportation facilities have mostly changed their economy from *jhuming*, hunting, fishing and food gathering to timber extraction. The villagers of Jamiri, Khuppi, Bhalukpong, Tippi, Palizi, Bana, etc. became more or less timber contractors and heavily depended on it. Sawmills have been established at Bana, Palizi, Tippi. A plywood mill was also established at Tippi. In addition to timber extraction the other business activities were the production of cane, herbs for medicine and dye extraction, etc. But, the ban on timber operations by the Hon'ble Supreme Court in the year 1996 has badly affected the economic life of the people. This ban has created such a situation for them that neither they can continue their business nor they can go back to their traditional economy. For a period of about 5–10 years they had forgotten their traditional economy and shifted themselves towards timber business. The ban on timber extraction had slowed down the pace of timber extraction to certain extent but still it is going on for personal use. Along with the timber operations, the extraction of cane, herbal/medicinal plants, etc. are also gaining momentum. The abundance of different species of trees, bamboo groves and plantation leaves play important role as these provide them variety materials for house-building, hunting/fishing implements, agricultural tools, dress and ornaments, domestic utensils and other implements.

10.4.5 Changing Biodiversity

The area is enriched by its great diversity of plant and animal resources. By virtue of this, they are largely dependent on the surrounding forest resources for their livelihood. The inherent Indigenous Knowledge System (IKS) allows them to interact with surrounding forests in differential ways. The Akas as a forest people have age-old tradition of conservation, preservation, protection, management and optimal utilization of the forest resources. They worship the forests as the feeder and rearer to the human beings. These include the protection of sacred-groves that are not intruded on, as plucking of even single leaf from such areas is strictly prohibited. They do not kill some animals and birds, which maintains the ecological balance. A number of social restrictions and customs have been in vogue, which every man and women has to abide by. They do not fell some trees because there is a strong belief behind the felling of these trees e.g. felling of the banyan tree causes illness to the person. Their belief systems, along with concept of sacred groves play significant role in conservation of plants and animals (Fig. 10.4). For example, the source of drinking water is believed to be an abode of a deity locally known as *Hugyein ubrao/ubram*, who helps to retain the purity of the water. They worship the Peepal tree (*Ficus religiosa*), which is considered to be the home of deity known as *Wiinjigi*. Mountains are also worshiped by them and are locally known as *Vojophu, Liiriiphu*, *Jyopsinfo*, *Nyejowoh*, etc. These areas are believed to possess magical powers and if any one extracts materials from the area, they will get lost in the forest. The forests are the source of vegetables, fruits, barks, edible leaves, wood, bamboo, cane and thatch. Agricultural implements, bows and arrows, handles, cloths, hair oil, musical instruments, ornaments and ethno- medicines (Table 10.2) are also met out from forest resources. Some parts of animals such as antelopes, bear, etc. and fish are used for orthopedic and stomach-ache problems. However, the penetration of scientific knowledge has caused deep impacts on their linkages with forest resources leading to loss of biodiversity in the recent past in the form of large scale deforestation and the loss of important plants and animal species. The survey reveals that some important species of plants are on the verge of extinction and certain wild animals like *Muntiacus muntjak*, *Selenarctos thibetanus laniger*,

Fig. 10.4 Sacred places, beliefs and rituals. Photo: **a** *Nyezowo* (An isolated hill believed to be sacred at Jamiri village), **b** *Jyopsinfo* (Shining stones believed to be sacred near Prizin village), **c** *Vojophy* (A mountain believed to be sacred at Palizi village), **d** An elongated stone believed to be sacred near Kararamu village, **e** *Hugien* (Source of drinking water with puja altars at Yayong village), **f** *Nyetrii Psigha* (A sacred place where meetings and festivals are conducted at Yayong village, **g** *Suin* (Erection of stones as a mark of agreement between two parties), **h** Altars for the sacrifice of *Mithun*, **i** Puja altars constructed inside the house

Table 10.2 Ethno medicinal plants used by the Akas

Scientific name	Local name	Therapeutic indication	Mode of use
Paederia foetida Linn. Rubiaceae	*Adraluhumbe*	Fire and hot water burns	10 gm of fresh leaves and stems are crushed in indigenous mortar and pestle, sieved with fine cloth and the juice is applied on the burns thrice a day
Macaranga denticulata (Blume) Muell. Arg.	*Liidzin*	Fire and hot water burns	Fresh resin is collected from the plant in bamboo tube and applied in the burns (anti-inflammatory) SOS

(continued)

Table 10.2 (continued)

Scientific name	Local name	Therapeutic indication	Mode of use
Euphorbiaceae			
Artemisia nilagirica (Clarke) pamp. Asteraceae	*Syowum*	Cough and fever	10 gm of fresh leaves and stems are crushed in indigenous mortar and pestle, sieved with fine cloth and the juice is applied on the burns thrice a day
Begonia sp. Begoniaceae	*Pelowo*	Boil	Tender leaves are steamed and taken for quick recovery and lessening of pain and also used as blood purifier
Ageratum conyzoides Linn. Asteraceace	*Pasong*	Cuts	Paste are prepared from leaves by hand squeezing method and applied in fresh cuts for clotting of blood and anti-microbial action
Curcuma sp. Linn. Zingiberaceae	*Kiistradu*	Stomachache	5 gm of clean rhizome are consumed twice a day for at least 3 days in severe stomach pain
Dendrocalamus hamiltonni Poaceae	*Si-emnyo*	Cuts	With the help of dao/peeler the culm is peeled and powdery peel off is directly applied to fresh cut and injury for clotting of blood
Zanthoxylum piperatum (L.) DC. Rutaceae	*Siina*	Labour pain	Dry fruits are fried in hot plate and taken with warm fermented local made alcohol *Tsii* (rice beer) mostly during labour pain and after delivery
Clerodendrom colebrookianum Walp. Lamiaceae	*Droloin*	Diarrhoea	The tender leaves are steamed and consumed during diarrhoea and other indigestions
Centella asiatica Linn. Apiaceae	*Tuffo*	Jaundice	Raw roots and leaves are taken along with routine food during jaundice for appetizer
Discorea sp. Dioscoreaceae	*Nyemumsi*	Dysentery	Cooked yam (both boil and burn) is consumed for curing dysentery. It is also applied to domestic animals for the same purposes
Ricinus cummunis Linn. Euphorbiaceae	*Migyim jyoksu*	Orthopaedic	10–20 gm of fresh leaves are made into paste and applied to fractured bones and joint pains. 2–3 leaves are slightly heated in fire and tied with a rope over the fractured part and kept for a week
Costus specious (Koenig) Smith Zingiberaceae	*Rumo-sana-dugo*	Jaundice, gastric and eye infection	The juicy beat is chewed just like sugarcane during jaundice and gastric. 1–2 drops of the extracted juice from the stem is applied during eye infections twice a day for immediate relief and recovery
Rhus javanica Linn. Anacardiaceae	*Subyutro*	Dysentery	In mild condition raw fruits are consumed directly. But during severe dysentery the juice is extracted from the fruit by squeezing method and taken 3–4 glass a day
Unidentified	*Mechme*	Diarrhoea	2–3 leaves are consumed directly with water for at least 3 days

Source Field survey 2006, 2008

Macaca munzala, *Sus scrofa*, etc. are endangered. The plants which were found within 1 km radius are now found in around 40 km radius from the villages. The existing modern judicial laws against illegal exploitation are hardly implemented. They deal with these problems through customary laws known as *Nobazio*.

The existing customary laws and knowledge base in relation to their ecosystem may facilitate a revolution in preserving and protecting huge forest resources in other areas. The forest policy makers/authority concerned with the Anchal forest may invite villager's views and their participation to avert the adverse impact on the forest ecology and adopt mitigation measures to prevent certain problems in the surrounding forests. The planning for tribal development must take forest resources as the base on which the tribal economy can progress with greater confidence. The tribal people should become not only the co-sharers in the new wealth created in those areas but should also have an active role in its management. Their existing knowledge base, which is popularly termed as 'Indigenous Knowledge System' (IKS) in relation to their ecosystem may facilitate a movement to preserve and protect huge forest resources.

10.5 Conclusion

The Akas are also one of the tribal groups of the state, who have been exposed to changes in their social, economic and cultural life. Societal changes seem to be more prominent among the Akas. The society has experienced remarkable changes with the passage of time. The marriage institution is one of the aspects, where changes are taking place rapidly. The concept of marriage among the younger generation is becoming different than the meaning of the traditional marriage. The traditional marriage practices are loosing their effectiveness in the present context. The traditional village council is also losing its importance these days in the society due to the introduction of the Panchayati raj system. The literacy rate of the people was very low in the past, but now it has increased to 63.57% being 71.26% for males and 55.74% for females respectively as per the findings of the field survey 2006. But, the level of education is disproportionate as the majority of the literates belong to the primary and secondary level of education. Nowadays the rituals are declining due to conversion of religion coupled with developmental activities. In recent days, there is a rapid decline in the influence of priests. In some villages it has depleted so fast that even a single priest is not available throughout the village. Lack of interest among the younger generation is one of the important causes for the rapid decline in the number of priests. The penetration of the scientific knowledge and technology into the land of this forest people had caused deep impacts on the lives of these people. Due to the assimilation of knowledge and the early adaptation of the local people, there are lots of changes in the sphere of religion as well as in the festivities of the people. People are converting to Christianity, Hinduism, Buddhism, etc., which may be the outcome of either the local people's inclination towards new faiths or for getting better facilities from recognized religions.

However, in the Aka society the oldest religion, which prevails from the earliest times is only the *Nyezino* or *Nyezi aou, No ain*, sky the father and earth the mother. Though, any form of statue or idol worship like other religions is nonexistent, the word *Nyezino* has such a place in their minds that they use the name of *Nyezino* at every moment of grief, problems or sufferings.

The area has not progressed impressively during the post independence period. They represent one of the oldest recognized tribe of Arunachal Pradesh. But, the region is still lagging behind in terms of the present day indicators of development. There are many factors behind such underdevelopment. The lack of efficient & effective transportation system and the inadequate means of communication are the major hurdles of development. The whole area is located in between the two district headquarters i.e. East Kameng and West Kameng. Both the headquarters are linked with metaled roads. Excluding the settlements situated along this metaled road, most of the settlements are not linked by road. Low literacy rates as well as low per capita income are the other factors responsible for the underdevelopment. Their social customs, beliefs, faith, tradition etc. reflects a deep imprint of the nearby forests. They worship the forests as gift of god to them. Their folktales, folklores, myths and festivities are highly influenced by the nearby forests. Besides, they practice traditional mode of conservation and management of plant and animal resources like protection of sacred-groves. The economic organization, material culture, food habits, social, cultural and religious practices are largely determined by the existing forest resources. There is an urgent need for the formulation of a sustainable forest management plan for the area. The community forestry system of Nepal has performed well over the decades where the forest user groups are primarily responsible for forest management plans. Hence, management plans in line of other successful management systems could prove fruitful for the area in future. In view of the above mentioned changing aspects of indigenous communities there is a need of developing strategies to cope up with the process of globalization. Further, an assessment of healthy and unhealthy changes in the society is highly essential for balanced development of the people. Therefore, the indigenous people must be taken into confidence for preservation of rich cultural practices, livelihood and biodiversity in future though participatory approach.

Acknowledgements The authors are thankful to the Rajiv Gandhi University, Rono Hills Itanagar for providing basic facilities for carrying out their research. The authors also acknowledge the grant received from Indian Council of Social Science Research (ICSSR), New Delhi in the form of a major research project. Here presented chapter is a piecemeal work from the findings and analysis of the project.

References

Balan, K. (1992). *Socio-economic change in India* (p. 73). New Delhi: Ashish publishing house.

Choudhary, S. D. (1996). *Arunachal Pradesh district gazetteers, East Kameng, West Kameng and Tawang districts* (p. 272). Shilong: Gazetteers Department, Government of Arunachal Pradesh.

Geology and Mineral Resources of Arunachal Pradesh. (2010). Miscellaneous Publication No. 30 (Part–IV, Vol. I(i)). Shillong: Geological Survey of India, Government of India.

Ghosh, G. K. (1992). *Tribals and their culture*. New Delhi: Ashish publishing house.

Gohain, B. K. (1994). *Continuity & change in the Hills of Assam* (p. 22). New Delhi: Omsons publications.

Joshi, R. C. (2006). Physiography of Arunachal Pradesh. In N. Nagaraju & B. Tripathy (Eds.), *Cultural heritage of Arunachal Pradesh* (pp. 19–31). New Delhi: Indus Publishing Company.

Kholey, R. N. (1997). In P. C. Dutta & D. K. Duarah (Eds.) *Aspects of customary laws of Arunachal Pradesh* (p. 16). Naharlagun: Arunachal Government press.

Nimachow, G. (2008). *Tribals in transition: A geographic study of society, economy and culture of the Aka tribes of Arunachal Pradesh*. Unpublished Ph.D. thesis submitted to Rajiv Gandhi University, Itanagar.

Nimachow, G. (2011). *The Akas: Land and people*. New Delhi: Commonwealth Publishers.

Parihar, V. (1989). *Society in transition* (p. 36). Jaipur: Printwell Publishers.

Riba, T. (1997). *The tribes and their changing environment: A case study of Galos (Adi) of West Siang district. Project report*. Rono Hills Itanagar, Arunachal Pradesh: Department of Geography, Rajiv Gandhi University.

Sinha, R. (1962). *The Akas* (p. 143). Itanagar: Directorate of Research.

Soumen, S. (1988). *Between tradition and change* (p. 16). New Delhi: B. R. Publishing Corporation.

Statistical Abstract of Arunachal Pradesh. (2011). *Directorate of economics and statistics*. Itanagar: Government of Arunachal Pradesh.

Verrier, E. (1968). *Myths of the North-East Frontier of India* (pp. 116–117). Shillong: Research Publications.

Chapter 11
The Changing Socio-economic Profile of the Shaukas and Rangs Vis-à-Vis Loss of Agro-Diversity, and Yartsa Gunbu: A Case Study of the Askote Conservation Landscape, District Pithoragarh, Kumaun Himalaya

Chandra Singh Negi

11.1 Introduction

The Askote conservation landscape is located in eastern Kumaun, in the state of Uttarakhand of India and lies between the coordinates 80°15′–81°5′ E longitude and 29°5′–30°32′ N latitude, and is bounded by the international border with the Tibetan Autonomous Region of the Peoples' Republic of China in the north. The Kali River forms the boundary in the south-east of the landscape till Jauljibi, which also constitutes the border of India with the Kingdom of Nepal. The western and north-western boundary of the landscape run along the true left bank of Gori River, till close to the latitude 30°11′ N, and then follows the ridge on the true right of the Poting sub-basin to Silwa Dhura (Fig. 11.1). Encompassing an area of 4463 km^2, the entire project landscape is about 120 km long and in average, about 51.5 km wide. There is a great altitudinal range within the landscape, from 560 m amsl at Jauljibi, to 7434 m at the summit of Nandadevi East, and this constitutes the Centrum of the bio-geographic elements of the Western Himalaya, the Central Himalaya and the Tibetan plateau (Negi 2010). As a result, the landscape is rich in terms of biodiversity, containing reportedly 2607 species of vascular plants, 265 species of birds and 37 types of mammals (FES 2003). The landscape contains 129 villages spread from sub-tropical to alpine altitude zone, falling within the Dharchula and Munsiari Tehsils.

The altitudinal range of the valley in the landscape is as follows: 12% of the villages lie in the sub-tropical altitudes, 55% in the warm temperate zone, 18% in

C.S. Negi (✉)
Department of Zoology, L S M Government Postgraduate College,
Pithoragarh 262501, Uttarakhand, India
e-mail: csnsacred1@gmail.com

R. Chand et al. (eds.), *Societies, Social Inequalities and Marginalization*,
Perspectives on Geographical Marginality,
DOI 10.1007/978-3-319-50998-3_11

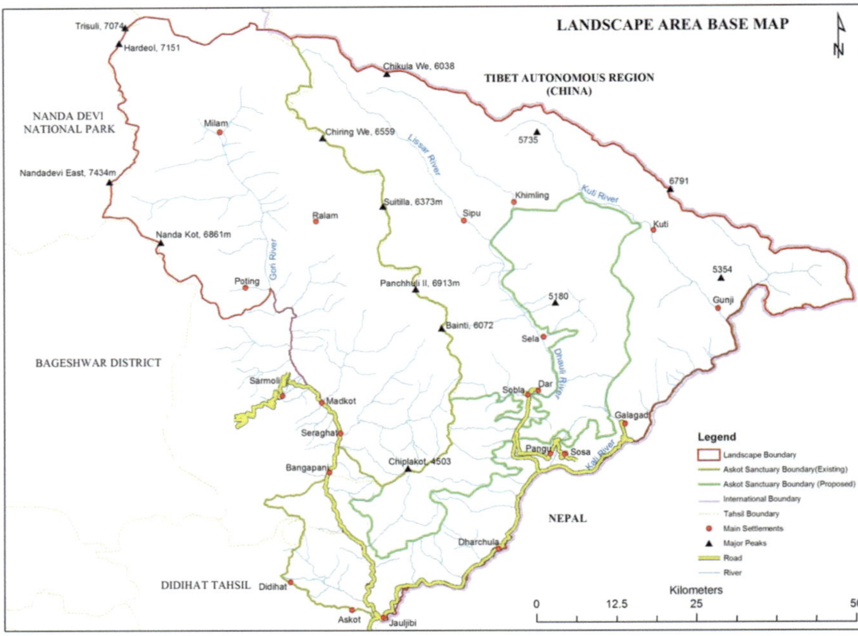

Fig. 11.1 The Askote conservation landscape

the cold temperate zone, 7% in the sub-alpine zone and 8% in the alpine zone. Overall population density for the landscape is 21 people per square kilometer and the per capita availability of agriculture land in the landscape is 0.13 ha. Permanent human habitations occupy the lower altitudes, up to the temperate zone, which are more densely populated. In comparison, there are only a few settlements in the cold-temperate area and villages in the sub-alpine zone are only seasonally occupied, and form part of the transhumant land use practices common in high mountain areas. The landscape being described renders little land suitable for agriculture. Of the total land area of about 4463 km^2, only about 103 km^2 or just 2.31% of the area is cultivated land. For about 68,939 people in 14,010 households, who live in 129 villages in the landscape, with highly fragmented holdings, this 2.31% of land does not suffice even for subsistence. An estimation of the proportion of food grain produced locally in 80 villages in the Gori basin indicates that only close to half the food grain requirements of human and their livestock are met from local agriculture (FES 2003). Again, when a mere 2.3% of the total land is cultivable, the means of sustenance remains not agriculture, but the collection of the wild resources, including medicinal and aromatic plants (MAPs), and Yartsa Gunbu. This fact alone becomes all the more accentuated, when one analyzes the actual land use configuration within the landscape. Reserve forests cover only 8%, the Van Panchayats or village forests, managed by village forest councils covering 34.48% (and encompasses some of the most pristine areas of highest biodiversity values), while Civil and Soyam revenue land comprises the rest 54.59%.

Loss in the crop biodiversity has taken place over the years, which principally and inadvertently relates to the changing lifestyle with the closure of the traditional trade with Tibet, aftermath the 1962, Sino-India war. Added to the phenomenon is the growing demand of certain MAPs, principally Jambu (*Allium stracheyi*) and Caraway (*Carum carvi*), whose acreage has increased with the concomitant decline in area under the traditional crops. In addition the recent discovery of Yartsa Gunbu (*Ophioc ordyceps sinensis*) in the alpine zones of their summer homes have dwindled their interest in the sowing, rearing and caring of their traditional crops, viz. beans, potato, amaranths, etc., whose growing season coincides with the harvesting season of Yartsa Gunbu (Negi 2007a, b).

It needs to be emphasized here that the traditional crop varieties and races adapted or acclimatized to the local high altitude climatic regimes and conditions, not only meet the nutritional and food security of these traditional communities, but more importantly formed a major commodity that was traded with Tibet in the immediate past. This traditional crop diversity has not only withstood the rigours of time, survived the pest infestations, disease, and so forth, but more importantly possesses the desired agronomic and genetic traits needed to develop resistant varieties (Altieri 1991; Anonymous 1989). Loss in crop biodiversity or the loss in the productivity of the traditional crops or the general prevalence of raising monocrops through replacement of the traditional crops with high yielding varieties (HYVs), or the effect of green revolution on the traditional crop biodiversity, or even the effect of global warming and the loss of traditional crop biodiversity, are some of the topics that has been dealt with (Maikhuri et al. 1999; Altieri 1991; Maikhuri et al. 1991). However, in the present landscape, the resultant closure in trade ties with the Tibetans post 1965, and the resultant changes in the socio-economic profile, infact remains the prime reason behind the resultant loss of traditional crop diversity.

11.1.1 The Mode of Sustenance

Bhotiya is a generic term derived from *Bodh,* the ancient name for Tibet, and is commonly used to describe the border tribes who conduct Trans-Himalayan trade. However, the people themselves resent being referred to as *Bhotiya* and prefer to be known by their indigenous names—*Shauka* and *Rang*. Primarily Mongoloid in appearance, with occasional Aryan and aboriginal traits still perceptible, the Shaukas and Rang are actually an agglomeration of independent and mutually exclusive sub-groups like the Tolchchhas, the Marchchhas and the Jads, residing mainly in the frontline areas bordering on Tibet (Hoon 1996). They are a loosely knit indigenous community, having their permanent abode in the seven river valleys —Bhagirathi-Jahanavi-Ganga in Uttarkashi district; Alaknanda and Dhauli Ganga in Chamoli district: Gori Ganga, Darma or Eastern Dhauli, Kali and Kuti-Yangti Rivers in Pithoragarh district.

Transhumance

The migratory folks practicing the transhumance, restricted to the annual movement from the winter homes to their high altitude summer residences are termed as *Kuncha.* Transhumance enables the *kunchas* to gather rare Himalayan herbs, chiefly *Kutki* (*Picrorhiza kurrooa*), *Gandrayan* (*Pleurospermum angelicoides*), *Atis* (*Aconitum heterophyllum*) and *Hathazari* (*Dactylorhiza hatagirea*) and to utilize agricultural land in the summer villages by making use of the growing season to raise specialized high altitude crops, such as buckwheats-*phaphar* (*Fagopyrum esculentum*) and *palthi* (*Fagopyrum tataricum*), which are adapted to grow in poor soils. This agricultural endeavour provides them with almost 50% of their dietary needs. Yet another class—the *anwals* (the traditional shepherds), practice pastoral nomadism as an appropriate technology, which allows them to utilize marginal resources unsuitable for other uses, and at the same time maintain a breed of sheep and goat that is much larger and more productive than those raised by sedentary peasants. Additionally, the livestock population can be harnessed as energy converters in environments, which would otherwise be less productive to human population. Notwithstanding, the economics of these two governing principles of sustenance, a marked decline in the transhumance has been observed over the last two decades, born out of (i) the severance of trade with Tibet, post the 1962 Indo-China war and (ii) lure of government services with extension of the reservation (Negi 2010).

11.1.2 Trade: Past and Present

The Shaukas and Rang had a barter trade relationship with the Tibetans and were masters of the trans-Himalayan trade. Sheep and goats transported the trade merchandise. They transported the grain over the passes from June to October, and returned with borax, salt and wool. People from the lower hills bartered their excess grain at any one of the trading depots of the Shaukas and Rang, for salt and wool. The Bhotiya traders jealously guarded their trade routes and maintained a monopoly over the trans-Himalayan trade with Tibet. The main items exported were food grain, sugar, gur (unrefined sugar), spices, tobacco, cotton cloths, hardware, corals, beads, etc., while the main items of imports were wool, gold, chirbi, Thi (a sweet), Yak's tail, Yak's skin, corals, precious stones, salt, etc. Western Tibet depended on the Bhotiya trade for its food grain supply. This apart, since pastoralism was closely related to the Bhotiya trade economy, the Bhotiyas' livestock population comprising of sheep, goats, ponies, yak, and jhuppu, were all primarily of Tibetan stock (Hoon 1996).

The fortunes of the Shaukas and Rang, however, changed drastically with the Chinese assumption of control over Tibet in 1959 and the Sino-India war in 1962. In 1965, the Indo-Tibetan border was sealed and trans-Himalayan trade came to a grinding halt. The livelihood of at least two-thirds of the Bhotiya population was bond exclusively with income from trade. In fact, it can be said that the transhumance lifestyle of raising sheep and goats for pack animals and the family movement from winter village to summer village had developed around the need to maintain trade relations. The disruption of the trade not only deprived the professional traders of their livelihood, but it also had an adverse effect on the livelihood of several other professions, which were indirectly dependent on the continuance of trade. With the closure of the trade, for example, there was no longer a demand for the freight shepherds, who used to carry merchandise for the Bhotiya traders *to and fro* from Tibet. Traditional crafts also suffered, since as long as Tibetan wool was imported in sufficient quantities, weaving flourished and the Shaukas and Rang produced a variety of woolen cloth for their own use as well as for sale in other parts, including neighbouring Nepal.

However, with the re-opening of trade in the mid-90s, agricultural produce, have now become the chief form of export, from these valleys. For example, potatoes are sold at 6 Yuan/kg, radish at 12 Yuan/kg, grated and dried radish for 30 Yuan/kg and hill tobacco for 12 Yuan/kg (1 Yuan = 6 Re) (Hoon 1996). This is all the more significant, since more than 2/3rd of all the export or trade items are locally produced. However, at present the Shaukas and Rang believe that the government would do best by lowering the prevailing custom duties (which are very high). A duty free structure with less government interference would ensure prosperity and thus help in sustaining this unique lifestyle and culture, they believe in. Additionally, it needs to be emphasized here that if trade potential is to be enhanced, then one will have to make a study of the commodities that could be traded at present with the Tibetans, for the same reason that the commodities of the past, viz., say *Fagopyrum* spp. and the products made out of the same acreage of the said crop has drastically reduced since then, with whatever produce being procured out of cultivation, barely meeting the requirement of the resident population (Negi 2010). Also, it needs to be seen that even if the stakeholders go in for the cultivation of the erstwhile crops, will the commodities still find a market in the Tibet? These are the issues that need to be urgently taken care of, if we do sincerely think of planning to revitalize the economy around trade.

11.1.3 Agro-Diversity

Although agriculture land covers just 2.3% of the project landscape and though landholdings are marginal and fragmented, the landscape harbours rich crop diversity. 211 different local varieties of food crops have been identified as being grown in the landscape, of which 105 can be classified as cereals or pseudo cereals, 21 pulse crops, 10 oil-yielding plants, and 10 species (FES 2003). 41 varieties of paddy (*Oryza sativa*) are grown here of which 31 are upland varieties, not requiring field flooding. 20 varieties of wheat and barley (*Triticum* sp.) are grown and 14 varieties of finger millet (*Eleusine coracana*) apart from 5 other millets from the *Panicum* sp. and *Pennisetum* sp. The list also includes 31 different vegetables and

31 fruits plants. However, the villages located above 1800 m do not cultivate more than 7–8 varieties of vegetables in a year due to the short growing season but make use of 10–12 different wild plants as vegetable. The equivalent of staple cooked as rice are paddy, setaria, and amaranth, while chapatti or its equivalents are made from wheat, maize, barley, millets and buckwheat. The narrowing genetic base of traditional agriculture crops in the landscape is a matter of concern, since fewer varieties of *Triticum sp.*, beans and peas and of paddy is grown now. Agriculture in the lower (sub-tropical to sub-temperate zone) in the landscape is characterized by subsistence cultivation of crops like Phaphar (*Fagopyrum esculentum*), Palthi (*F. tataricum*), barley (*Hordeum vulgare*), Mandua (*Eleusine coracana*) mixed with cheena (*Panicum miliaceum*), been (*Phaseolus vulgaris*), mustard (*Brassica campestris*), and potato (*Solanum tuberosum*). The traditional system of crop management here involves a mixed range of crop rotation and multiple cropping patterns, involving inter-change of crops and mixture of varieties and different levels of input for each of numerous small plots of land (Negi 2007a). However, in the high altitudes, due to the short growing season (April–October) there is only one crop season in a year. Water availability is often a constraining factor, for example in Johaar valley, 14 out of the 19 migratory villages, lack adequate supply of water.

Bottlenecks in agricultural production

Heavy monsoon rains do not reach the Darma and Vyas valleys due to the rain shadow of the high Himalayan peaks. Ambient temperature is closely related to elevation and frequently is a determinant of the crop selection made. Low temperatures during winter months may force the choice of frost tolerant crops in many environments and set the time needed for a crop to reach maturity. In general, the potential for agricultural intensification decreases with altitude in mountains. The per capita availability of agriculture land is poor, mountain soils are poor in nutrients and holdings are severely fragmented. Even though, a high degree of micro-climatic variability based on slope aspect, humidity, rainfall, temperature and the resulting diversity at different altitude and aspect gradients in species and varieties being cultivated in quite a few villages; the local inhabitants, in general have failed to optimize the different niches by growing variety of crops, or minimize risk by harnessing different production systems.

In low elevation, the traditional methods of farming here incorporate some techniques to maintain and improve the fertility of the field, which include the following—(i) a high input of compost manure, (ii) cultivation of legumes or nitrogen fixing crops, (iii) fallowing fields for a season, and (iv) changing crops in rotation (taking a low nutrient demanding crops after a high nutrient demanding crop). The productive capacity of different field are kept in mind e.g. prime or irrigated fields are used for growing wheat and paddy and the relatively degraded and gravely land is used for growing pulses and oil seed crops. However, in higher elevations (summer homes), agricultural practice is defined by the prevalent physical limiting factors, as also by the short duration, for which cultivation is possible.

11.2 Agriculture in Johaar Valley: A Case Study

Here the per capita availability of agriculture land is comparatively poorer, defined by soil, which is poor in nutrients, and compounded by severely fragmented landholdings. On an average around 0.11% (or about 1.31 kms^2) of the landscape comprises of agricultural land, of which around 16% is devoted to the crops (or actually cultivated). Some of the most characteristic features of the agriculture in these areas (modified from Asher et al. 2002) are (i) agriculture is more for cash cropping than to meet the subsistence food requirement, with the use of large quantities of organic manure, and virtually no use of chemical fertilizers, (ii) single cropping—because of the short growing season, and thus no mixed cropping. Today, one comes across three basic forms of agricultural fields (Negi 2007a):

1. Traditional fields that continues to be used, and are big, flat with mostly sandy loam soil with no irrigation facility. Usually cereals, mustard, barley, buckwheat and some durum wheat are grown in these fields.
2. Vegetable gardens, usually very close to the homes or within the home precincts, often very rich in organic matter- those procured from the kitchen waste, or livestock manure or even leaf litter from the milch animals. However, vegetable gardens form a very small proportion of the total area under cultivation.
3. Fields within the village ruins. As most of the villages encountered bear a very desolate look with very few families actually residing in the villages, most of the dwellings have broken down, with crumbled walls protecting the crops being grown (chiefly *Carum carvi* and *Allium stracheyi*) within.

Since, the mountain soils by nature are poor in nutrients, and to produce any crop requires a heavy input of manure, which is made possible by procurement of nutrient cycling from the surrounding grasslands. Often one encounters fallow land, where the livestock are left in the night or the anwals (the transhumant herders) are told to arrange their stay accompanied by their stock of sheep and goats; all principally with an aim that a mass of the livestock manure could be procured, without any wastage, whatsoever. Manuring or nutrient enhancement is primarily done through these following means (i) application of composted manure: Phicchi grass (*Danthonia sp.*), often used as bedding for the cattle, is mixed with manure to form the compost, (ii) crop residue: The stalks of crops such as Thoya (*Carum carvi*) and mustard are spread on to the fields and left to decompose. In addition some of the grass species growing in the wild too are collected and spread in the similar fashion to decompose and rot.

Polyculture: The preferred norm

The crop management involves a mixed range of crop rotation and multiple cropping patterns, involving the inter-change of crops and the mixture of crop varieties and different level of input for each of numerous small plots of land. Crops, such as Phaphar (*Fagopyrum esculentum*) are grown in all villages within the landscape (but not to the extent as observed in Darma valley), and often mixed with lentils, bean, sesame or mustard. Potato fields are seldom changed, due to its consistency in the production. As long as the regular rotation between buckwheat and potatoes is observed, the yield of these crops does not go down and there is no need for extended periods of fallow. Subsistence crops cover more than 90% of the total area under cultivation. The short cultivation season available to these transhumant migrants dictate the cropping system to be followed, and thus only those crops, which would ripen within a very short span of 3 and half months to 4 months, are grown. Around 21 crops or vegetables are currently grown in these villages, all of which are seasonal (Table 11.1).

Out of the total number of crops cultivated, mention must be made of two intensively grown remunerative crops—Jambu (*Allium stracheyi*) and Caraway (*Carum carvi*). *Allium* remains a highly priced spice, which sales for 150–200 Rs. in the local market at Munsiari. This crop, along with the caraway covers the greater

Table 11.1 Some of the crops encountered in the Johaar valley

Crop species	Vernacular names	Area (ha) 2002	Percent of total area	Production (kg) 2001
Fagopyrum esculentum	Oggal/palthi	0.020	0.09	18
F. tataricum	Phaphar	4.380	21.03	2975
Hordeum himalayens	Uwa-jau	0.166	0.80	60
Solanum tuberosum	alu	3.950	18.97	18,556
Brassica campestris	Sarson/mustard	6.960	33.42	3683
Triticum himalayens	Nappal/durum wheat	0.218	1.04	40
Lens esculentum	Masoor	0.086	0.41	19
Pisum sativum	Kalon	0.004	0.02	7
Other vegetables	–	0.012	0.06	Unavailable
	Total food crops	15.795	75.85	
Allium stracheyi	Jambu/chives	3.499	16.80	2431
Carum carvi	Thoya/caraway	1.328	6.38	1121
	Total spices	4.827	23.18	
Angelica glauca	Gandrayan	0.137	0.66	15
Aconitum heterophyllum	Atis	0.024	0.12	5
Rheum australe	Dolu	0.005	0.02	6
Saussurea costus	Kuth	0.030	0.15	–
Picrorhiza kurrooa	Kutki	0.005	0.02	–
	Total MAPs	20.823	100.00	

Source Asher et al. (2002)

proportion of the cultivable land in Johaar. An additional feature of this crop is that on per unit area basis, it gives a higher yield as the same is harvested twice in the very limited growing season. *Allium* happens to be the perennial crop (i.e. the plants regenerate each year from the same vegetative stock every growing season without having to resort to fresh seeding for propagation). Of the three different species of *Allium*: *Allium stracheyi*, *A. wallichii* and *A. humile,* it is only *A. stracheyi*, which is actually cultivated. In comparison, caraway is a singular biennial crop. According to Asher et al. (2002), the Johaar valley alone yields around 1100 kg of caraway seeds annually, contributing as much as 1.76 lacs towards income from agriculture. Since the crop is remunerative, selling for 250 Rs. per kg, villagers are devoting a greater portion of their cultivable land to this crop, even though the crop is harvested after two years.

A comparative study of the different valleys brings forth the salient aspect of crop productivity, as well as the preferred crops being grown in the valley. Secondarily, crop productivity directly correlates with the population size of the villages as well as, as emphasized with their duration of stay. Choudas valley comprises of completely resident population and hence the crop productivity is many times more than the rest of the three valleys under study. Also the crop species being grown are tied up with income generation in the form of export. Rajma (*Phaseolus vulgaris*), potato (*Solanum tuberosum*) and maize (*Zea mays*), are all commodities, which are traded. More over the economy is completely agricultural based, with no cultivation of MAPs. One can thus contrast the findings with other three valleys, where MAPs cultivation (a garb, but never the less the know-how of cultivation need to be acknowledged) is a norm, with the productivity of the agricultural crops being marginally very small. Potato remains the most preferred crop species, not just in Choudas, but throughout the studied villages. The only fruit species—apple (*Malus pumilo*) is grown in Choudas valley only.

11.2.1 Yartsa Gunbu—Boon or Bane

Ophiocordyceps sinensis (Berk.) G.H. Sung et al., commonly referred to as ‘Yartsa Gunbu’, which translates from Tibetan as ‘winter worm, summer grass’, or locally in Kumaun and Garhwal Himalaya as ‘*Keera ghaas*’—referring to the larva (syn. *Keera*) and the emergent fruiting body that appears as sprouting grass (syn. *ghaas*), is a genus of mostly entomophagous flask fungi (Pyrenomycetes, Ascomycotina) belonging to family Ophiocordycipitaceae (Fig. 11.2). The most determining factor that invariably relates to the future availability of caterpillar fungus is the ever increasing demand in the international markets. Greater the demand, higher the price of the commodity being traded, greater would be the intensity to harvest the fungus and longer the duration of stay in the habitat sites. In fact, during the last five years, the price of the commodity has multiplied six-fold, from around US$ 3333 per kg in the year 2008 to US$ 20,000 per kg in the year 2012. The unbelievable increase in the sale price has only accentuated the pressure on Yartsa Gunbu.

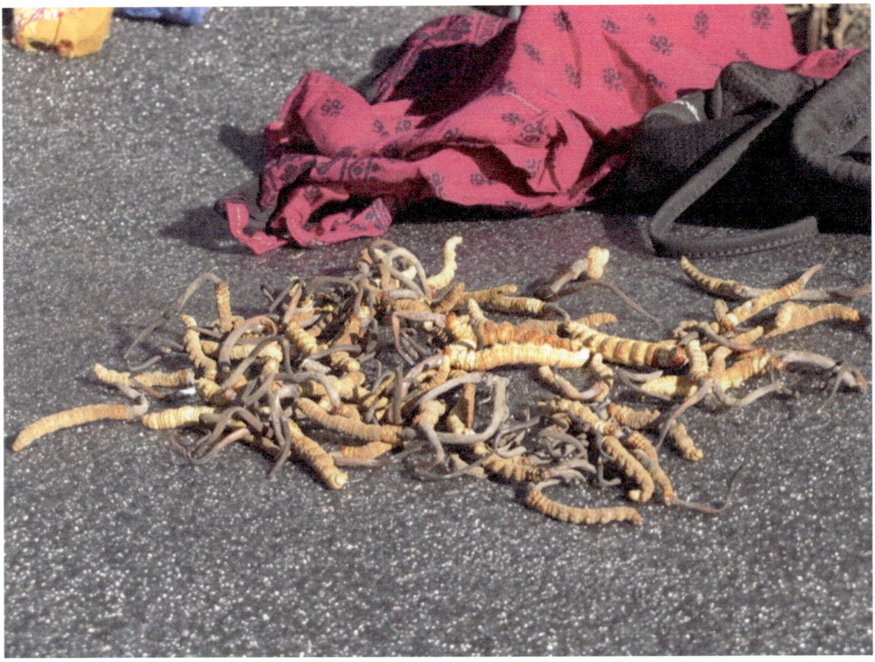

Fig. 11.2 While the yield per hectare has declined perceptibly, the price of the commodity has increased from a mere 20,000/kg to astounding 1.2 million at present; has obviously led to its overexploitation and concomitantly raised concerns to its future availability

While production has remained more or less constant over the last three years (2010–12), the income earned has more than doubled (Fig. 11.3). In fact, the income generated by the Yartsa Gunbu, in a short period of one and half months in a year far exceeds the traditional sources of income combined (Fig. 11.4).

Invariably on account of the two pulling forces—ever-increasing demand and concomitantly its astounding price, has resulted in not just the rampant exploitation of Yartsa Gunbu but also the degradation of the prime habitat, thus endangering not just the future viability of the prized species, but also of the associated flora. Again, it's not just the biodiversity aspect per se, but most importantly (and probably having lasting long-term consequences), the drastic changes in the socio-economic and cultural fabric of the cultural landscape that should be a major concern for the policy makers, and of the interest to the academia. In a survey lasting 2 years and conducted across Askote Conservation Landscape the author gathered the following vital information (Negi et al. 2014);

1. Habitat destruction by the ever increasing harvesters is a major cause of concern. Lately, the lower reaches of the habitats have shown signs of degradation (with no yield whatsoever as per the informants), primarily on account of the space being occupied by the harvesters during their stay in the harvesting season.

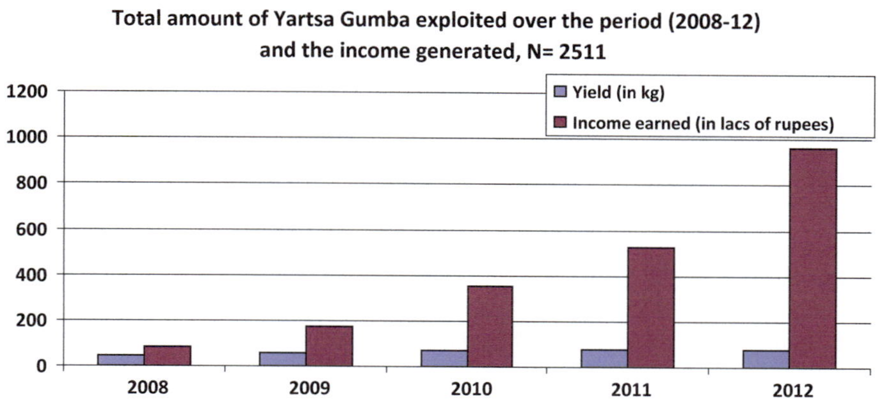

Fig. 11.3 While the yield of Yartsa Gunbu has remained more or less static over the last three years, the cost per kg has multiplied 6 folds in the mere 5 year period

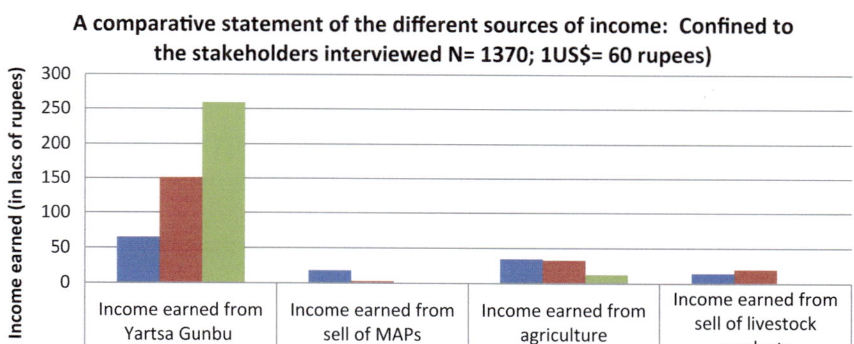

	Income earned from Yartsa Gunbu	Income earned from sell of MAPs	Income earned from agriculture	Income earned from sell of livestock products
Johaar	64.88	17.9	34.49	14.29
Darma	150.84	2.67	32.55	19.78
Gori Paar	258.77		12.25	

Fig. 11.4 Income earned from the sale of Yartsa Gunbu offshoots the income earned collectively out from traditional crops, livestock or livestock products and sale of medicinal and aromatic plants. This fact alone is a cause of concern for the simple fact that long-term sustenance of the inhabitants has to be bound with the sustainable harvesting of the Yartsa Gunbu, which at present is not

2. The healing power of Yartsa Gunbu is believed to be concentrated in the caterpillar, filled with *Ophiocordyceps* mycelium. Firmness of the larva and the precise ratio of the size of the stroma, the fungal fruit body growing out of the head of the larva, versus the size of the insect larva, determine the price of the specimen. Smaller the ratio, better the price. The harvesters are thus forced to collect caterpillar fungus early in the season. In fact, 70–80% of the harvested lot consists of immature specimens. The exorbitant price hike only makes the practice more detrimental.

3. The anthropogenic pressure undoubtedly has increased on account of two major factors—(i) womenfolk (previously tabooed and barred from entering into the sacrosanct landscape) accompanying the males, and (ii) hiring of contract labourers from outside by the local powerful villager, who makes provision for their lodging in the alpine meadows, in return for 50% of their harvested lot.
4. For the conservationists, the overall viability of the Yartsa Gunbu has to be seen vis-à-vis (a) the population dynamics of the host insect (*Thitarode*)—the obligate out-crosser. A significant, 14.81% decline in the population of the host insect has occurred within just two years; (b) the elders raising concern to near complete decimation of the woody species—*Rhododendron campanulatum, R. anthopogon, Juniperus communis,* as well as the exploitation of the MAPs, primarily of *Picrorhiza kurrooa, Dactylorhiza hatagirea, Chaerophyllum villosum,* and *Aconitum heterophyllum,* and of hunting of wildlife, during the stay.
5. Even though the collated yield throughout the broad landscape show a steady increase, with 372 kg of Yartsa Gunbu exploited in the year 2012; in Johaar, Darma and the Gori Paar valleys, the yield has shown a downhill trend over the last five years (Fig. 11.5).
6. Dilution of traditional conservation practices, inclusive of taboos, has occurred. The fact remains that until a few years back, the womenfolk were tabooed to enter into the alpine meadows (considered sacred), which obviously restricted the number of the harvesters. However, with the increase in the price of the commodity, the age-old taboo is now grossly discarded, and the womenfolk are actually encouraged to go for collection.

One of the fortunate outcomes of the Yartsa Gunbu phenomenon is that the Bhotiya tribesmen, who had long lost touch with their villages, principally their

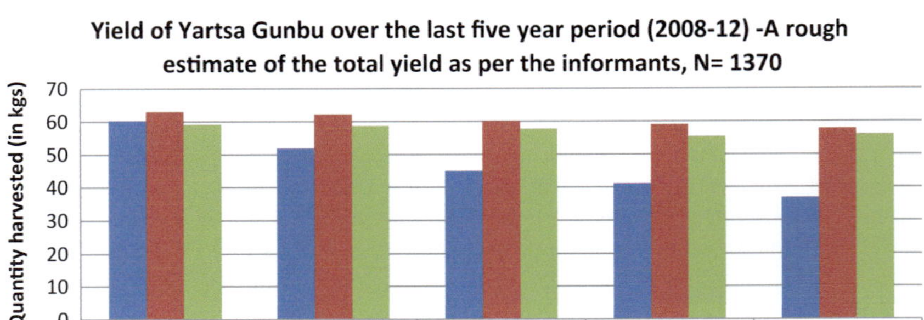

	2008	2009	2010	2011	2012
Johaar	60.31	51.96	44.97	41.04	36.83
Darma	62.99	62.18	60.04	59.05	57.8
Gori Paar	59.11	58.68	57.74	55.38	56.01

Fig. 11.5 Johaar, Darma and the Gori Paar represent the landscapes, where the Yartsa Gunbu exploitation started nearly two decades back. With no additional habitats left out, the yield has shown a downhill trend over the last five years. The same fate awaits the rest of the recently discovered landscapes, too

summer homes, are now returning back to their roots. In fact, claimants are busy tracing their roots to a village with Van Panchayat harbouring Yartsa Gunbu. One could thus envision that the resultant longer duration of stay in summer homes will in turn encourage the villagers to go-in for cultivating the crop species, such as *Nappal* (durum wheat), erstwhile main crop of commerce traded with the Tibet. In zist, any effort on part of the policy makers or the enforcement agencies bent on prohibiting the exploitation of the Yartsa Gunbu, need to appreciate the above-mentioned facts; and thus endeavour not to bar the locals from harvesting of Yartsa Gunbu per se, but more importantly make them aware of the declining yields of the produce, and how best they can stop the trend.

11.3 Discussion and Recommendations

11.3.1 Crop Diversity and Sustenance

It is not just the reduced period of stay in their summer homes, which has resulted in reduction in crop diversity, principally of species requiring relatively longer period to mature. Among the traditional crops with negligible acreage at present, *Parilla frutescens, Macrotyloma uniflorum, Vigna sp., Hordeum himalayens* and durum wheat (*Triticum himalayens*) are under threat in the region, due to lack of interest in growing them. It is important to take note of the fact that, presently in the region, where the population is sparse, dispersed in small hamlets and villages, where the principal mode of agricultural practice is subsistence type, and where the principle crops are still to a major extent-millets, pseudo-millets and course cereals (having negligible market value, added to their little palatability as compared to rice and wheat), little attention has been paid by the policy makers as well as by the agricultural scientists to focus on increasing their yield and at the same time increasing their palatability, too. This fact alone remains the prime factor behind the decline in the acreage of the traditional crops, not only in the region but throughout the Central Himalaya (Maikhuri et al. 1999).

11.3.2 Poor Marketing Facilities Coupled with Inefficient and Inadequate Credit Systems

The increasing demand for cash income from farming has made marketing facilities fundamental to rural development. Mountain areas in general, have limited markets for their produce, owing primarily due to sparse population, difficult terrain and dearth of transport links, makes the same non-profitable, due to high transportation cost. Very often, the prices of their produce are undervalued, compared to their true social cost. Thus, inappropriate prices for inputs and outputs, force the stakeholders to pay least hid towards the conservation of the agricultural base (primarily, the soil) by making unsustainable practices towards procuring profitable margins! This could be exemplified by the example offered in the Choudas valley, where few of the villages, who were cultivating apples, have forsaken the same, due to lack of market and undervalued price offered, and thus have shifted to cultivating potato, Rajma, and other cash crops, eventually leading to increased soil loss during rainfall.

Given the variability of mountain areas in terms of agro-ecological potential (soil fertility, rough terrain, fragility, among others) and market access, different types of development strategies need to be adopted based on agro-ecological potential and market access (Table 11.2). In areas with better agro-ecological conditions and access to markets and infrastructure, the strategies should be promoting land use intensification, crop diversification, and growing of cash crops, which offer higher incomes. Similarly, where market access is constrained and infrastructure is poor, less perishable-high volume and non-perishable agricultural commodities, such as

Table 11.2 Developmental strategy based on agro-ecological potential and market access

Agro-ecological potential	Market access			
	High	Potential villages	Low	Potential villages
High	• High-value cash crops • Horticulture, commercial dairy, intensive food crop production, etc. • Investments in irrigation, land management, and the agro-processing cottage industry with institutional support base	Sirkha, Sirdang, Pangla	• High-value, low volume crops such as NTFPs and medicinal plants • Subsistence food crops • Infrastructure development to utilize the mountains high agro-ecological potential	Baling, Dugtu, Daantu, Bon, Goe, Sipu Budi, Napalchhu, Gunji, Nabi, Rongkong, Kuti, Milam, Martoli, Burfu, Laspa
Low	• Commercial agro-forestry, farm forestry, livestock, pastoralism, off-farm employment • Crafts and services for markets • Promote technologies that enhance agricultural potentials and utilize local niches	Sosa, Sirkha, Sirdang Dar, Sela, Bungling Budi, Paanto	• Agro-forestry, tree farming for timber and NTFPs, medicinal plants • Subsistence agriculture with zero tillage, mixed cropping, livestock production • Tourism and recreation • Local Institutional based conservation programmes and strengthening of local institutions	Jipti, Gala, Simkhola Nangling, Dhakar, Marchha, Tedang Garbyang Tola, Rilkot, Lwa, Bilju

Adapted from Ruben and Pender (2004)

honey, mushrooms, dry fruits and nuts, medicinal, aromatic and dye plants that suffer lower post-harvest losses should be promoted (Ruben and Pender 2004). In general, emphasis should be given to harness the locally available mountain's comparative advantages. For example, climatic and ecological variations in hills and mountain areas provide opportunities for ecological niche products such as fruits, vegetables, medicinal plants, herbs, spices, agro-forestry, tree farming, seed production, and many other high value-low volume crops. Thus, a part of action research and extension services should be oriented towards mountain areas to take into account mountain specificities, diversities, niches, socio-economic and cultural factors, and market access. In selecting crops and tree species along with bio-physical suitability, distance from roads and accessibility factors should be taken into account, as they influence the performance of land use systems, which determine farmer's land use preferences.

11.3.3 Sustainable Harvesting of Yartsa Gunbu

As relates to sustainable means of harvesting of Yartsa Gunbu, several scholars have argued that community-based management practices can be effective mechanisms to ensure sustainability of Yartsa Gunbu harvest (Cannon et al. 2009; Stewart 2009; Weckerle et al. 2010; Shrestha and Bawa 2013). Again, measures for sustainable resource management would prove to be effective, when stakeholders are formally or informally integrated into policy making and control mechanisms (Dietz et al. 2003; Robbins et al. 2009), but before that the local communities must appreciate the problems arising out of the commercial exploitation, and thereby institute mechanism/s at the local level itself for the sustainable harvesting of the Yartsa Gunbu. Bhutan offers one salient example of how Yartsa Gunbu harvesting should be managed. These include, (i) relaxation of laws on gathering Yartsa Gunbu in order to provide locals with an incentive to police their areas and protect natural resources; (ii) delegating the powers to restrict the number of harvesters to just a few members per household; (iii) bringing forth a legislation that (a) Yartsa Gunbu can only be sold at authorized auctions by authorized collectors, and (b) buyers must be Bhutanese nationals only. The government imposes a 4.9% levy on sales to cover the expenses of auctions and to support environmental protection programs (Cannon et al. 2009).

Similarly, in Nubri's Samagaun, Nepal, local VDC (Village Development Council) has taken responsibility for devising and implementing management practices and regulations that governs the exploitation of Yartsa Gunbu. These involve (as per Childs and Choedup 2014)—(i) setting a date for the commencement of the harvest; (ii) village leaders have the authority to even postpone collection dates, when conditions warrant; (iii) the right to gather Yartsa Gunbu is held by any bonafide resident of the village, a status defined through participation in a household taxation system (Childs 2005). Each household must register its collectors with the village administration and pay a Yartsa Gunbu tax of NRs 100 ($1.20) for the first household member and NRs 4500 ($53) for each additional member. These success stories as relates to the sustainable harvesting of Yartsa Gunbu could be replicated through Biodiversity Management Committees (BMCs), which are being constituted at village level under the ambit of National Biodiversity Authority of India (NBAI).

11.4 Conclusions

A significant proportion of people in this landscape are poor. This is more so in case of those still practicing transhumant mode of sustenance, arriving in their summer homes, primarily to harvest the MAPs from the wild; and anwals, who still carry out their traditional mode of lifestyle with dwindled numbers of livestock. Three decades of lost contact with their traditional trade partners across the border—the *gamgya*, and most thereafter making best use of the extension of reservation; the left out generation are bereft of the knowledge as well as interest to reinstall or restrengthen the former trade. It is in this background that efforts to diversify and strengthen the various options and assets, currently available within the landscape, be made. Strategies that could be explored include the following;

1. *Enhance livelihoods security*-With regard to sustainable forest resource use, access with regulation to the collection of certain medicinal plants from Van Panchayat area grazing rights, and meeting subsistence needs.
2. *Improve infrastructure and capacities for livelihoods*—Opportunities through adventure tourism, eco-tourism, and pilgrim Yatra exist and needs to be capitalized upon. Greater potential of trans-border trade, horticulture produce, wool-based cottage, can be realized.
3. *Biological research outputs*—While opportunities for MAPs cultivation exist, efforts should be made to remove the bottlenecks, viz. subsidies of 40–50% have been envisioned by the National Medicinal Plants Board, New Delhi, to be shared on an equal basis with the State Herbal Research and Development Institute, Gopeshwar (Uttarakhand), and farmers interested in the cultivation of MAPs. However, there is a clause stating that to benefit from the subsidy, a minimum of 20% of the cost of cultivation must be procured from a bank. In view of the terrain and the remoteness of these villages, banks are reluctant to provide such loans.
4. As relates to agriculture, diversification and better utilization of the crop diversity across the landscape in terms of agro-ecological potential (soil fertility, rough terrain, fragility, among others) and market access, needs to be explored.

Acknowledgements The authors gratefully acknowledge the financial help received from Dr. Naseem Ahmad, Director, Ministry of Environment, Forests & Climate Change, Paryavaran Bhawan, New Delhi. The study however, would not have been possible without the help of the village residents for reposing their faith and sharing the information.

References

Altieri, M. (1991). Traditional farming in Latin America. *The Ecologist, 21,* 93–96.

Anonymous. (1989). *Lost crops of the Incas: Little known plants of the Andes with promise for worldwide cultivation.* National Academy of Science, Washington, DC: National Research Council.

Asher, M., Bhandari, P., Ramnarayan, K., & Theophilus, E. (2002). *Livelihood in transition* (Working Paper 5). Anand, Gujarat, India: Foundation for Ecological Security.

Cannon, P. F., Hywel-Jones, N. L., Maczey, N., Norbu, L., Tashi Samdup, T., & Lhendup, P. (2009). Steps towards sustainable harvest of *Ophiocordyceps sinensis* in Bhutan. *Biodiversity and Conservation, 18*(9), 2263–2281.

Childs, G., & Choedup, N. (2014). Indigenous management strategies and socioeconomic impacts of Yartsa Gunbu (*Ophiocordyceps sinensis*) harvesting in Nubri and Tsum, Nepal. *Himalaya, The Journal of the Association for Nepal and Himalayan Studies, 34*(1), 8–23.

Childs, G. (2005). How to fund a ritual: Notes on the social usage of the Kanjur (bKa'-'gyur) in a Tibetan village. *Tibet Journal, 30*(2), 41–48.

Dietz, T., Ostrom, E., & Stern, P. C. (2003). The struggle to govern the commons. *Science, 302,* 1907–1912.

Foundation for Ecological Securities (FES). (2003). A biodiversity log and strategy input document for the Gori River basin, Western Himalaya region.

Hoon, V. (1996). *Shaukas and Rang of the Kumaun Himalayas (Living on the move series).* New Delhi: Sage publications.

Maikhuri, R. K., Nautiyal, M. C., & Khali, M. P. (1991). Lesser-known crops of food value in Garhwal Himalaya and a strategy to conserve them. *Plant Genetic Resources Newsletter, 86,* 33–46.

Maikhuri, R. K., Rao, K. S., Saxena, K. G., & Semwal, R. L. (1999). Traditional crops in the central Himalayas. *Plant Genetic Resources Newsletter, 120,* 1–7.

Negi, C. S., Pant, M., Joshi, P., & Bohra, S. (2014). Yar tsa Gunbu [Ophiocordyceps sinensis (Berk.) G.H. Sung et al.]: the issue of its sustainability. *Current Science, 107*(5), 882–887.

Negi, C. S. (2007a). Changing face of polyculture in the Darma and Johaar valleys, Pithoragarh, Kumaun Himalayas. *International Journal of Sustainable development and World Ecology, 14,* 428–436.

Negi, C. S. (2007b). Declining transhumance and subtle changes in livelihood patterns and biodiversity in the Kumaun Himalaya. *Mountain Research and Development, 27*(2), 114–118.

Negi, C. S. (2010). *Askote conservation landscape: Culture, biodiversity and economy* (Vol. xviii, p. 590). Connaught circus, Dehradun: Bishen Singh Mahendra Pal Singh.

Robbins, P., McSweeney, K., Chhangani, A. K., & Rice, J. L. (2009). Conservation as it is: Illicit resource use in a wildlife reserve in India. *Human Ecology, 37,* 559–575.

Ruben, R., & Pender, J. (2004). Rural diversity and heterogeneity in less-favoured areas: The quest for policy targeting. *Food Policy, 29,* 303–320.

Shrestha, U. B., & Bawa, K. S. (2013). Trade, harvest, and conservation of caterpillar fungus (*Ophiocordyceps sinensis*) in the Himalayas. *Biological Conservation, 159,* 514–520.

Stewart, M. O. (2009). The 'Himalayan Gold' rush: Prospectors' practices and implications for management. *In* T. B. Dotson, et al. (Eds.), *Contemporary visions in Tibetan studies: Proceedings of the first international seminar of young tibetologists* (pp. 1–26). Chicago and Bangkok Serindia Publications.

Weckerle, C. S., Yang, Y., Huber, F. K., & Li, Q. (2010). People, money, and protected areas: the collection of the caterpillar mushroom *Ophiocordyceps sinensis* in the Baima Xueshan Nature Reserve, Southwest China. *Biodiversity Conservation, 19,* 2685–2698.

Chapter 12
Brokpa Yak Herders of Bhutan: A Study in Pastoral Livelihood Patterns, Transhumance and "Drukor"

Raghubir Chand

12.1 Introduction

Marginal societies are undergoing unprecedented changes all over the Himalayan high lands with changing local development policies and forces of globalization reaching their doorsteps. While most of the Himalayan states retain the colonial blueprint of economy, promoting commercial agriculture through the increased labour force in the service sector economy reflects the forces of globalization and modernization much earlier than in the pastoral societies. The pastoral people surviving on the transhumance and trade depending on the extensive use of alpine grazing lands at the margins of the Himalaya preserved their self-reliant mode of economy and defended autonomy even during the colonial era. The Tibet-Himalayan Highland has nowadays attracted more and more attention for its increasing geo-political, environmental, and socio-economic importance (Tsering et al. 2010, 191). However, the material base of the pastoral economy and the social fabric of pastoral life eroded gradually with the increasing intervention of a commonly perceived development planning imposed from the political powers centered on the agricultural societies in the plain areas of the country. The livelihood pattern is broadly similar in all pastoral areas in the Himalaya, but in case of Bhutan, it is significantly different from other Himalayan states in terms of its size, social structure, economic base, nature of governance, market situations and resource use. During the twentieth century the pattern and impact of development was different in Bhutan as its approach to development and policy of interaction to outside world was markedly different than other states in the Himalaya. Bhutan in fact followed the policy of self-imposed isolation and thus remained cut-off from the rest of the world in terms of foreign direct investment and market links. One important aspect

R. Chand (✉)
Department of Geography, DSB Campus, Kumaun University,
Nainital 263002, Uttarakhand, India
e-mail: raghubir.seri@gmail.com

R. Chand et al. (eds.), *Societies, Social Inequalities and Marginalization*,
Perspectives on Geographical Marginality,
DOI 10.1007/978-3-319-50998-3_12

of Bhutan's development philosophy is that it pursued the concept of Gross National Happiness (GNH) in place of Gross National Product (GNP) which was introduced by the fourth King of Bhutan Jigme Singye Wangchuk who opened Bhutan to the outside world in 1972. The Dragon Kingdom was thus lead to the path of modernization with environmental friendly, and a more sustainable approach of development based on Buddhist spiritual values (Ura et al. 2012). The issue of development gained significant momentum in Bhutan after this Dragon Kingdom has evolved into a new era of democracy in the year 2008. In particular, the modernizing processes accelerated through increased access to education, health care, facilities such as transportation and communication and phenomenon of migration leading to urbanization introduced at a much larger scale in Bhutan at the dawn of twenty first century. There is therefore a need to have a proper understanding of these processes to relate them to the process of transformation taking place in Bhutan especially in the context of tribal mode of living. The pastoral communities in Bhutan were never in direct conflict with the agricultural communities as they had larger areas of their grazing lands and were freely moving up and down following transhumance and trade. In Brokpa land, there was no clear demarcation of their pasture land before the state introduced the laws of environmental protection and brought their land under the control of national parks, nature reserves and wildlife sanctuaries. The factor of climate change proved more disastrous in the pastoral world. The Brokpa pastoral community is facing newer challenges due to the dwindling population of yak, degradation of high altitude pastures, and subsequently shortage of feed and fodder (Mait et al. 2014). The Brokpa economy and the pastoralist way of life in Bhutan is at the crossroad of change though not completely out of its traditional hold. The present chapter is thus based on a brief overview of pasture economy and associated pastoral life and the changes taking place in the Brokpa society in Bhutan. The focus is on the new generation Brokpas and how they can adapt their livelihood strategies to the new democratic setup within Bhutan and changing global context.

12.2 The Brokpa Yak Herders of Bhutan

Brokpas are perceived as a distinct segment of Bhutanese society, though they belong to the same racial stock and pursue animistic moulds. They are part of the same Tibetan cultural world and have been dropouts of the Tibetan history; they were forced to leave Tibet due to political reasons, as other segments of society in Bhutan (Chand 2000, 48). In fact, all highlanders residing in the upper valleys of western, central and eastern Bhutan above 3000 m are Yak herders known by different names such as Bjops in western Bhutan, Laps in Central Bhutan (to the east of Black mountain) and Brokpas from the Tang valley to Valleys of Merak and Sakteng in eastern Bhutan (Fig. 12.1).

The Brokpa is a professional and occupational name without any reference and bias to race, blood and regional affinity. Brokpa is derived from the Tibetan word

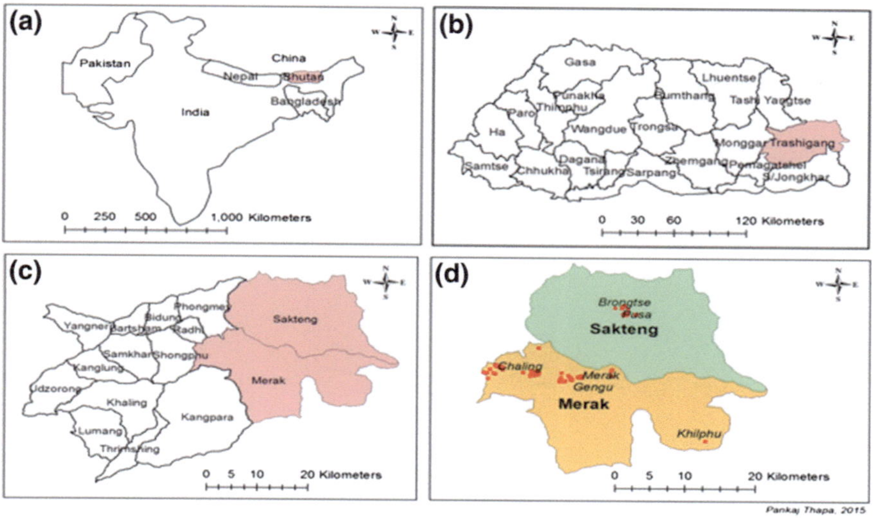

Fig. 12.1 The Brokpa habitat **a** Location of Bhutan in South Asia. **b** Location of Trashigang Dzongkhag in Bhutan. **c** The Brokpa home land Merak and Sakteng Gewog within Trashigang Gewog. **d** Important Settlements of Brokpa home land

Brogpa, brog meaning pasture and *pa* stands for inhabitants. Brokpa is synonymous with herdsman, shepherds or people who herd cattle even for their owner (Chand 2004, 29). Like all other Himalayan highlanders of India, Nepal, Bhutan and Tibet, Yak is the heart and soul of Brokpas (Wangchuk et al. 2013; Tshering 2004; Chand 2004; Ura 2002; Chettri 2008; Karchung 2011; Gyamtsho 2000; Wangmo 1990). It is believed that yak was domesticated by the Qiang people, the Qiang inhabitants of the Tibetan Plateau dates back to the Pleistocene period about 5000 years ago during the Longshan cultural era (Wiener et al. 2003, 3; Rhode et al. 2007 as cited in Wangchuk 2013, 189). Yaks were then disseminated to other parts of the world (Wiener et al. 2003, 3; Rhode et al. 2007 cited in Wangchuk et al. 2013, 190). Qiang, which literally means 'shepherds' in Tibetan language are the true ancestors of the yak herding tribes in the Himalayas and still possess a very rich yak-herding culture (Wangchuk et al. 2013, 190). The Brokpa of the Merak and Sakteng villages are believed to have fled from Tibet during the fifth and sixth centuries along with their yaks and settled within the rugged terrain of Eastern Bhutan about 14th and 15th Centuries (Dompnier 2007, 12–15 as cited in Wangchuk et al. 2013, 190). Historically the Merak and Sakteng valleys were founded by Lady Aum Jomo and Lama Jarappa during the reign of King Sangtsen Gampo (627–49 A.D.) of Tibet (Chand 2004, 36). With the immigration of Brokpas from Tibet in Merak and Sakteng Valleys of Bhutan, Their territorial political organization (Fig. 12.2) was independent of neighbouring states because the type of religion followed by Brokpas is a Gelugpa sect of Tibetan Buddhism. They were ruled by the Kushuguru and his incarnations before this Brokpa territory became part of Bhutan. Still today, the Brokpa community practices the Gelugpa sect and

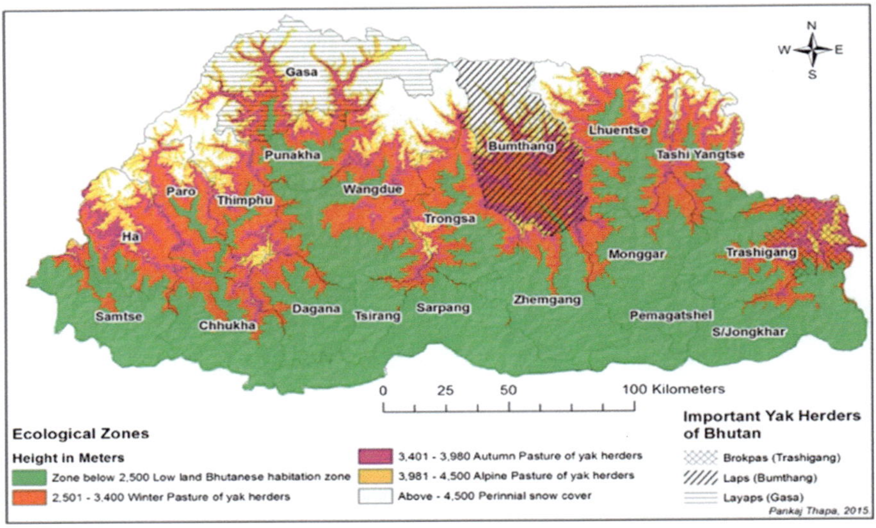

Fig. 12.2 Ecological zones and important Yak Herders of Bhutan and their Home land

maintain a sophisticated religious practice which they claim dates back to Lama Jarappa during the reign of King Sangtsen Gampo of Tibet (Chand 2004, 63).

Brokpas thus represent a truly yak herding community in Bhutan occupying the Merak and Sakteng *gewogs* (administrative blocks) of Trashigang *Dzongkhag* (district) in eastern Bhutan. They also have a separate sub-district level administrative unit called *Dunkhag in dzongkha* language. Sakteng Dunkhag is located between 27°3′ N. and 27°28′ N. latitude and 91° 42′E. to 92° 8′E. longitude which encompasses an area of 910.9 km^2. It shares the border with Arunachal Pradesh (India) in its north and east. Phongme, Songphu and Kangpara *gewogs* of Trashigang *Dzongkhag* lie in the west and the southern boundary is shared by Lauri gewog of Samdrup Jongkhar Dzongkhag. There are 12 villages in Sakteng gewog with a total population of 2072 of which 1078 are males and 994 are females (PHCB 2005). Sakteng is the biggest village with the Dunkhag (sub-district level administrative unit) head quarter, other six villages in upper Saketengchhu valley are Tengma, Borongma, Borongtse, Murbee, Thrakthri and Pusa. The lower half of the Saktengchhu valley is warmer where the remaining 6 villages namely Jonkhar, Tholong, Bumpalock, Moenlamdung and Yongbazor are located. Merak, Gengu, Khilphu and Kharsithang are four villages under Merak Gewog with a total Population of 1621 of which 876 are males and 745 are females (PHCB 2005). All villages in Sakteng *gewog* lie along the Saktengchhu river while Merak and Gengu are in Nyeramachhu valley and other two villages of Merak gewog namely Khilphu and Kharsithang are located in Jorangrichhu valley. Village Merak and Sakteng are 25.5 km apart being separated by the Nakchungla (4140 m) pass (Chand 2000). The local geography provides the vast expanse of alpine meadows and grassland as well as forest meadows and bamboo grasslands which gives rise to three sets of pastures in Brokpa habitat.

This study is based on the primary information at household level for which data have been collected from randomly selected 210 households from 13 villages of Brokpa habitat in January 2010. It comes to 25.73% of the total 816 households (PHCB 2005) living in about 16 Brokpa villages in both Merak and Sakteng gewogs. Out of four villages of Merak gewog, 45 households from village Merak and 27 households from Gengu were selected for the present study. Similarly, out of 12 villages of Sakteng gewog, 11 villages were selected taking a maximum of 38 households from village Sakteng followed by 12 from Borongma, 11 from Tengma, 9 from Pusa, 6 from Borongtse, 5 from Manidungjur and 2 from Thrakthri, thus in all 83 from upper Sakteng valley and remaining 55 households were selected from lower Sakteng valley. From lower Sakteng valley, a maximum of 20 households were surveyed from Murbee followed by 18 from Jonkhar, 9 from Tholong and 8 from Bumpalog (Table 12.1). Information related to livestock population, pasture ownership and coverage, animal products, the population engaged in transhumance and "*drukor*"—*a* barter system, practice of "*nepos*"—a village host who helps Brokpas to sell their milk products; were collected from each sample households. The sample size is thus fairly representative and likely to contribute to our understanding of the Brokpa society as it stands in the twenty first century.

12.3 A Brief Overview of Brokpa's Pastoral Life and Livelihood Pattern

The life and livelihood of Brokpas is very simple and sparse. Collectivism and cooperative living is the sole character of the Brokpa society and they maintain a homogeneous socio-cultural category. Brokpas still have a very tribal way of living. They remained isolated to maintain their tribal polity. The rhythms of life in this remotest corner of Bhutan are vibrant with a profound sense of its own intrinsic worth (Chand 2000). Animal rearing has all along been the primary occupation of the Brokpas. Weaving of woolen goods and the bartering of milk products with lower valley inhabitants are a part of their livestock economy. Herding and trading together account for 66.17% of the total work force. About 67.30% of Brokpa families practice transhumance with a maximum of 91% of families in Merak (Chand 2000, 63). "Drukor"—a barter trade link between Brokpas and their Sharchop "nepos" (the village hosts as known in Eastern Bhutan) is their integral economic practice which has been followed for centuries. Products such as cheese, butter homespun blankets and floor mats are bartered by Brokpas for rice, barely, chilies and other items. Agriculture adds to their normal animal husbandry activities. A summer crop of buckwheat and muster in the upper valleys, and maize, wheat and barley are cultivated in the lower Brokpa villages. While Merak is purely a pastoral village, upper Sakteng valley is also marked by the dominance of yak

Table 12.1 Villagewise distribution of population in sample households, 2010

S. no.	Villages	Households	Population			Male per 1000 females	Average household size
			Male	Female	Total		
1.	Tengma	11	35 (53.0)	31 (47)	66 (100)	1129	6.1
2.	Borongma	12	32 (52.5)	29 (47.5)	61 (100)	1103	5.1
3.	Sakteng	38	94 (49.8)	95 (50.2)	189 (100)	989	4.97
4.	Manidungjur	5	18 (62.1)	11 (37.9)	29 (100)	1636	5.8
5.	Borongtse	6	11 (47.8)	12 (52.2)	23 (100)	916	3.83
6.	Pusa	9	31 (58.5)	22 (41.5)	53 (100)	1409	5.88
7.	Thrakthri	2	9 (64.28)	5 (35.72)	14 (100)	1800	7.00
8.	Jonkhar	18	49 (49.5)	50 (50.5)	99 (100)	980	5.5
9.	Murbee	20	63 (56.76)	48 (43.24)	111 (100)	1312	5.55
10.	Tholong	9	24 (47.06)	27 (52.94)	51 (100)	888	5.66
11.	Bumpalog	8	28 (60.86)	18 (39.14)	46 (100)	1555.55	5.75
Sakteng gewog (total)		138	394 (53.10)	348 (46.90)	742 (100)	883.25	5.37
12.	Merak	45	147 (51.4)	139 (49.6)	286 (100)	1058	6.35
13.	Gengu	27	86 (56.21)	67 (43.79)	153 (100)	1284	6.04
Merakgewog (total)		72	233 (53.07)	206 (46.93)	439 (100)	1131	6.09
Grand total		210	627 (53.09)	554 (46.91)	1181	1132	5.62

Source Field work, 2010, Figures shown in parenthesis are percentage

herding over the summer cultivation of buck wheat. However, in the lower Sakteng valley, maize, wheat and barley are cultivated with a minimal practice of yak herding. The formal education which was introduced in 1987 (Chand 2004) has achieved 39.26% literacy as recorded by the first Population and Housing Census of Bhutan in 2005. The male literacy rate is 47.94% leaving behind females with 22.78% (PHCB 2005). Occupational change is very minimal for the Brokpas being recorded at only 9.19% from 1999 to 2010 (Chand 2000) as compared with the first study conducted by the author in 1999 with the present sample survey of 2010. Merak has witnessed however the maximum change of 17.83% during the same period which was noted just 1.82% change in the year 1999 (Chand 2000, 63). For the Brokpas cattle rearing seem to be still perceived and favored as the best occupation.

12.4 Livestock Composition with Yak Dominance

The entire gamut of Brokpa's pastoral life is woven around the yak. Yaks are reared in 10 out of 20 districts in northern belt of Butan. Total number of yak in Bhutan is recorded as 43,144 which constitute only 5.2% of the total livestock population of Bhutan (Livestock statistics 2011). The Brokpas have the largest share of the yak population as compared to other yak herders in Bhutan. Although the yak population is increasing in recent years because of yak being used in high altitude trekking and other tourist purposes mainly in western and central Bhutan, the number of yak herding households are decreasing in Bhutan in general.

The livestock description requires first the introduction of crossbreeds of yak. The term yak is applied only to the male and female of the same species (bos grunniens), known as 'chuk', or 'di'. The crossbreed animals from yak bulls are described quite differently. The cross between Golung (a good quality Tibetan yak bull) and the chuk is called dzo (male) and dzomo (female). The cross between the dzomo and a yak bull is called the koi (both male and female). The cross between the koi female and the yak bull is the shingolengma; the cross between shingolengma and a yak bull, is the dagolengma; and the progeny between the dagolengma and the yak bull is the bagolengma. The seventh generation cross between the bagolengma and the yak bull is a Golung or Golengma. Similarly, a cross between a yak bull and a siri cow is known as a dzomo but the next generation crossbreeds are described as Tui, Gar, Kyuk, and Zen and in the seventh generation again a yak or di is produced. In the present yak survey 2010, a new breed called dremo was identified which is a cross breed of the female yak called Shug and a brown Swiss bull. Dremo or bremo is considered to be the best breed.

The wealth of Brokpas is expressed in terms of head counts of the cattle. In 210 sample households, a total of 7321 cattle heads are reported with a majority of 66.07% in the two sample villages of Merak gewog followed by remaining 33.93% in 11 sample villages of Sakteng gewog (Table 12.2). Within the Sakteng gewog, more cattle are found in upper Sakteng valley (77.70%) and the remaining 22.30%

Table 12.2 Livestock population of sample households, 2010

Animals	Merak gewog	Sakteng gewog			Grand total
		Upper Sakteng valley	Lower Sakteng valley	Sakteng gewog total	
Yak	557	191	28	219	776
Dremo	804	451	38	489	1293
Dzo	457	194	27	221	678
Dzomo	680	294	34	328	1008
Mithun	2	1	0	1	3
Koi	275	101	32	133	408
Cow	303	123	214	337	640
Sheep	1418	258	71	329	1747
Goat	10	3	3	6	16
Dog	114	68	66	134	248
Horses	217	246	41	287	504
Total	4837	1930	554	2484	7321
Average cattle per household	67.18	23.25	10.07	18.00	34.86

Source Field work, 2010

were found in Lower Sakteng valley. Out of the total cattle population, yak stood highest among all cattle with 56.79%. Merak has also the highest yaks (66.55%) in the total yak population of both gewogs followed by one third (33.39%) in Sakteng gewog. The main herding grounds of yak population is Merak along with sheep (80.18%). Other categories of cattle such as cows (640), followed by horses (504), dogs (248), goat (16) and mithun (Bas frontalis, a large semi-domesticated bovine) (3). Horses and dogs constitute an important part of the livestock economy due to their role in transhumance and pasturing. The data is shown in Table 12.2. The average cattle per households is 34.86 in general with a highest of 67.18 in Merak followed by 18.00 in Sakteng gewog. The average cattle per households is more than double (23.25) in upper Sakteng valley than the lower Sakteng valley (10.07). The agro-pastoral mode of economy is more dominant in lower and warmer Sakteng valley.

Over the years, the yak population has not changed much as it was compromised by the increasing intervention of developmental activities from the outside. The economic profitability of livestock production in these remote Brokpa villages is undisputable even in the wave of commercialization of livestock production system on one hand and increasing encroachment of national park and wildlife sanctuaries in the traditional Brokpa grazing lands on the other. In the total sample households of 210, the majority of households (41.43%) have reported the same situation of their livestock population in the near past. Interestingly the percentage of sample households reporting the increase in their livestock herds is one quarter (25.24%) of

Table 12.3 Changing livestock situation in sample households, 2010

Sample villages	No. of cattle increasing	No. of cattle decreasing	Same	Households with no cattle	Total house holds
Merak gewog	18 (25.00)	15 (20.83)	31 (43.06)	8 (11.11)	72 (100)
Upper Sakteng valley	29 (34.94)	27 (32.53)	18 (21.69)	9 (10.84)	83 (100)
Lower Sakteng valley	6 (10.91)	11 (20.00)	38 (69.09)	–	55 (100)
Sakteng gewog total	35 (25.36)	38 (27.54)	56 (40.58)	9 (6.52)	138 (100)
Grand total	53 (25.24)	53 (25.24)	87 (41.43)	17 (8.09)	210 (100)

Source Field work, 2010. Figures shown in parenthesis are percentage

the total households with a same matching figure of 24.24% with those reporting decrease in the number of their livestock thus making for a balance at household level. The poor Brokpa families without any cattle are 8.09% in the total sample who usually work as a yak herder for rich families. There are more sample households in the increasing category in upper Sakteng valley with 34.94%, whilst the lowest is in the lower Sakteng valley (10.91%). Likewise, there are more households within decreasing category (32.53%) in upper Sakteng valley with a lowest (20.00%) in lower Sakteng valley. The situation in Merak is more or less stable. The livestock situation remaining the same (Table 12.3) as witnessed by the field data suggest that yak herding is still a way of life in Brokpa community.

12.5 Pasture Ecology and Pasture Ownership Pattern

Pastures are the main domains of the Brokpas. Three main areas, as dictated by seasonal control, make up their domain: summer, autumn and winter pastures. Brokpas have therefore a defined seasonal cycle of movement within these three pasture regimes located between 2500 and 4500 m and categorized by a harsh climate with snow in winter and abundant rain in summer (Chand 2002). Summer pastures are found in the upper reaches of the region and Brokpas move to these pastures in April and stay till the end of June—that is, as long as the green pastures last. The herders drop lower and lower in pursuit of the fast disappearing grasses and reach the autumn pastures which are normally located on the mid-mountain slopes and they stay there until the end of August. As cold weather sets in the upper reaches, they slowly move towards the winter pastures that they use from September through to December. Winter pastures are located on snow-free areas where grasses are available in the cold season. May and October constitute the peak months of their summer and winter migration, when most of the households are set out to move. The yak herders live in stone houses in summer and in thatched huts in winter, and they also camp in the pastures.

There are more pastures in Sakteng gewog. Sakteng has pasture coverage of about 43,075.15 ha. The maximum pasture area belongs to the winter pastures (32,016.66 ha) followed by summer pastures (8410.77 ha). The area under autumn pastures is the lowest and accounts for 2611.34 ha (Chand 2004). On the contrary, in the Merak gewog, out of a total of 9619.35 ha of pastures, a maximum of 6152.69 ha is under summer pastures followed by 2163.49 ha under winter pastures and a minimum of 1303.17 ha under autumn pastures (Chand 2004). The autumn pastures act as transit camps between winter and summer pastures and are spread over a distance of five to ten days' walk.

The pasture ownership pattern of sample households presented in Table 12.4 reveals that about 60% of total surveyed households own pasture lands. In Merak, 90.28% households own pastures while in Sakteng gewog, the share is 43.47%. The situation in upper Sakteng valley is better with 67.47% households owning the pastures. There are just 7.27% of households who own the pastures in the lower sakteng valley. Those who own the pasture land also own pastures (42.82%) in all three pasture areas including summer, winter and autumn. Those who have pastures in only the autumn and winter rank next (7.62%) followed by 4.76% households with only summer and winter and 4.28% with only summer pastures. Sample households with pastures at three locations are maximum in the upper sakteng valley (59.04%), households with only summer pasture are at maximum in Meark (9.72%) because of the scale and the extent of the landscape conditions. The percentage of households without pasture land is also considerably high (40.08), ranging from a minimum of 9.72% in Merak to 92.73% in lower Sakteng valley. More than half (56.52%) families have no pasture land in Sakteng gewog. They survive on the pasture lands of their kind neighbours and, in return, pay them with cheese and butter.

12.6 Yak Products

Animal production systems (or livestock herding systems in extensive yak breeding areas) are generally studied in terms of the complex interaction between three components (animals, humans and resources), which are linked by a set of interdependent and interrelated outputs (e.g. animal products, increase of herd, wealth, social position, sustainable land resource management etc.) (Jest and Bonnemaire 2002). All yak products are extremely valuable food materials (meat, milk, fat) and non-food items (wool, leather, horn, blood, bones etc.). Yak is an important natural resource in Brokpa economy and they utilize it sufficiently. Only a few mountainous countries have yak breeds outside Tibet and Bhutan is one important yak habitat. A typical Brokpa hut in a summer pasture is full of yak milk stored in bamboo canes. Yak milk is churned into butter, which keeps a Brokpa family busy all day and night. It is difficult to churn stored milk into butter and empty the bamboo canes to use them for the next milking session. After churning, the cheese is kept in bags made of yak hide over the fireplace for six months. This fermented

Table 12.4 Pasture ownership pattern in sample households, 2010

Sample villages	Households with all three pastures	Households with only summer pastures	Households with only summer and winter pastures	Households with only autumn and winter pastures	Households without any pastures	Total households
Merak gewog	37 (51.40)	7 (9.72)	6 (8.33)	15 (20.83)	7 (9.72)	72 (100)
Upper Sakteng valley	49 (59.04)	2 (2.41)	4 (4.82)	1 (1.20)	27 (32.53)	83 (100)
Lower Sakteng valley	4 (7.27)	–	–	–	51 (92.73)	55 (100)
Sakteng gewog total	53 (38.40)	2 (1.45)	4 (2.90)	1 (0.72)	78 (56.52)	138 (100)
Grand total	90 (42.86)	9 (4.28)	10 (4.76)	16 (7.62)	85 (40.48)	210 (100)

Source Field work, 2010. Figures shown in parenthesis are percentage

cheese; known as '*yitpa*' by *Sharchops* (people of eastern Bhutan) is a delicacy for the Brokpas. The Brokpas here do not make '*chugos*', a hardened cheese produced in western and central Bhutan. However, the livestock extension centers in Merak and Sakteng have conducted training for them in making *chugos*. Yak meat is a luxury food for many and people generally do not kill productive yaks. It is only old yaks and sheep that are slaughtered for meat and skin. Yak hair and sheep wool provide important raw materials for Brokpa clothes, caps, blankets, bags, ropes and their dress. Regrettably, woolen manufacturing has still not acquired a central place in the Brokpa economy. Clothes from wool are woven mainly for their own domestic use.

Primary information collected about the yak products is less reliable. Brokpas always fear to tell the real number of their yak herds and also the milk products. In the present sample, 54.76% households responded that they sell cheese and 53.81% sell butter. Merak has the highest percentage (84.72%) of sample households selling the cheese and butter while in Sakteng gewog only 39.13% households sell cheese and 37.68% sell butter. Yak meat is not sold at a larger scale and is mostly used within the village. Sample households trading yak meat are 29.04% of the total and it is higher in the upper Sakteng valley (55.42%) because of the larger number of villages and population (Table 12.5).

12.7 Transhumance and "Drukor"

Yak herding and transhumance is the traditional way of supplementing livelihood system in all highland pastoral communities. Transhumance, the seasonal migration of herds between different but complementary ecological points, summer pastures in the mountains and winter pastures in the lowlands, is key grazing pattern and it is a recurrent feature of indigenous grazing management systems (Chetri et al. 2011). In the truest sense of the word, the Brokpas cannot be identified as nomads. It would be more correct to call them transhumant as they have a well-defined seasonal cycle of movements. The migratory pattern of their life is an economic necessity (Chand 2004). The extreme climate of the place makes it hard to live there throughout the year. As the cold weather sets in, entire families have to keep moving with their herds across several passes, dropping lower and lower to the warmer regions for frost-free pastures. They move slowly over a period of four to six weeks along with their cattle down up to 2500 m in different directions (Fig. 12.3).

The Brokpas of Merak follow four main routes. The first group marches towards Khiliphu and the second towards Shingkhar along the river Jomori in the south. The third group follows Nyeramachhu valley and then Brongzongla ridge up to Danglingtsho diverting into two main branches: one to Kangpara and the other towards Khaling and Womrong. The fourth group roams around Mindula pass between Galing and Phongme villages. The winter pasture movement of the people of Sakteng breaks up into two general directions. Some of them go up to Shingkhar

Table 12.5 Sample households involved in the trade of Yak products, 2010

Sample villages	Cheese		Butter		Yak meat		Total sample household
	Trading	Not trading	Trading	Not trading	Trading	Not trading	
Merak gewog	61 (84.72)	11 (15.28)	61 (84.72)	11 (15.28)	14 (19.44)	58 (80.56)	72 (100)
Upper Sakteng valley	36 (43.37)	47 (56.63)	36 (43.37)	47 (56.63)	46 (55.42)	37 (44.58)	83 (100)
Lower Sakteng valley	18 (32.73)	37 (67.27)	16 (29.09)	39 (70.91)	1 (1.82)	54 (98.18)	55 (100)
Sakteng gewog total	54 (39.13)	84 (60.87)	52 (37.68)	86 (62.32)	47 (34.06)	91 (65.94)	138 (100)
Grand total	115 (54.76)	95 (45.24)	113 (53.81)	97 (46.19)	61 (29.04)	149 (70.96)	210 (100)

Source Field work, 2010. Figures shown in parenthesis are percentage

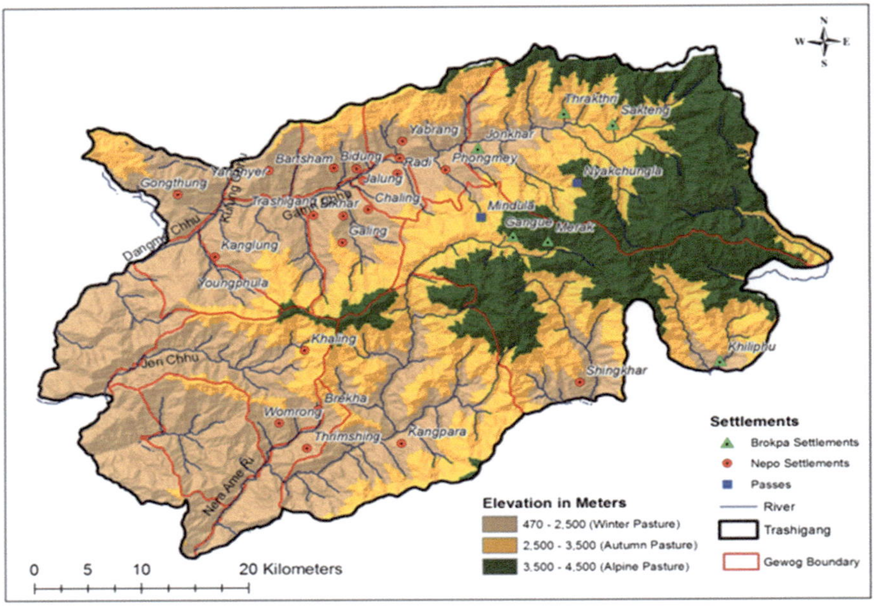

Fig. 12.3 Pasture ecology and economy of Brokpa Yak Herders of Trashigang, Eastern Bhutan

and Khiliphu, crossing the Nyakchungla pass. The majority of them move down to along Gamrichhu on both sides. The summer movement of the Sakteng people is confined to Zangphu Sum range bordering Arunachal Pradesh of India, while the people of Merak go up to Sherphu, Jomokunkhar, Dangzong and the region around the Nyakchungla pass.

Brokpas need more hands in the same households to look after yak, sheep and other cattle in different pastures and also to trade with the low lying valleys. Table 12.6 presents the nature of transhumance as practiced by Brokpas.

In the present sample, about 54.76% families were practicing transhumance while the remaining 45.24% did not migrate. Merak had the highest number of transhumant families (88.89%) while in Sakteng gewog the migratory households were 36.39%. There is a tendency of migrating more with other cattle (26.19%) followed by yak and other cattle together (16.66%) and families moving with purely yak herds are just 11.90%.

While this migration is a climatic compulsion and an example of the ecological adaptation of the Brokpas, it also provides them opportunity to go for "drukor" during winters. "Drukor" or "brukor" roughly means moving around for grains to exchange with their dairy products. "Drukor" is thus a unique trading relationship that exists between Brokpas and their Sharchhop (people of eastern Bhutan) hosts and friends. These strong social ties bind together the highlanders and the valley-dwellers underlying their economic necessities. While some men and children tend the yaks in nearby pastures, other members of the family come down to

Table 12.6 Transhumance practice in sample households, 2010

Sample villages	With Yaks	With other cattle	Both Yaks and other cattle	Households practicing Transhumance	Households not practicing Transhumance	Total house holds
Merak	12 (16.67)	34 (47.22)	18 (25.00)	64 (88.89)	8 (11.11)	72 (100)
Upper Sakteng valley	13 (15.66)	21 (25.30)	13 (15.66)	47 (56.63)	36 (43.37)	83 (100)
Lower Sakteng valley	–	–	4 (7.27)	4 (7.27)	51 (92.73)	55 (100)
Sakteng gewog total	13 (9.42)	21 (15.22)	17 (12.32)	51 (36.96)	87 (63.04)	138 (100)
Grand total	25 (11.90)	55 (26.19)	35 (16.66)	115 (54.76)	95 (45.24)	210 (100)

Source Field work, 2010. Figures shown in parenthesis are percentage

Table 12.7 Population engaged in "drukor" in sample households, 2010

Sample villages	Male	Female	Total
Merak gewog	102 (55.13)	83 (44.87)	185 (100)
Upper Sakteng valley	97 (55.11)	79 (44.89)	176 (100)
Lower sakteng valley	4 (80.00)	1 (20.00)	5 (100)
Sakteng gewog total	101 (55.80)	80 (44.20)	181 (100)
Grand total	203 (55.01)	163 (44.99)	369 (100)

Source Field work, 2010. Figures shown in parenthesis are percentage

the villages and barter their products. Merakpas go up to Singkhar Lauri across Samdrup Jongkhar Dzongkhag and up to Kangpara, Thrimshing, Brekha and Wamrong. A majority of them visit Radhi, Phongme, Yabrang, Saling, Bidung, Bartsham, Ranjung, Chaling, Changme, Galing, Shongphu, Bikhar, Kanglung and Yongphulla and around Trashigang. Some go across Dangmechhu up to as far as Gongthung, Yangnyer and Jamkhar villages. Table 12.7 presents Brokpas who are engaged in bartering the yak products visiting the warmer valley villages of eastern Bhutan. A total of 369 Brokpas were found engaged in Brukor in 2010. There is no gender preference for "drukor" as this requires skills as well as social contacts to perform trade with their trading partners. There are 55.01% Brokpa males followed by 44.99 female Brokpas in the total sample households. In the lower Sakteng valley, however it is male dominated (80.0%) practice as they have few yaks to graze and simultaneously, they are also engaged in agricultural pursuits.

Almost all the Brokpa families have 'nepos' or 'nygepo' (village hosts). In summer months, the Brokpas bring down butter and cheese and their "nepos" sell it for them in the village for them in exchange for rice, maize and chillies. The drukor practice starts by late October and continues for two to three months. They move from one village to other collecting grains that have been bartered for butter and cheese by their "nepos". "Nepos" also help the Brokpas in gathering their grains from various families. It is imperative for a Brokpa to eat and lodge at his "nepo's" house, during his stay in the village. The "nepo" in return receives cheese and butter as gift. A single Brokpa family will have one or more "nepo" families. This tradition has been sustained through several generations. The Brokpas from Sakteng usually trade with Tawang and Bomdila regions of Arunachal Pradesh in India. Brokpas still trade with their Monpa "nepos" across the border and visit as far as Udalguri in Assam to buy raw silk, cooking pots, salt, gumboots etc.

The Drukor tradition and practice is still very much an essential economic ingredient for the Brokpa society. According to an estimate, one Brokpa family traded close to 300 kg of butter and about 60 kg of cheese and yitpa. In exchange, the family received 35 quintals of unhusked rice and 30 quintals of maize grains in the winter of 1999. Another family bartered two hand-woven blankets for 280 changdeys (one changdey is about 2 kg) of maize and dried chilly. About 14 changdeys of maize was bartered for 12 kg of butter. In cash, one blanket is sold for Nu. 5000, butter for Nu. 120 per kg and yitpa for Nu. 90 per kg (Kuensel 1999, 4).

Table 12.8 represents a village wise distribution of "nepo" families serving Merak, upper Sakteng valley and lower Sakteng valley. In the present sample a total of 253 "nepo" families are in trading contacts with the Brokpas of Merak and Sakteng gewogs. The maximum "nepo" families (131) are serving Sakteng gewog followed by Merak (122). There is almost no tradition of "drukor" in lower Sakteng valley lying below 2500 m of altitude which is not favourable for yaks. Radhi and Phongme are two main villages with a maximum 49 and 41 "nepo" families serving both Sakteng and Merak. Third important "nepo" village is Yabrang with 35 "nepo" families followed by Chaling (19) and Songphu (13) and Changme (12). As such Brokpas have six main sharchhop villages (in eastern Bhutan) with a maximum of 169 "nepo" families covering 66.78% of the total "nepo" households. The Brokpas of Sakteng also go to Arunachal Pradesh of India to sell their dairy products, for which data was not collected. On an average, sample households in Merak maintain 1.70 "nepos" per household. Brokpas of Sakteng have 1.05 "nepo"s per household. In general Merak and Sakteng gewog have their "nepos" distributed equally in 12 villages. The main "nepo" village of Merak is Radhi with 17.21% "nepo" families and Yabrang is the main "nepo" village of Sakteng with 26.78% "nepo" families. Although Radhi is the central "nepo" village to both Merak and Sakteng gewogs, it is gradually being transformed into a small rurban center with gewog headquarter and concentration of development activities. In the recent years, Some Brokpas buy rice from government subsidized shops and prefer to sell butter and cheese in cash. Mr. Ngangong, the former gup (the village head) of Merak, was of the opinion that the un-husked rice, which they received from villagers, is less profitable and demands more labour. However, they need maize to feed the herds so they need to do the "drukor". Increased access to market centers facilitates monetary transaction practices. One dissatisfying factor to this effect has been that urban population does not seem to like the butter and cheese prepared by the Brokpas. On the other hand, "drukor", too, has lost its original significance and value that was once attached to it.

Table 12.8 Distribution of "nepo" families serving sample households, 2010

Sl. no.	Nepo villages	From upper Sakteng gewog valley	From lower Sakteng valley	Sakteng gewog total	From Merak gewog	Grand total
1.	Yabrang	33	2	35	–	35
2.	Radhi	26	2	28	21	49
3.	Phongme	24	–	24	17	41
4.	Chaling	–	–	–	19	19
5.	Songphu	–	–	–	13	13
6.	Changmey	–	–	–	12	12
7.	Saling	9	–	9	–	9
8.	Khardung	8	–	8		8
9.	Phimsong	8	1	9	–	9
10.	Khaling	–	–	–	8	8
11.	Galing	–	–	–	7	7
12.	Jangyen	6	–	6	–	6
13.	Kangpara	–	–	–	6	6
14.	Bidung	–	–	–	6	6
15.	Wamrong	–	–	–	5	5
16.	Bikhar	–	–	–	4	4
17.	Bartsham	–	–	–	4	4
18.	Pakling	4	–	4	–	4
19.	Shokang	3	–	3	–	3
20.	Gazari	3	–	3	–	3
21.	Rolam	–	1	1	–	1
22.	Dongrom	–	1	1	–	1
Total "nepo" families		124	7	131	122	253

Source Field work, 2010

Nevertheless, "drukor" tradition is still essential to the Brokpas, at least as long as they continue with yak herding and occupy their original homes in the upper reaches of the mountains.

12.8 Conclusion

Yak herding is central to Brokpa pastoral livelihood. It is not just their economy but their socio-cultural heritage. Surviving on transhumance, trade and herding, they represent the initial mode of human adjustment to nature. Brokpas have developed a symbiotic relationship with their pasturelands, which provides for them a perennial source of livelihood and sustenance. The Brokpa pastoral system is unique and rich in biological diversity. They have three sets of pastures which they are still maintaining even in the wake of the biodiversity conservation movement taking away their traditional pasturing rights. Even in this conflicting situation, their occupational patterns as yak herders and sellers of milk products, their social magnitudes and their persistence as distinct entities in Bhutan are immensely significant. Still living in harmony with the organic and inorganic entities in nature and preserving the bio-cultural diversity they are away from the evils of hierarchical lives of peasant caste societies. Moving from one pasture to another with the change in season and bartering with the settled villagers, Brokpas like all other indigenous tribes in the world are now getting trapped in the web of modernization and development. Today Brokpa families are facing labour shortage owing to increasing participation of their children in education and other developmental activities. The issue of climate change is another threat impacting on the availability of pastures in three ecological settings unique to Brokpa region of Bhutan. How they perceive climate change scenario and how they cope with negative impacts of climate change will determine the future of their survival as a semi-nomadic community. At large it also poses a question of the survival of bio-cultural diversity of Bhutan.

Acknowledgements The author is deeply grateful to Mr. Sangey Khandu and Mr. Leki Khandu, B. A. (Hons) Geography student of batch 2009–2010, Sherubtse College Kanglun, Royal University of Bhutan, who collected field information for the present research in 2010. This work could not have been undertaken without the field support extended by them during his tenure as the Colombo Plan Professor from India to the Royal Government of Bhutan from 2008 to 2010. His sincere thanks to Dr. Pankaj Thapa, Assistant Professor, Department of Geography and Regional Planning, Sherubtse College Kanglun, Royal University of Bhutan for cartographic assistance.

References

Anonymous. (2011). *Livestock statistics 2011*. Department of Livestock, Ministry of Agriculture and forest, Royal Govt. of Bhutan, Thimphu.

Anonymous. (2005). *Population and housing census of Bhutan 2005*. Office of the Census Commissioner, Royal Government of Bhutan, Thimphu.

Chand, R. (2000). Brokpas in transition—A study of tribal self reliance in Bhutan. *Sherub Doenme, 5,* 47–74.

Chand, R. (2002). Pastoral economy & pasture ecology of Brokpas. In T. Gyeltshen & Sadruddin (Eds.), *Environment and life support system of Bhutan Himalaya* (Vol. 1, pp. 233–249). Kanglung, Bhutan: Sherubtse College.

Chand, R. (2004). *Brokpas: The hidden highlanders of Bhutan, PAHAR (People's Association for Himalaya Area Research)* (164 p). Nainital.

Chettri, N. (2008). Local and indigenous practices on adaptation: An experience from herder's life of western Bhutan. *Mountain Forum Bulletin, 8*(1), 19–21.

Chetri, D. K., et al. (2011). Transhumance effect on husbandry practices and physiological attributes of Chauri (Yak-Cattle) in Rasuwa District. *Our Nature, 9,* 128–137.

Dompnier, R. (2007). *Traditions of the Brokpa of Merak and Sakteng*. Thimphu: Ministry of Home and Cultural Affairs.

Gyamtsho, P. (2000). Economy of Yak herders. *Journal of Bhutanese Studies, 2*(1), 90–135.

Jest, C., & Bonnemaire J. (2002). Towards a global approach to the study of yak research and development. In H. Jianlin, et al. (Eds.), *Yak production in the central asian high lands, Proceedings of the third international congress on yak held in Lhasa, P.R. China* (pp. 5–10), September 4–9, 2000. Nairobi, Kenya: ILRI (International Livestock Research Institute).

Karchung, G. (2011). Diminishing cultures of Bhutan: Costume of Merag community. *SAARC Culture, 2,* 17–43.

Kuensel. (1999). The National News paper of Bhutan, 4.

Mait, S., et al. (2014). Determinants to climate change adaptation among the Brokpa Pastoral Nomads of Western Arunachal Pradesh. *Indian Research Journal of Extension Education, 14* (2), 36–40.

Rhode, D., Madsen, D. B., Brantingham, P., & Dargye, T. (2007). Yaks, yak dung and prehistoric habitation of the Tibetan Plateau. *Developments in Quaternary Sciences, 9,* 205–224.

Tshering, C. (2004). Rangeland of Bhutan. In *Proceedings of the International Congress on Yak, Chengdu, Sichuan, P.R. China*. IVIS publication.

Tsering, R., et al. (2010). The Brokpa and their social development: The work of M/S Dunkarpa welfare association at Dirang Circle of West Kameng District, Arunachal Pradesh, India. *Himalayan Study Monograph, 11*, 191–195.

Ura, K. (2002). The Herdsmen's dilemma. *Journal of Bhutanese Studies, 7*(2), 1–43.

Ura, K., et al. (2012). *A short guide to gross national happiness index*. The Centre for Bhutan Studies (96 p), Post Box No. 1111, Thimphu, Bhutan.

Wangchuk, D., et al. (2013). The Yaks: Heart and soul of the Himalayan tribes of Bhutan. *Journal of Environmental Research and Management, 4*(2), 0189–0196.

Wangmo, S. (1990). The Brokpa: A semi-nomadic people in Eastern Bhutan. In N. K. Rustomji & C. Ramble (Eds.), *Himalayan environment and culture* (pp. 141–158). Shimla: Indian Institute of Advanced Study.

Wiener, G., Jialin, H., & Ruijun, L. (2003). *The Yak* (2nd ed.). Bangkok: Regional Office for Asia and Pacific, Food and Agriculture Organization of the United Nation.

Chapter 13
Occupational Diversification and Changing Marginality Conditions of Brokpas of Bhutan in Twenty First Century

Raghubir Chand

13.1 Introduction

The main conceptual base of this study is that the Brokpas of Bhutan were socially and culturally marginal not only to other Himalayan communities in general but to the Bhutanese population. Although, there is no tribal and non-tribal demarcation adopted officially or otherwise in Bhutan and from anthropological viewpoint, all segments of Bhutanese societies belong to the same racial stock of Tibeto-mongoloids. However due to the marginal locations of Brokpa habitat in Bhutan and also from the interaction pattern, as observed across the communities boundaries, Brokpas are numerically in the minority. Curiously, the Bhutanese call them Brokpas, a name that is derived from the Tibetan word Brogpa, Brog means pastures and pa means inhabitants, literally meaning nomads or herdsman. Even the Sharchop (people of eastern Bhutan) call them Brami which means other people (Chand 2004, 29). This understanding about Brokpas may be somewhat limited or even distorted, but one thing is clear that they are not among the majority people of Bhutan. Brokpas are minority people in Bhutan as in most societies and they are by all standards considered as marginal people which also involves the question of identity as perceived by others in Bhutan. Keeping in consideration their own and unique ways of living, The Royal Government of Bhutan preserved and protected their cultural and socio-religious identity, they were allowed to wear their own distinct dress in particularly and practice their own sect of Buddhism.

Marginal societies are undergoing unprecedented changes all over the Himalayan highlands with changing local development policies and forces of globalization reaching at their doorsteps. While most of the Himalayan states retain the colonial blueprint of economy, as already stated in the previous chapter, pro-

R. Chand (✉)
Department of Geography, DSB Campus,
Kumaun University, Nainital 263002, Uttarakhand, India
e-mail: raghubir.seri@gmail.com

R. Chand et al. (eds.), *Societies, Social Inequalities and Marginalization*,
Perspectives on Geographical Marginality,
DOI 10.1007/978-3-319-50998-3_13

moting commercial agriculture, and the increased labour force in the service sector, were linked to the forces of globalization and modernization much earlier in farming societies than in pastoral societies. The Pastoral people surviving on the transhumance and trade depending on the extensive use of alpine grazing lands at the margins of the Himalaya preserved their self-reliant mode of economy and defended their autonomy even during the colonial era. The Tibet-Himalayan Highland has nowadays attracted more and more attention by its increasing geo-political, environmental, and socio-economic importance (Rinchin et al. 2010). Wangchuk et al. (2013a) have analyzed the impact of globalization on the Drogpas (Brokpas are also known as Drogpas in Tibet) and other tribal people and ultimately portrayed their future outlook. However, the material base of the pastoral economy and the social fabric of pastoral life eroded gradually, with the increasing intervention of a commonly perceived development planning imposed from the political powers centered on the agricultural societies in the plain areas of the country. While the livelihood pattern though is broadly similar in all pastoral areas in the Himalaya, Bhutan, is significantly different from other Himalayan states in terms of its size, social structure, economic base, nature of governance (Ura et al. 2012), market situations and resource use.

13.2 The Brokpa Habitat

As mentioned in the previous chapter the Brokpas are a truly yak herding community in Bhutan occupying Merak and Sakteng gewogs (administrative blocks) of Trashigang Dzongkhag (district) in eastern Bhutan (Fig. 13.1). They are believed to have fled from Tibet during the fifth and sixth centuries along with their yaks and settled within the rugged terrain of Eastern Bhutan in about the 14th and 15th Centuries (Dompnier 2007, 12–15 as cited in Wangchuk et al. 2013b, 190). Their territorial political organization was independent of neighbouring states because they followed the Gelugpa sect of Tibetan Buddhism which they still practice to this day.

This study is based on primary information collected at the household level from a randomly selected 126 households from 7 Brokpa villages in January 2010. It was a sample of 15.44% of the total of 816 households (PHCB 2005) living in the 16 Brokpa villages in both the Merak and Sakteng gewogs. Out of four villages in the Merak gewog, 45 households from the village of Merak were selected. Out of the 12 villages of Sakteng gewog, 6 villages were selected, sampling 38 households from the village of Sakteng, 12 from Borongma, 11 from Tengma, 9 from Pusa, 6 from Borongtse, and 5 from Manidungjur. Thus in total 81 households from the Sakteng valley and 45 households from Merak were sampled, giving a total of 126 households in all were selected with a total sample population of 707 persons (Table 13.1). In this chapter, only seven sample villages were considered so that they may be compared with the same number of villages surveyed in 1999.

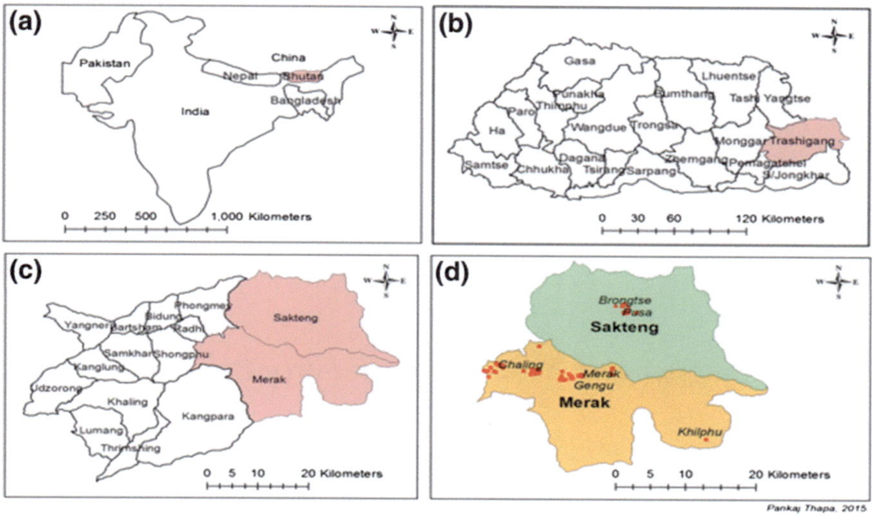

Fig. 13.1 The Brokpa habitat **a** Location of Bhutan in South Asia. **b** Location of Trashigang Dzongkhag in Bhutan. **c** The Brokpa home land Merak and Sakteng Gewog within Trashigang Gewog. **d** Important Settlements of Brokpa home land

Table 13.1 Village wise distribution of population in sample households, 2010

S. N.	Villages	Households	Population			Male per 1000 females	Average household size
			Male	Female	Total		
1.	Tengma	11	35 (53.0)	31 (47)	66 (100)	1129	6.1
2.	Borongma	12	32 (52.5)	29 (47.5)	61 (100)	1103	5.1
3.	Sakteng	38	94 (49.8)	95 (50.2)	189 (100)	989	4.97
4.	Manidungjur	5	18 (62.1)	11 (37.9)	29 (100)	1636	5.8
5.	Borongtse	6	11 (47.8)	12 (52.2)	23 (100)	916	3.83
6.	Pusa	9	31 (58.5)	22 (41.5)	53 (100)	1409	5.88
Sakteng gewog (total)		81	221 (52.49)	200 (47.51)	421 (100)	905	5.19
7.	Merak	45	147 (51.4)	139 (49.6)	286 (100)	1058	6.35
Grand total		126	368	339	707	982	5.76

Source Field work, 2010, figures shown in parenthesis are percentage of total population

Information related to work force participation and dependent population, occupational structure and change, income sources and occupational diversification and structural occupational changes were collected from each sample household in both years. The sample size is thus fairly representative.

13.3 Brokpa's Occupations and Income

The Brokpa's pastoral life and livelihood pattern was examined in Sect. 12.3. As stated there, the life and livelihood of Brokpas is very simple and sparse. They are still close to a tribal way of living and they remain isolated. Animal rearing has all along been their primary occupation. Woolen goods and bartering of milk products with lower valley inhabitants are a part of their livestock economy (Ura 2002; Wangmo 1990).

13.3.1 The Changing Nature of Work Force Participation

The economic status of any population is best explained by looking at the working and dependent population. These show the size of the labour force and its growing impact on economic gains. Primary information collected about the work force participation and dependent population in sample households in 1999 and 2010, is presented in Table 13.2. There was an overall increase of +8.9% in the working population in the age group of between 15 and 59 in the Brokpa villages since the first survey conducted in 1999 and a gradual decrease of −5.4% in the childrens' age group of below 14 years and also in the old age persons group of above 60 years of −3.5%. The impact of changing age compositions has already occurred because of the gradual fertility decline that has been underway in Brokpa society due to increasing use of family planning measures since the late 1990s.

The biggest increase in the working population is found in the village of Borongtse (+32.6%) with a least increase of +5.8% in village Manidungur in upper Sakteng valley. The corresponding decrease may be seen in the same villages in the

Table 13.2 Work force participation and dependent population in sample households, 1999–2010

S. N.	Villages	Children below 14 years in %			Work force 15–59 years in %			Old age persons >60 years in %		
		1999	2010	Change	1999	2010	Change	1999	2010	Change
1.	Tengma	35.3	24.3	−11	48.0	63.6	+15.6	16.7	12.1	−4.6
2.	Borongma	33.3	29.5	−3.8	58.3	67.2	+8.9	8.4	3.3	−5.1
3.	Sakteng	36.7	31.2	−5.5	56.0	64.6	+8.6	7.3	4.2	−3.1
4.	Manidungjur	36.2	34.5	−1.7	56.3	62.1	+5.8	7.5	3.4	−4.1
5.	Borongtse	35.0	4.4	−30.6	50.0	82.6	+32.6	15.0	13.0	−2.0
6.	Pusa	21.3	39.6	+18.3	70.2	50.9	−19.3	8.5	9.5	+1.0
Sakteng gewog (total)		34.1	29.7	−4.4	55.4	63.9	+8.5	10.5	6.4	−4.1
7	Merak	40.1	34.3	−5.8	52.9	61.9	+9.0	7.0	3.8	−3.2
Grand total		36.9	31.5	−5.4	54.2	63.1	+8.9	8.9	5.4	−3.5

Source Field work, 1999 and 2010

childrens' age group below the 14 years. However, the village of Pusa has demonstrated a decline of −19.3% in the labour force with the corresponding increase of 18.3% in the childrens' age group below the 14 years. There was a moderate decrease in old age persons above 60 years ranging between −2.0% in Borangtse to a maximum of −5.1% in Borongma village between 1999 and 2010. Age is the foremost criteria for assessing the potential of economic growth through the labour force participation. The dependency ratio has already decreased in all six sample villages, except in the case of Pusa. A way of looking at the changes is to review the changes in absolute figures. In general, the work force recorded in 2010 was 63.1% compared with 54.2% in 1999. Likewise the percentage of children and old age persons has reduced from 36.9 to 31.5% and from 8.9 to 5.4% during the period 1999 to 2010.

13.3.2 Occupational Change

The study of occupational change provides a baseline set of measurement for the comparison of past and potential future trends of occupational mobility within Brokpa society. There has been no comprehensive effort to monitor changes that were taking place in Brokpa economy so far. This investigation provides such a comparison, resulting from the two surveys conducted by the author. The data presented in Table 13.3 demonstrate the occupational change and the reasons behind such changes. It is interesting to note that Brokpas show remarkable stability in their occupational behaviour (Chand 2000, 2002, 2004). The hope of bringing about changes through the social development programmes in the late 1990s was not proven by the results. It seems that it is too early to establish the lasting effects of such development programmes on the new generation of Brokpas.

The data set reflects that in 90.81% of cases there were no occupational changes. In Sakteng gewog, the occupational change is recorded as averaged a mere 3.33% while in Merak, it is 17.83%. As such, the occupational change ranges from a minimum of 1.52% in the village Tengma to 13.04% in the village Borongtse and 17.83% in Merak. Limited educational changes and less man power are the two major factors behind such a low rate of occupational change. The influence of schooling on occupational change is found to be at a maximum in the village Merak in which 62.75% of the total 51 of persons who changed their occupation, did so for education related reasons. In Sakteng gewog, education ranks second (35.71%) to less man power which is the reason 42.28% of those persons who changed their occupation. The contribution of other factors such as climate change (Mait et al. 2014) behind occupational change was found to be the explanation in 18.19% of cases.

Table 13.3 Occupational change in sample households, 1999–2010

Sl. N	Villages	1999–2010			Reason for change			
		Change	No change	Total	Education	Less man power	Other	Total
1.	Tengma	1 (1.52)	65 (98.48)	66 (100)	0	1	–	1
2.	Borongma	2 (3.28)	59 (96.72)	61 (100)	1	0	01	2
3.	Sakteng	4 (2.12)	185 (97.88)	189 (100)	2	1	01	4
4.	Manidungjur	2 (6.90)	27 (93.10)	29 (100)	1	1	–	2
5.	Borongtse	3 (13.04)	20 (86.96)	23 (100)	0	01	2	3
6.	Pusa	2 (3.80)	51 (96.20)	53 (100)	1	1	0	2
Sakteng gewog (total)		14 (3.33)	407 (96.67)	421 (100)	5 (35.71)	6 (42.86)	3 (21.43)	14 (100)
7.	Merak	51 (17.83)	235 (82.17)	286 (100)	32 (62.75)	13 (25.49)	6 (11.76)	51 (100)
Grand total		65 (9.19)	642 (90.81)	707 (100)	37 (56.92)	19 (29.23)	9 (13.85)	65 (100)

Source Field work, 2010, figures shown in parenthesis are percentage of total population

13.3.3 Occupational Structure

Occupational structure is a key component and manifestation of population composition which gives a proper illustration of the economic base of a society and engagement of any population in an area or a country. Occupational structure also influences and determines the pattern of the socio-economic development of an area. The village wise distribution of the occupational structure of Brokpas is presented in Table 13.4 on the basis of data collected in the year 2010. The proportion of workers engaged in various occupations highlights the economic and socio-cultural features of the Brokpa society. In Bhutan, where socio-cultural moorings have strong bearing on a persons livelihood, livestock herding, trading and weaving are integral parts of this livestock economy (Chand 2004, 109). When the respondents were asked to give their main occupations, they were mostly referred to engaging in herding which accounted for 39.8% of responses on average. In Merak, herding contributes 48.7%, while in Sakteng, it is 36.5% of all occupations. The second most important occupation was agriculture (18.3%) followed by weaving (15.4%). As per the survey of 1999, weaving was second to herding with almost the same number of people i.e. 15.63% but agriculture was being practiced by only 4.65% in 1999 (Chand 2004, 109). It seems that people from herding backgrounds have gradually moved into agriculture over the years. In 2010, agriculture accounted for 5.7% of jobs in Merak, while in Sakteng, 23.3% people are practicing agriculture. Gomchens (lay priest) and monks constitute 5.8 and 3.2% of the total occupations being adopted by Brokpas respectively. They get food grains and dairy products for their services (Chand 2004, 109). Government jobs which were alien to Brokpas in 1999, now engage 5.3% people and rank in fourth place. There were two shops, one each in Merak and Sakteng in 1999 (Chand 2004, 109) which has now grown to 13 (and 3.5% of jobs). Other categories of occupation include: masons and carpenters who share 3.2% of the total occupations. Those who have no cattle earn their livelihood from labour, road work during the winter months and also work as herders with merchant Brokpas. They account for 5.6% of the total occupations.

13.3.4 Occupational Diversification and Structural Occupational Changes

Occupational diversification refers to a continuous adaptive process whereby households add new activities, maintain existing ones or drop others, thereby maintaining diverse and changing livelihood portfolios (Ellis 2004, p. 1). Diversification takes place mostly in order to overcome risk and seasonality in natural resource-based livelihoods. In the case of Brokpa economy, the situation is different, where the risk factor is less important and educational attainment plays a significant part in occupation diversification. Traditionally, the Brokpas were not

Table 13.4 Occupational structure of sample households, 2010

Villages	Herding	Weaving	Agriculture	Govt. job	Gomchen	Monk	Shop-keeper	Mason & carpenter	Others	Total
Tengma	13	10	9	1	4	1	6	1	3	48
Borongma	15	8	9	1	1	2	0	0	0	36
Sakteng	36	19	31	8	5	3	4	5	4	115
Manidungjur	11	3	2	1	1	0	0	0	1	19
Borongtse	14	2	2	0	1	2	0	0	1	22
Pusa	10	7	10	2	1	0	0	1	1	32
Sakteng gewog total	99 (36.5)	49 (18.1)	63 (23.3)	13 (4.8)	13 (4.8)	08 (2.5)	10 (3.7)	7 (2.6)	10 (3.7)	272 (100)
Merak	51 (48.7)	9 (8.5)	6 (5.7)	7 (6.7)	9 (8.5)	4 (3.8)	3 (2.8)	5 (4.8)	11 (10.5)	105 (100)
Grand total	150 (39.8)	58 (15.4)	69 (18.3)	20 (5.3)	22 (5.8)	12 (3.2)	13 (3.5)	12 (3.2)	21 (5.6)	377 (100)

Source Field work, 2010, figures shown in parenthesis are percentage. People are engaged in more than one occupations

keen on enrolling their children in school. The admission of children has not been voluntarily forth coming. The local administration exercised power to generate enrollment in the beginning (Chand 2004, 54). It all happened because the man-power requirement for yak herding had an adverse impact on education of Brokpa children as they are retained in the pastures to help the parents. Of late Brokpas realized that the undue reliance on pastoral activities has hampered the generation of cash resources. There is a barter tradition still practiced by Brokpas and the milk products of Brokpas once a sole source for low landers Bhutanese are now competing with Indian butter and cheese in the urban centers. At the same time, Brokpas are undergoing constant changes in the socio-economic setup and the impact of market forces and globalization are reaching their doorsteps. The failure of pastoralism to deliver improved livelihoods in the post-liberalization era in Bhutan, particularly after 2008, is another important factor encouraging occupational diversification. The data related to percentage change in various Brokpa occupations is presented in Table 13.5.

The average rate of change in both the gewogs (administrative blocks) has been found to be the most in relation to herding. About −11.16% people of the total sample households have left herding. Sakteng has witnessed −18.61% shift in herding, while Merak had just −3.72% loss of the total sample households population. Those engaged in other occupations like labour and road work etc. have also recorded equally important shifts (−10.52%). All other categories of Brokpa occupations have gained workers with a maximum by agriculture (+9.26%) followed by government jobs (+4.96%), Gomchens, the lay priest (2.94%), Shopkeepers (+2.83%), Mason and carpenters (+2.61%), weaving (+2.51%) and Monks (+2.51%). At the gewog (administrative blocks) level maximum changes were recorded in Sakteng with herding (−18.61%) followed by agriculture (+15.64%) and weaving (+4.67%). In Merak gewog (administrative block), the highest change was found in government jobs (+6.30%) followed by mason and carpenters (+4.40%) and herding (−3.72%).

Table 13.5 Occupational diversification and structural occupational changes (in %) in sample households, 1999–2010

Sl. n.	Occupations	Sakteng gewog total	Merak	Total
1.	Herding	−18.61	−3.72	−11.16
2.	Weaving	+4.67	+0.36	+2.51
3.	Agriculture	+15.64	+2.88	+9.26
4.	Govt. jobs	+3.63	+6.30	+4.96
5.	Gomchen	−1.42	+7.3	+2.94
6.	Monk	+1.62	+3.40	+2.51
7.	Shop-keeper	+3.26	+2.40	+2.83
8.	Mason and carpenter	+0.83	+4.40	+2.61
9.	Others	−10.52	–	−10.52

Source Field work, 1999 and 2010

13.3.5 Main Source of Income and Occupations

The sample households were asked to furnish the information about the sources of their income in their preferential order which was a critical part of the survey. In the case of Brokpas, fear of taxation impacts on potential responses. The data presented in Table 13.6 however provide a some approximation of the situation. There is a marked conformity in the sense that dairy has been the prime source of income for 57.9% of the sample households in both the gewogs (administrative blocks). The income from dairy is highest in Merak (66.6%) since Merak is a purely pastoral village. Dairy is a main source of income in Saktend as well, but for 53.1% of total sample households. In a Buddhist societies like the Brokpas, the gomchens or the lay priests, earn the second most sizable proportion of income in the villages (10.93%) which is just below that of dairy. The possibility to earn cash income from the labour and road work stands is at third place (9.5%) followed by government jobs (6.3%) and agriculture (6.3%). Shopkeepers (4.8%) and mason and carpenters (4.8%) earn equally from their respective occupations. However dairying is the main source of income of Brokpas.

Looking at agriculture, it is a source of income for 8.6% of sample households in Sakteng while it is a source of income by merely 2.2% in Merak. The government jobs are also important for 7.5% sample households in Sakteng, owing to the presence of administrative offices.

13.4 Conclusions

Yak herding is the main occupation of Brokpas and they depend heavily on dairy products, which also provides them with the opportunity to go for "drukor" during winters. "Drukor" is thus a unique trading relationship that exists between the Brokpas and their Sharchhop (people of eastern Bhutan) hosts and friends. Strong social ties bind together the highlanders and the valley-dwellers based on economic necessities. The "drukor" tradition and practice is still very much an essential economic ingredient for the Brokpa society. Although "drukor" has lost its original significance and value that was once attached to it, nevertheless, "drukor" tradition is still essential to the Brokpas, at least as long as they continue with yak herding and occupy their original homes in the upper reaches of the mountains. Today Brokpa families are facing labour shortage owing to increasing participation of their children in education and other developmental activities. They need more hands in the same households to look after yak, sheep and other cattle in different pastures and also to trade with the low lying valleys. As a result, the average rate of occupational change in both the gewogs (administrative blocks) has been found be most distinctive in terms of decreased herding. The occupational engagement in herding is gradually shifting towards other occupations with the rate of −11.16% of the total sample households. All other categories of Brokpa occupations have

Table 13.6 Sample households classified according to their main source of income, 2010

Sl. n	Villages	Dairy	Agriculture	Shop-keepers	Govt. jobs	Mason & carpentry	Gomchen	Others	Total families
1.	Tengma	5	1	1	0	1	2	1	11
2.	Borongma	7	2	0	1	0	1	1	12
3.	Sakteng	16	4	2	5	2	4	5	38
4.	Manidungjur	3	0	0	0	0	1	1	5
5.	Borongtse	6	0	0	0	0	0	0	6
6.	Pusa	6	0	0	0	1	1	1	9
Sakteng gewog total		43 (53.1)	7 (8.6)	3 (3.7)	6 (7.5)	4 (4.9)	9 (11.1)	9 (11.1)	81 (100)
7.	Merak	30 (66.6	1 (2.2)	3 (6.6)	2 (4.4)	2 (4.4)	4 (4.4)	3 (6.6)	45 (100)
Grand total		73 (57.9)	8 (6.3)	6 (4.8)	8 (6.3)	6 (4.8)	13 (10.3)	12 (9.5)	126 (100)

Source Field work, 1999 and 2010

gained workers with a maximum in agriculture (+9.26%) followed by government jobs (+4.96%), Gomchens (lay priest), and monks (+2.51%). The occupational change impacted on 9.19% of jobs during 1999–2010 in Brokpa society. Despite this, livestock forms the basis of Brokpa economy without which the Brokpa identity seems indistinguishable. So long as Brokpas continue with their livestock economy, Brokpas will retain their unique place in Bhutan. However, the image of Brokpas as a fringe dwellers and marginal people is changing and The Royal Government of Bhutan is paying special attention to the need to preserve and protect their cultural and socio-religious identity.

Acknowledgements The author is deeply grateful to Mr. Sangey Khandu and Mr. Leki Khandu, B. A. (Hons) Geography student of batch 2009–2010, Sherubtse College Kanglung, Royal University of Bhutan, who collected field information for the present research in 2010. This work could not have been undertaken without the field support extended by them during his tenure as the Colombo Plan Professor from India to the Royal Government of Bhutan from 2008 to 2010. His sincere thanks to Dr. Pankaj Thapa, Assistant Professor, Department of Geography and Regional Planning, Sherubtse College Kanglung, Royal University of Bhutan for cartographic assistance.

References

Anonymous. (2005). *Population and housing census of Bhutan 2005*. Thimphu: Office of the Census Commissioner, Royal Government of Bhutan.

Chand, R. (2000). Brokpas in transition—A study of tribal self reliance in Bhutan. *Sherub Doenme, 5*, 47–74.

Chand, R. (2002). Pastoral economy & pasture ecology of Brokpas. In T. Gyeltshen & Sadruddin (Eds.) *Environment and life support system of Bhutan Himalaya* (Vol. 1, pp. 233–249). Kanglung, Bhutan: Sherubtse College.

Chand, R. (2004). *Brokpas: The hidden highlanders of Bhutan, PAHAR (People's Association for Himalaya Area Research)* (p. 164). Nainital.

Dompnier, R. (2007). *Traditions of the Brokpa of Merak and Sakteng*. Thimphu: Ministry of Home and Cultural Affairs.

Ellis, F. (2004). *Occupational diversification in developing countries and the implications for agricultural policy, programme of advisory and support services to DFID (PASS), overseas development group (ODG)* (Project No. WB0207). Norwich, UK: University of East Anglia.

Mait, S., et al. (2014). Determinants to climate change adaptation among the Brokpa pastoral nomads of Western Arunachal Pradesh. *Indian Research Journal of Extension Education, 14* (2), 36–40.

Tsering, R., et al. (2010). The Brokpa and their social development: The work of M/S Dunkarpa welfare association at Dirang circle of West Kameng District, Arunachal Pradesh, India. *Himalayan Study Monograph, 11*, 191–195.

Ura, K. (2002). The Herdsmen's Dilemma. *Journal of Bhutanese Studies, 7*(2), 1–43.

Ura, K., et al. (2012). A short guide to gross national happiness index. *The centre for Bhutan studies* (p. 96). Post Box No. 1111. Thimphu, Bhutan.

Wangchuk, D., Dhammasaccakarn, W., & Tepsing, P. (2013a). Survival of Drogpa cultural and traditional memes and the threat of modern development. *Asian Social Science, 9*(15), 142–155.

Wangchuk, D., et al. (2013b). The Yaks: Heart and soul of the Himalayan tribes of Bhutan. *Journal of Environmental Research and Management, 4*(2), 0189–0196.

Wangmo, S. (1990). The Brokpa: A semi-nomadic people in Eastern Bhutan. In N. K. Rustomji & C. Ramble (Eds.), *Himalayan environment and culture* (pp. 141–158). Shimla: Indian Institute of Advanced study.

Chapter 14
Socio-cultural and Nutritional Environment of a Marginal Community of Uttarakhand, India—A Case Study of the Tharu Tribe

B.R. Pant and Surendra Pal

14.1 Introduction

Tribal communities and their habitats constitute very significant parts of the marginal populations of India. They comprise about 18% of country's land and 8.6% of its population (Census of India 2011). There is a great variation between the varieties/type of the Scheduled Tribe in one state/UT (Union Territory is a region govern by the Lieutenant Governor) to another. Indian tribes are multi-cultural and multi-racial. It is estimated that the total number of tribal communities in India are more than 1000 (See notes at the end). In the Indian context, a tribe is defined as a group with traditional territory, specific name, common language, strong kin relations, association with clan structure, tribal authority, and rigid inclination to religion and belief. Functional independence, homogeneity, primitive means of exploiting resources, economic backwardness, rich culture and tradition and least (minimum) desire to change are some of the other characteristics dominant among the tribes (Majumdar and Madan 1970).

About half of the total tribal population of the world lives in India. The aborigines in Indian language are known as "Adibasi"-Adi and basi standing for "original" and "inhabitants" respectively. Constitutionally these human groups are known as Scheduled Tribes (S.T.), "Anusuchit Jati", "Vanjati", "Vanbasi", "Pahari", and "Adimjati" etc. According to article 342 of the constitution of India, the Scheduled Tribes are the tribes or tribal communities, which may be notified by the President of India. The census enumerates only such tribal population as are scheduled under the relevant constitutional order in force at the time of the census.

B.R. Pant (✉)
Department of Geography, M. B. Govt. P. G. College, Haldwani, Nainital, Uttarakhand, India
e-mail: brpant123@rediffmail.com

S. Pal
G. I. C. Ramak, Champawat, Uttarakhand, India

R. Chand et al. (eds.), *Societies, Social Inequalities and Marginalization*,
Perspectives on Geographical Marginality,
DOI 10.1007/978-3-319-50998-3_14

The main characteristic features of Indian tribes are variation in size of the population, inhabiting all climatic zones. They belong to all four races-Negrito, Proto-Austroloid, Mongoloid and Caucasoid and speak Indo-Aryan, Dravidian and Tibeto-Burman language. India has the largest tribal population of the World. According to the 2011 census, the tribal population of India is enumerated 104,281,034 persons residing in about 18% of the country's geographical area (Anon 2011). As per the 2011 census, country's largest tribal population is found in Madhya Pradesh (14.7%) with 46 tribes (main and sub-tribes). The second largest tribal population is enumerated in Maharashtra (10.1%) followed by Orissa (9.2%). It is clear that the states which have more inhospitable geographical areas registered more tribal population such as Madhya Pradesh, Jharkhand, Orissa, Chhattisgarh, Rajasthan and the north eastern mountainous states of India. Contrary to this, the states having more fertile land, irrigation and other infrastructural facilities have less tribal population and two states namely Haryana and Punjab, and three UTs-Delhi, Chandigarh and Pondicherry, have not notified tribes. According to the share of the tribal proportion compared to the total state population (Anon 2011), Mizoram (94.5%) ranks at the top followed by Nagaland (86.5%), Meghalaya (86.1%), Arunachal Pradesh (68.8%) Manipur (35.1%) Chhattisgarh (30.6%), Jharkhand (26.2%) etc. Among the Union Territories (UTs) Lakshadweep (94.8%) ranks first followed by Dadra & Nagar Haveli (52.0%) (Anon 2011).

14.2 Tribes of Uttarakhand[1]

As per the 2011 census, Uttarakhand has 291,903 tribal people which constitute about 2.9% of the total population of the state. Uttarakhand state contributes only 0.83% to the total population and 0.3% to the total tribal population of India (Anon 2011). The habitation of a particular tribe is confined and specific to a particular location in the Uttarakhand, Tharu and Buksa are confined to once waterlogged land of Tarai and Bhabar region while Bhotia (Shauka, Tolcha, Marcha and Jad) are confined to the higher altitudes of Dharchula, Munsyari, Kapkot, Joshimath, Bhatwari and Dunda blocks and Raji is a tribe confined to Didihat, Dharchula and Champawat blocks. Likewise, the Jaunsari resides in Chakrata and Kalsi block of Dehradun district. Despite their habitation in different zones, their prevails a few

[1]There is confusion on exact number about the varieties/types of tribes in the country, the reason being there is no single list at the central level and each state/UT has its own list. Just adding the numbers of tribes that have been scheduled in each state/UT does not serve the purpose and the total number thus arrived should not be taken as total number of tribes in the country. For example Santhals have been notified in Orissa, Bihar, West Bengal, same case with the Bhills, which has been notified in Gujarat, Rajasthan, Maharastra, Madhya Pradesh etc. Likewise Bhotia, Buksa, Tharu, Jaunsari, Gonds, and many other tribes have been notified in more than one state. Therefore, there is lot of multiple counting. There are five state/UTs like Haryana, Punjab, Chandigarh, Delhi and Pondicherry, where no tribes have been notified as scheduled. However, there are about 250–300 major tribal groups in the country.

common features in the economic and social life of all the tribal communities with minor variations necessitated to maintain a harmonious coordination between the resource availability and needs of the society. Agriculture forms the basic livelihoods of the Tharu, Buksa and Jaunsari tribes while sheep rearing and trade is the main source of livelihood of the Bhotia tribe. However, the Raji (*Vanrawt)* whose population is lower than 600 persons (2001) are mainly dependent on subsistence agriculture, animal husbandry and daily wages.

The tribal population of Uttarakhand is concentrated in the remote forest areas of Tarai and Bhabar and higher reaches of Pithoragarh, Chamoli, Uttarkashi districts and the Jaunsar-Bawar area of Dehradun district. Among all 13 districts of the state, Udham Singh Nagar with a tribal population of 43.03% is at the top followed by Dehradun (38.3%), Pithoragarh (6.7%) and Chamoli (4.2%). In fact these four districts of Uttarakhand contribute 91.25% of total tribal population. The districts with lowest tribal population in the state are Rudraprayag, Tehri, Champawat, Almora, Pauri, Bageshwar, Uttarkashi, Hardwar and Nainital having respectively 0.1, 0.3, 0.5, 0.4, 0.8, 0.7, 1.2, 2.1 and 2.6% of the population. Udham Singh Nagar also ranks highest with 7.5% tribal population of its total district population followed by Dehradun (6.6%), Pithoragarh (4.0%) and Chamoli (3.1%). The proportion of the tribal population to the total population of the remaining districts is found less than 1% (Table 14.1).

As per 2011 Census, Jaunsari (30.37%) and Tharu (31.29%) contribute 61.66% tribal population of the state. Bhotia and Buksa both the tribes have respectively about 13.4 and 18.51% and only 0.24% tribal population belongs to Raji (*Van Raut*) while 6.19% population of 4314 househols have not disclosed their sub-tribe and finally they are enumerated as unclassified (Anon 2011b). As per the 2001 Census, Tharu (33.9% and Jaunsari (32.9%) contribute more than two thirds of the tribal population of the state. On the other hand, in Bhotia and Buksa tribes constitute respectively about 14.4 and 18.6% of the population leaving behind only 0.2% of the tribal population of the Raji tribe while about 0.50% of the population is enumerated as unclassified. The population of these tribes was respectively 34.24, 33.29, 15.8, 15.52, and 0.6 and 0.55% population is registered as unclassified in 1991.

14.3 Tharu Habitat

14.3.1 Geo-Historical Background of the Tharu Habitat

The overwhelming proportion (>95%) of the total Tharu population of the state resides in the Sitarganj and Khatima tehsils of Udham Singh Nagar district. It stretches from the Sharda River on the east to the Kichha tehsils of the same district on the west and lies between 28°44′ to 29°0′30″ North latitude and 79°28′ to 80° East longitude. It is bounded by the Nainital district on the north, the districts of

Table 14.1 Distribution of scheduled tribe population in Uttarakhand, 2011

Sl. no.	District/ state	Scheduled tribe population			% of state's S.T. Population	% of total population	Sex ratio F/1000 M	Growth	Urban population
		Total	Male	Female				2001–2011	% of total S.T.
1	Uttarkashi	3512	1651	1861	1.2	1.1	1127	30.8	3.9
2	Chamoli	12,260	6021	6239	4.2	3.1	1036	16.9	26.2
3	Rudraprayag	386	217	169	0.1	0.2	779	107.5	19.9
4	Tehri	875	459	416	0.3	0.1	906	26.6	28.0
5	Dehradun	111,663	58,264	53,399	38.3	6.6	917	12.4	9.1
6	Pauri	2215	1174	1041	0.8	0.3	887	39.0	11.9
7	Pithoragarh	19,535	9558	9977	6.7	4.0	1044	1.3	18.5
8	Bageshwar	1982	971	1011	0.7	0.8	1041	2.0	5.4
9	Almora	1281	633	648	0.4	0.2	1024	45.9	41.5
10	Champawat	1339	777	562	0.5	0.5	723	80.9	19.0
11	Nainital	7495	3801	3694	2.6	0.8	972	51.1	22.9
12	U. S. Nagar	123,037	61,758	61,279	42.1	7.5	992	11.6	4.6
13	Hardwar	6323	3385	2938	2.1	0.3	868	101.4	17.0
	Uttarakhand	291,903	148,669	143,234	0.2*	2.9**	963	14.0	9.3
	India	104,281,034	52,409,823	51,871,211	100	8.6	990	23.7	10.0

Source Census of India, Uttarakhand, 2011. *–% of India's population
**–% of total population of Uttarakhand

Bareilly and Pilibhit to the south, Kichha Tehsil on the west and Nepal on the east. It comprises a narrow strip of the country running for 40 km long and 25 km wide along the foothills of the Himalaya. The reporting area of the Tharu habitat is 68,206 km^2 with a total of 358,498 persons of which 82,385 populations belong to the Tharu tribe as per the census of 2001. The Tarai region is a vast alluvial land lying at the southernmost part of the Outer Himalaya. It is separated from the Bhabar by the spring line in the north and densely populated Gangetic plain in the south. It is characterized by a deposit of fine sediments carried by the rivers flowing down from the Bhabar region. Tarai is a zone of super abundance of water because the underground water of the Bhabar zone reappears on the surface at the junction of the Bhabar and Tarai known as the spring line belt. The Tarai belt is the lowest part of Uttarakhand with an altitude of 200 m or less. There is a general decrease of elevation towards the south. The surface of the area is formed of a fine alluvial material of sand and clay. Presently, due to the construction of artificial water reservoirs and other land reclamation plans, once swampy Tarai well known for its malarial climate and luxuriant growth of broad leaved vegetation cover has now turned into a zone of fertile agriculture.

The drainage system of the study region consists of the streams and rivers which flow directly from the Himalaya. The Sharda River is the main river flowing in the extreme east of the study region. From east to west there are the Saniha, Chauka, Khakra, Lohia, Jagbura, Kamin, Deoha, Kailash, Bahgul and Dhaura streams flowing in the study region. The drainage of the region as a whole ultimately leads to water logging resulting in the breeding of mosquitoes and an unhealthy climate, as most of these streams are subject to heavy floods during the rains. The main reservoir is Nanaksagar which is used for irrigation and fishing. Khatima hydro-power project is constructed in Sharda River in the east of the region. From the earliest time, the Tharu inhabitants of the region were using the streams and rivers for irrigation and fishing purposes. The most common method employed by them was to construct earthen dams wherever required across the small streams and rivers. This had caused a ruinous effect both to the land and climate.

In general, the study region experiences extreme climate, summer being severely hot and winter very foggy cold. It receives most of its rain in June, July, August and September; winter has only a few showers. The average annual rainfall of this region is 150–200 cm. The Savanah type of forest occurs within an altitudinal range of 150–350 m. The vegetation mainly consists of tall grasses and mixed tree species such as Amla, Bahera, Bamboo, Dhank, Gutel, Haldu, Jamun, Jhingan, Kachnar, etc. together with newly introduced, eucalyptus, popular and teak which are either found along the road sides or as a plantations inside forest areas. The forest areas have been encroached by the in-migrating population for expansion of cultivation and habitation. The habitat of the Tharu is a land of numerous sluggish stream and water locked morasses, full of forests and swamp with scattered patches of cultivation. Owing to the fertility of the soil cultivation has made a rapid progress in the study area. It is known as a leading rice producing area in the country. The principal crops grown in this area, are rice, sugarcane, wheat, sunflower, gram, peas, mustered, lentil and red pepper.

The means of communication in the study area so poor that it is not at all economical to transport field produced to distant market. The only railway in this region is that of N.E. Railway which connects the main habitat of the Tharu. The nearest railway stations are Kichha (25 km) and Khatima (30 km). In this study region metalled roads are rare except between Haldwani to Khatima—Tanakpur via Kichha, and via Chorgalia, though unmetalled roads are numerous. The low lying parts of the region are not easily accessible during the rainy season. The Udham Singh Nagar district was carved out from Nainital district in 1994. It is generally spread over the Tarai region of the state. The U.S. Nagar district is administratively divided into seven tehsils some are development blocks, namely, Jaspur, Kashipur, Bajpur, Gadarpur Rudrapur (Kichha), Sitarganj and Khatima, of these seven tehsils, the last two, that is Sitarganj and Khatima are inhabited by the Tharu whose socio-economic and nutritional status is the main subject of the study. Tharu is concentrated in these two tehsils of U.S. Nagar in Uttarakhand, but they are also settled in the districts of Gonda, Gorakhpur, Bahraich, Lakhimpur, Kheri and Pilibhit of Uttar Pradesh and Nepal. According to various census figures, the population of Tharus have increased in recent years. U.S. Nagar (older Nainital Tarai) rose from 15,397 in 1891 to 82,385 in 2001 recording an increase of 435% in a steady manner except in the decade between 1911 and 1921 when influenza was responsible for considerable mortality and in 1941 when an epidemic of cholera was responsible for a decrease of Tharu population (Pant 2010). Thus the Tharus, survive in the unhealthy climate of the region.

The anthropologists, sociologists, anthropo-geographers and historians who have traced the origin of Tharu tribes in the Tarai region hold conflicting views on the matter. The word Tharu is derived from '*tahre*' which means 'they halted' after their alleged flight into the Tarai forests; or to '*tarhuwa*' which means 'became wet' or '*thatharana*', signifying 'trembling' or 'quaking' during their flight from Hastinapur to the Tarai after a fierce battle between the Rajputs and the Muslims. Knowles traces the origin of the Tharus to the word *tharua* (Knowles, quoted by Srivastava 1958) *of* hill dialect, meaning a 'paddler'. The *Tharu* name also denotes a wine bibber (Crooke 1896). Dr. Buchnan introduced another idea when he made a reference to the expulsion of the Gurkhas from Magadh by a people called the Tharu. These are said to have descended from the hills and spread over every part of the north of the Ganga River. But, this again is a controversial point, because the Tharus themselves claim Chittor as their original home and to have descended from Jaimal, Fateh Singh and Tarran Singh of Sisodiya clan of Rajputs (1303 AD.). They say that they were driven out, apparently in the third siege of Chittor by Akbar in 1567 AD and they were originally Rajputs who lost caste by using intoxicating liquor and rearing fowls. They never claim Gurkha and hill origin (Srivastava 1958).

According to H.R. Nevill, the Tharus are an aboriginal race who claims royal descent on the female side. H.R. Nevill gives a story that 'once upon a time in the remote past when the king of these parts was defeated by the forces of an invader, the women of the royal palace, rather than fall into the hands of the foe fled into the jungles with the Saise and Chamar (lower category people- servants/ workers of the royal family) belonging to the palace" (Srivastava 1958). From these sprang the two indigenous races of the Tharu and Buksa, the former is said to have descended from the Chamar and the later from the Saise. This tradition is further explained by the dominance of women among the Tharu. Mahalanobis, Majumdar and Roa have found the Tharu a Mongoloid tribe and they could not be placed in any other constellation of tribes and castes of the Province, Indo-Aryan or Australoid. In features they possess, more or less, oblique eyes, mostly brown or yellow-brown complexion, very scanty and straight hair on the body and the face, thin noses of medium size; while in other features they resemble the Nepalese more than any of the Australoid or pre-Dravidian castes and tribes. Furthermore, Majumdar (1942) has contested the supposed Rajput origin of the Tharu on the basis of the blood group tests carried by him in connection with the Census Operations of the United Provinces, 1941. On the basis of the studies done by the various people, we can conclude that the Tharu are a Mongoloid people, or predominantly so, who have successfully assimilated non-Mongoloid physical features as well.

14.3.2 Cultural Set-up of the Tharu Habitat

Tharus are very bold and strenuous. They have done a lot of struggle for their existence in the study region. As a result they are more or less successful in their efforts to exploit the natural resources to meet their primary requirements like food, shelter and clothing. But at the present time their mode of living is changed because of their increasing contact with the people of other castes and cultures. The Tharu villages are generally scattered and often located at a distance of three to four kilometers from one another. They are always in search of a good location for founding their villages. A good location means it must be on land at a high level with proximity to a river or some water supply yet it has to be safe from water logging and inundation during the rainy season. It must have a fertile soil and safety from the constant attacks of wild animals. The practice of deserting the villages and shifting the field cultivation off and on has brought to the Tharus additional responsibilities and labour demands and insecurity and instability of life. Now, they have permanent settlements which has materially benefited them and enabled them to lead a more settled and regulated life. The Tharu have a separate mandir (temple) for their village diety-Bhumsen.

In Tharu village agricultural fields separate one house to another. A Tharu house is generally surrounded by vegetable garden (*bari)* and small fields. The farmer grows seasonal vegetables in his garden and maize, mustard, tobacco, potatoes and onion in his fields for his personal consumption. On either side of the lane the houses are arranged in a lane on a higher level. It is noticed that there is a neat and clean arrangement of the houses in the Tharu hamlets (Villages). A Tharu village represents a closely knit society.

Tharu houses are surrounded by agricultural fields and vegetable gardens. They usually construct their houses in the months of April and May before the rainy season. The house must face to east to bring prosperity. Every house must consist of at least two partitions. The roof is thatched with *bharai* or *kans* grass which grows in abundance in the region. The plastering of the walls on both side and decorations on them are done by women. The home-made plaster for walls is made of clay, cow dung, water and chaff of wheat (*Tatar*). Economically well off Tharu families have made pucca houses (strong permanent houses) with iron, cement, khaprail (just like bricks which are used in the roof of the house and window) and bricks material. They have a separate cattle sheds.

14.3.3 Social Set-up of Tharu Habitat

The Tharu tribe can be divided into two groups, the higher and lower. The higher group comprises the Batha, Birlja, Badvait, Dahait, Mathum and Ranathakur, and the lower group, the Buksa, Khunka, Rajia, Sansa, Jugia and Dangra. The former was hypergamous (hypergamy is a term used in social science for the act or practice of marrying someone who is wealthier or of higher caste or social status than oneself.) to the latter but over the course of time the higher group stopped taking brides from the lower one (Turner 1931). Gotra (kuri) is generic name which is given after ancestors or preferably a sage or a holy person. One of the main functions of this name is that it serves as very important distinguishing category between the higher and lower castes. Thus, they prefer to marry the members of the same gotra. In the sample villages, Rana Thakur forms the bulk of Tharu population while other groups constitute a negligible proportion of the total population. It is also noticed that Tharu of lower castes do not want to be known as lower groups. All these sections restrict themselves in terms of inter marriage and inter dining. Tharu as a whole may be said to have a common culture. The nucleus of social organization in the Tharu society is the family. Both the joint and single family system is found among the Tharu. At present there is a tendency towards the breaking up of the joint family. The structure of joint family varies in form and size from village to village and from family to family.

The village community is a compact social unit which consists of a number of different families living in a common settlement and performing certain functions together. It is also responsible for maintaining low and order in Tharu society. The Panchayat is the only powerful tribunal for setting numerous disputes in their community. The Tribal Panchayat is recognized as the custodian of low and morals. Women of the Tharu tribe enjoy a very dominant position in their community. They have considerable freedom and wield great influence in the domestic sphere. The majority of the women do all the marketing while their counterparts only assists them in bringing the articles home purchased by their wives. Tharu women help men in manifold economic activities like, plastering the house, cleaning the cattle sheds, sowing and harvesting, winnowing, storing, pounding and grinding the grain, selling and purchasing the articles of their daily use. In addition to this, tribal customs and conventions also provide for their right of ownerships of property and possessions of the family. It is noticed that all women in the family do not have equal status. Their status is determined by the positions they hold in their own family.

The religion of the Tharu is based on a belief in ghosts, demons and the spirits of the dead whom they fear. They are worshipping Mahadeva known as Narayana who gives them sunshine, rain and harvest. The contact with the Hindus has enabled the Tharus to include the worship of Satyanarayana and a few other Hindu god and goddesses in their religion.

The Tharu tribe celebrates a number of religious festivals, such as holi, tij, janmasthami, anthai, diwali etc. The fundamental idea underlying all the religion rites, rituals, and ceremonies is to ensure the well being of an individual family, its cattle and the crop, the village and finally of the whole community and to protect them from unforseen dangers and hardship. There is a tendency growing among them to absorb the deities of Hindu religion on large scale and to respect the Hindu saints and pirs (Pir is a term used for the muslim saints) of other religions. It is observed that the educated and progressive Tharus have given up their faith in these gods due to the influence of other religions, particularly Arya Samaji. As a result they are weakening their faith in their old religions and spirits to some extent.

14.3.4 Demographic and Socio-economic Environment of the Sample Villages

This study takes into account a total of 907 households with a total population of 4992 persons (2593 males and 2399 females) living in of 22 sample villages (Table 14.2). On an average the sex ratio in the sample villages is 925 female per 1000 male which is lower than that of the study area—the district of U.S. Nagar, state and country averages (see Fig. 14.1). The sex ratio varies from a minimum of 808 in the village of Navinagar to the maximum of 1145 females per 1000 males in Baruwabag. Only 18% of villages have sex ratios more than 1000. The average sex ratio of 0–6 year age group is very alarming i.e. 853 females per 1000 males. It reflects the same attitude of Tharu society towards the female child, similar to that in the whole country. The sex ratio varies from a minimum of 333 females per 1000 males in Bharabhuria village to the maximum of 1357 in village Barkidandi. Due to the influence of Hindu society, the prominent factors for low sex ratio in the Tharu society are—son preference, unequal treatment given to boys and girls, female infanticide, neglect of female infants, early marriage; death during pregnancy and hard work. On the contrary, the average sex ratio in the hill districts of the state is high because of outmigration of the male population from the villages for employment and better education and is also due to the high male mortality. The average Tharu household size is 6 people per households.

Education is an important indicator used to measure the level of social progress in a society. On an average 69% of the Tharu population is registered as literate in

Table 14.2 Household size and sex ratio of tharu tribe in sample villages, 2006–2007

S. no.	Sample villages	Total tharu H.H. (no.)	Selected household		Total population (no.)			H. H. size (p/HH)	Sex ratio F/1000 M	
			(No.)	(%)	Persons	Male	Female		(Total)	(0–6 age)
1	Barkidandi	60	30	50	174	84	90	6	1071	1357
2	Baruwabag	46	23	50	133	62	71	6	1145	1222
3	Biriya	88	42	48	232	128	104	6	813	923
4	Bidaura	101	50	50	291	149	142	6	953	579
5	Bihi	43	21	49	96	53	43	5	811	500
6	Dhusari	66	31	47	173	90	83	6	922	875
7	Harraiya	64	33	52	187	95	92	6	968	1000
8	Navinagar	46	24	52	132	73	59	6	808	630
9	Pindari	71	35	49	181	90	91	5	1011	933
10	Sadhunagar	140	70	50	377	198	179	5	904	1222
11	Saronja	100	48	48	236	130	106	5	815	1176
12	Bharabhuriya	73	36	49	182	99	83	5	838	333
13	Charubeta	79	40	51	192	100	92	5	920	1222
14	Chhinki	63	32	51	170	92	78	5	848	500
15	Diya	87	43	49	251	130	121	6	931	636
16	Gauharpatia	121	60	50	356	183	173	6	945	957
17	Jogitharnagla	59	28	47	155	83	72	6	867	647
18	Khetalsanda	73	38	52	238	122	116	6	951	650
19	Kumraha	106	53	50	286	151	135	5	894	611
20	Mohd. Bhuriya	154	76	49	462	230	232	6	1009	1333
21	Uldhan	95	46	48	203	108	95	4	880	583
22	Umrukala	103	48	47	285	143	142	6	993	818
	Total/average	1838	907	49	4992	2593	2399	6	925	853

Source Field survey, 2006–07

the study area which is higher than the district but lower than the state average. Literacy varies from the lowest of 19.1% in the village Pindari to the highest of 79.5% in the village Khetalsanda. Among the total literates, 84.3% were males and 56.9% were females (see Table 14.3). The gender gap in literacy level was 27.4%. Out of total literates, a maximum 35.59% have an education level up to primary standard followed by 30.33% up to junior school, 24.36% up to high school, 9.72% intermediate and higher education. About 31% of the Tharus are still illiterate. Among them, illiterate females are three times more than males. About 49.95% of the Tharu are illiterate due to lack of access to schools followed by 23.34%% who are looking after the younger members in the family and 17.54% due to poverty. There are 9.17% illiterates who look after grazing animals.

The dropout rate of tribal girls and boys is more than the non-tribal children. Out of the total surveyed Tharu population, only 6.2% male Tharus are registered as

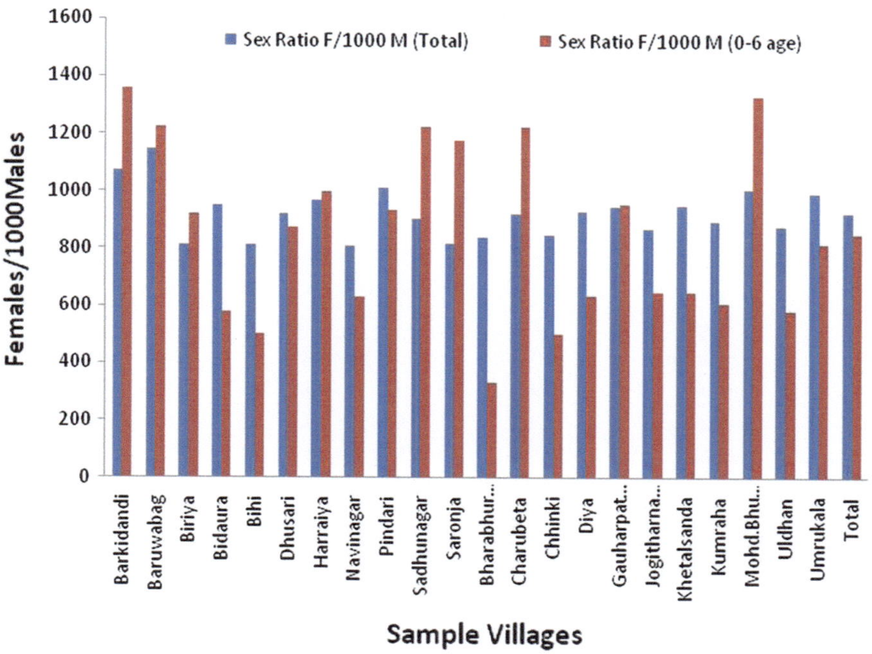

Fig. 14.1 Sex ratio of tharu tribe in the sample villages

out-migrants working as labourers outside the study region. Agricultural is the main occupation of the Tharu. But they were engaged in hunting, fishing, collecting forests fruits and roots in the past. Along with agriculture they have allotted sufficient time in their calendar to the supplementary pursuits of fishing, hunting, gathering of fuel and fruits from the forests and making nets, mats and ropes even today. The cycle of economic activities of Tharus regulates their daily routine work. During their sowing and harvesting seasons, they are busy in fields all the day. During other seasons, they employ their time in repairing and constructing house or engage themselves in the supplementary occupations. The Tharus have a division of labour according to sex, age and class. Children and aged people are mostly employed in various vocations requiring less physical strain. Certain occupations are exclusively assigned to a particular sex, for example, ploughing, leveling of the agricultural fields, hunting, building and repairing the houses etc. are done by men. Cooking meals, cleaning utensils, fetching of water, plastering of walls, collection of fuels and fodder, making mats, rearing livestock etc. are some of the duties allotted to women. There are number of tasks such as sowing of seeds, harvesting, gardening, fishing and poultry farming which are jointly done by them to help each other. It is observed that the number of old occupations (inherited profession/work) are no longer pursued by the Tharu and presently they are mainly engaged themselves in agriculture, government and semi-government services, business, animal husbandry, contract work etc. The awareness and education level of the Tharus has

Table 14.3 Literacy rates and gender gap in literacy of tharu tribe in sample villages, 2006–07

S. no.	Sample villages	Literacy rate (%)			Gender gap in literacy rate (%)
		Person	Male	Female	
1	Barkidandi	73.8	84.3	63.4	20.9
2	Baruwabag	69.9	83.0	58.3	24.7
3	Biriya	65.7	74.8	54.3	20.4
4	Bidaura	77.4	94.6	60.3	34.3
5	Bihi	74.1	93.0	52.6	40.4
6	Dhusari	73.4	86.6	59.2	27.4
7	Harraiya	73.2	84.2	61.6	22.6
8	Navinagar	50.0	91.3	59.5	31.8
9	Pindari	19.1	78.7	46.8	31.9
10	Sadhunagar	65.6	76.0	53.4	22.6
11	Saronja	63.8	77.9	45.3	32.5
12	Bharabhuriya	79.1	91.4	66.2	25.1
13	Charubeta	71.5	89.0	51.9	37.2
14	Chhinki	69.1	84.6	52.1	32.5
15	Diya	67.0	82.4	51.4	31.0
16	Gauharpatia	66.9	75.6	57.6	18.0
17	Jogitharnagla	67.7	80.3	54.1	26.2
18	Khetalsanda	79.5	92.2	67.0	25.2
19	Kumraha	70.4	82.0	58.1	23.9
20	Mohd. Bhuriya	72.7	90.6	54.1	36.6
21	Uldhan	72.8	86.5	58.0	28.5
22	Umrukala	75.9	86.8	65.3	21.5
	Total/average	69.0	84.3	56.9	27.4

Source Field survey, 2006–07

increased tremendously. It was also noticed that tribal people who have sold or rented their agricultural land (*rehan*) have become landless.

14.4 Dietary Habits

Food is the main source of nourishment for the human body. Health, nutritious food and a balance diet are interdependent. The diets of the Tharu people are mainly limited to cereals with little vegetables. Diets are monotonous and restricted in nature and variety. Generally the Tharu people of the sample villages take their food two/three times in a day. The main items of the food in any time are wheat, rice, pulses and vegetables. Taking together all the sample villages, the per capita per day cereals intake of a Tharu person is 404.7 g which is 8% less than the standard requirement of 440 g. The cereals intake of 80% of the sample villages was found to be less than the standard requirement. The per capita daily intake of pulse is only 27.4 g which is 39.1% less than the standard requirement of 45 g. The per capita daily consumption of all vegetable in Tharu villages was 60.4 g which is 68.2% lower than the standard requirement of 190 g (including leafy root and other vegetables). In the study region, the average consumption of ghee (homemade butter) and cooking oil per capita daily was found 9.2 g which is 77% less than the standard requirement of 40 g. Tharus are poor and have low purchasing capacity. The average intake of milk and milk products per capita was found to be 47.6 g which is 68.27% lower than the standard requirement of 150 g. The average fruits consumption per capita daily is 11.8 g. In the sample villages, average intake of the sweet (sugar and jagery) items per capita per day was 10.2 g which is 66% less than the standard requirement of 30 g. The sweet is generally consumed in tea and porridges. Fish is an important item in the traditional diet of Tharus. There are no permanent ponds for the fishing now due to agricultural encroachment by the immigrant population. Therefore Tharus buy eggs, fish and meat from nearby markets. Due to their poor purchasing capacity, the per capita daily intake of these items was very nominal being 11.12 g including flesh, fish and egg.

The average consumption of condiments and spices per capita per day is less than 10 g. It is however the same in all sample villages. It is clear from the above study that the per head per day intake of pulses, flesh, fish, milk, vegetables, sweet, ghee (homemade butter), cooking oil is inadequate when compared to the standard or minimum requirement as recommended by the National Institute of Nutrition, Indian Council of Medical Research. Tharus are not aware about the necessary essential intake of other food items such as pulses, vegetables, fruits etc. and consume more cereals.

14.4.1 Health and Nutrition Status

Health is a crucial attribute of human resources and a healthy human population is the most desired national asset. Adequate sanitation and safe potable water are necessary for the prevention of infectious and water borne diseases. There is no detailed study of the anthropometric measurements of tribal people. It is only during the last fifteen years that some information has become available through studies done by the National Nutrition Monitoring Bureau and National Family Health Survey of India. The mean height and weight of the sample people have been worked out in the present study. The mean height of the total surveyed people is 156.2 cm which is 160.0 cm for males and 152.5 cm for females. Similarly the mean weight of the surveyed tribal people is 42.3 kg while mean weight for male and female is respectively 45.1 and 39.4 kg. The average height of the Tharu tribe can be termed as short in stature in general and tribal women can be termed as low weight from a pregnancy point of view in particular. The Tharus are not very healthy due to their high work load on the one hand and low nourishment, including

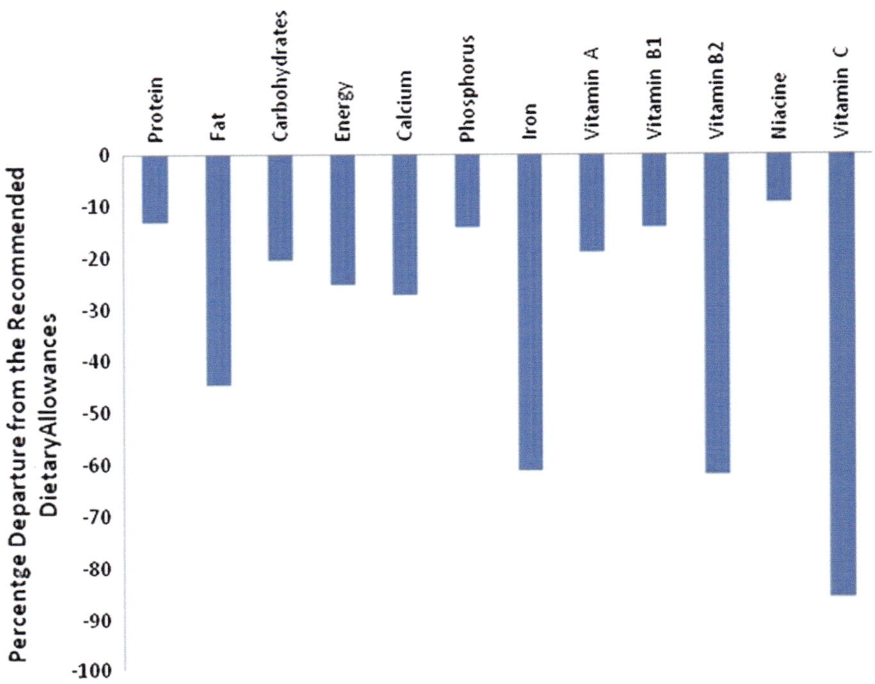

Fig. 14.2 Average nutrition availability pattern of tharu tribe in the sample villages, 2006–07

heavy drinking (liquor) on the other. Key statistics are shown in Fig. 14.2 and Table 14.4.

Body Mass Index (BMI) was calculated using the formula weight (kg)/ height^2(m) (James et al. 1988). Average BMI varied from 15.2 in the village of Bihi to 19.9 in village Charubeta. Similar trends were seen in the male and female Body Mass Index while the Body Mass Index is very low in the group of very low income and it shows an increasing trend as per increasing income levels. About one third of the villages of the Tharu belong to very low income groups with BMI below 18.5 which is the optimum criterion for Chronic Energy Deficiency (CED) grade I i.e. moderate type of malnutrition.

The per capita per day energy requirement of 2250 kcal has been taken as a standard requirement. In all sample villages, the average energy intake per capita per day was found to be only 1679.99 kcal which is 25.33% below the standard/ minimum requirement of 2250 kcal (Gopalan et.al. 2007). In about 72.72% of the total sample villages, the consumption of energy intake was found to be less than 1700 kcal. The average per capita per day protein intake was found 47.63 g which is 13.4% less than the standard requirement of 55 g. The protein intake in all sample villages was found lower than the standard. The average fat intake per capita per day was found to be 11.8 g which is 44.6 lower than the standard requirement of 20 g. In the study region, average carbohydrate intake was found 341.33 g per

Table 14.4 Nutrition availability pattern of tharu tribe in the sample villages (percentage departure from the recommended dietary allowances (R.D.A.))

S.no.	Sample villages	Protein	Fat	Carbohydrates	Energy	Calcium	Phosphorus	Iron	Vitamin A	Vitamin B1	Vitamin B2	Niacine	Vitamin C
	R.D.A.	55 gm	20 gm	430 gm	2250 kcal	400 mg	1400 mg	29 mg	600 µg	1025 mg	1.4 mg	14 mg	40 mg
1	Barkidandi	16.2	46.2	22.42	26.87	26.55	17.71	63.27	22.69	20.8	64.28	13.5	86.22
2	Baruwabag	15.18	45.85	15.77	21.58	27.46	10.48	63.72	20.61	11.2	60.71	5.57	85.4
3	Biriya	10.14	38.05	20.16	24.43	21.68	6.6	58.24	9.51	7.2	60.71	5.64	85.2
4	Bidaura	5.74	37.15	10.1	16.06	20.5	7.54	61.03	8.49	11.2	59.28	3.07	85.5
5	Bihi	18.05	44.55	3.01	33.39	29.44	16.74	57.86	27.66	6.4	71.42	11.14	86.4
6	Dhusari	11.7	40.09	19.39	24.13	24.39	12.63	61.24	12.49	13.6	61.42	8.21	85.2
7	Harraiya	12.78	46.05	20.44	25.48	28.07	13.61	60.62	18.4	11.2	62.85	7.78	85.12
8	Navinagar	3.36	41.15	9.71	15.87	20.44	4.45	57.37	21.2	4	58.57	2.21	86.4
9	Pindari	12.74	44.4	22.9	25.56	28.59	16.48	61.82	20.03	19.2	63.57	11.78	87.75
10	Sadhunagar	11.4	47.45	17.91	23.64	31.54	12.93	60.79	12.73	12	60.71	6.57	86.32
11	Saronja	16.52	46.85	2.2	27.18	31.21	19.16	63.06	14.39	20.8	30	14.92	84.75
12	Bharabhuriya	15.1	46.3	23	27.69	29.33	16.74	63.17	14.36	18.4	62.85	12.57	84.97
13	Charubeta	9.67	34.9	1.85	22.61	14.72	12.31	62.17	4.66	14.4	58.57	10.07	84.62
14	Chhinki	15.56	47.75	22.66	27.61	31.37	17.7	61.68	15.46	19.2	64.28	12.71	85.42
15	Diya	17.38	46.46	21.57	26.81	32.64	17.27	64.03	34.63	19.2	71.42	11.07	86.8
16	Gauharpatia	1.58	46.75	23.69	28.27	28.31	18.33	60.86	25.49	17.6	63.57	14.07	86.27
17	Jogitharnagla	15.41	46.25	22.23	25.74	30.7	17.5	62.68	23.4	17.6	66.42	12.42	86.4
18	Khetalsanda	12.67	46.9	19.3	24.9	29.27	13.31	60.41	37.61	11.2	61.42	8	87.47
19	Kumraha	13.65	44.3	2.11	25.95	28.01	16.03	57.51	8.95	12	63.57	12.14	84.75
20	Mohd.Bhuriya	16.67	45.55	23.59	28.29	30.65	17.96	63.65	31.48	19.2	65	8.5	86.27
21	Uldhan	15.61	48.1	2.53	29.74	29.59	16.72	61.86	16.42	15.2	64.28	12.57	85.5
22	Umrukala	13.32	48.3	21.49	26.53	30.02	12.05	61.03	16.92	14.4	64.28	11.71	89.72
	Total/average	13.4	44.6	20.62	25.33	27.48	14.28	61.27	19.15	14.4	62.14	9.64	86.02

Source Gopalan et al. (2007) and Field survey, 2006–07

head per day which is 20.62% below the standard requirement of 430 g. On an average all the villages were found to be below the standard requirement. Out of a total of 22 villages, the carbohydrates intake of 40.9 villages was observed to be below the regional average.

The average per capita daily calcium intake was found 290.08 mg in all Tharu sample villages which is 27.48% below the standard requirement of 400 mg. The calcium intake of 68.17% was found to be below the regional average. However all sample villages fall short varying from 14.72% in village Charubeta to 32.64% in Diya from the normal requirements. The average phosphorous intake per head/day was found to be 1199.95 mg which is 14.28% below the standard requirement of 1400 mg. In 54.54% of the total sample villages, phosphorous intake was below the regional average. The deficiency in the phosphorous intake is not as alarming as the main source of phosphorous is cereals which are the main food of the Tharu. The per head/day iron intake of Tharu was found to be 11.23 mg which is 61.27% below the standard requirement of 29 mg. The carotene (vitamin A) intake among the Tharus is very unsatisfactory being 485.07 μg per head per day which is 19.5% below the recommended dietary allowances (RDA) of 600 μg. Daily intake of carotene was found to be short by 8.49% in the village of Bidaura and 34.63% in village Diya. Thiamine (vitamin B1) intake in the sample villages was 14.4% above the normal requirement of 1.25 mg per capita daily. The intake of the riboflavin (vitamin B2) in the sample villages was 0.53 mg which was 62.14% below the standard requirement of 1.4 mg. The intakes of riboflavin in 95.45% of sample villages fell short by more than 58%. In the sample villages, the average niacin (nicotinic acid) intake was found 12.65 mg which is 9.64% below the standard requirement of 14 mg. On an average, the consumption of niacin was found below the standard requirement in 95.45% of the total sample villages.

Green vegetables and fruits are the main source of vitamin C or ascorbic acid. In the study region, its average intake was found to be 5.59 mg which is 86.02% below from the standard requirement of 40 mg. The per head/day intake of vitamin C in all the samples village fell short by more than 84% due to exclusion of fruits and fresh vegetables from the meal. There are many other factors that also influence the diet requirement. Much disparity was therefore seen in the nutritional category levels.

Out of the total sample population, 512 persons (10.25%) were found to be unhealthy in one way or other. Out of the total 512 ill people, 50% were females and 50% were males. Out of total indisposed persons, 43.36% were suffering from fever, 34.57% from stomach related problem, 19.92% eye related and 2.15% from water borne and other diseases. Out of total indisposed males, 50.0 and 38.67% were respectively suffering from fever and stomach while 36.72 and 30.47% females were also suffering from these two diseases. About 30.47 and 2.34% women were respectively suffering from eye and other water borne related diseases. Mortality in 1989–90 to 2003–04 periods was also recorded during the field study. A total 502 deaths occurred during this specific period. The annual number of deaths was decreasing due to increasing medical facilities and awareness among the Tharu. The study shows that a maximum of 81% of deaths occurred before the age of five. Infant mortality among the Tharu is still more due to early marriage of girls in one hand and poor nutrition intake on the other. Of the balance 4.2% deaths occurred above the age of 71 years. The longevity among Tharu is much higher than the average of India. Only 3% of deaths were recorded between the ages of 61–65 years (Pant and Chand 2009).

The prevalence of deficiency diseases among the sample of the Tharu population due to a particular deficiency of food nutrients has not been worked out in the present study. However a number of studies done by several authors Shafi (1967), Mohammad (1977), Mohammad (1978), Tiwari (1988, 1991), Pant (1988, 1994), Pant and Jalal (1991), Pant et al. (1991, 1992) in various parts of the country indicate the clear relationship between the nutrients intake and deficiency diseases.

14.5 Conclusion

Tharus (33.9%) and Jaunsaries (32.9%) are the dominant tribes in the study area and together they constitute about two thirds of the tribal population of the state. The Tharus and their habitat constitute very significant parts of the Tarai region of Uttarakhand. The Tharu women enjoy a very dominant position in their community. They have considerable liberty and wield great influence in the domestic sphere. The women are the sole owners of the domestic animals, and the producers of the fields and the gardens, while the men have their fishing and hunting nets and traps. The Tharus are passing through the stages of demographic transitions. The sex ratio is very low (853) in juvenile age groups due to the influence of Hindu societal values such as the son preference, unequal treatment given to boys and girls, female infanticide, neglect of female infants and early marriage etc. The average household size was 6 persons per household in the sample of Tharu villages. The Tharus are short in stature in general and women can be termed as low weight from a pregnancy point of view in particular. Tribal people are not very healthy due to over-work on the one hand and low nourishment with heavy drinking (liquor) habits on the other. Average BMI varies from 15.2 in the village Bihi to 19.9 in the village of Charubeta. A similar trend was seen in the male and female Body Mass Index. The Tharu belong to a very low income group with a BMI below 18.5 which is the optimum criterion for Chronic Energy Deficiency (CED) grade I i.e. a moderate type of malnutrition. On an average, 69% of the Tharu population is registered as literate this higher than the district (63.9%), state (63.2%) and national tribal population (47.1.8%) averages. About 56.9% of females and 84.3% of males are registered as literate.

The average energy intake per capita per day was found to be 1679.99 kcal in all the sample villages which is 25.33% below the standard requirement (2250 kcal). In in all of the sample villages, the consumption of energy intake was found to be less than 1900 kcal. The average per capita per day intake of important nutrients such as protein, fat, carbohydrate, calcium, phosphorous, iron, carotene (vitamin A), riboflavin (vitamin B2), was found less than the standard requirement.

The socio-economic and nutrition status of the Tharu population of Uttarakhand can be improved by adopting the following measures:

1. Implementation of Marriage Act (minimum age of girls and boys is 18 and 21 years respectively) strictly in Tharu society.
2. Special infrastructural facilities in the form of schools for tribal girls and health care centre for mother and infant child and priding career opportunities for tribal women nearer home should be established.
3. Eliminating all practices that discriminate against tribal people.
4. Eliminating all forms of exploitation abuse, harassment and violence against women and girls.
5. Promoting gender sensitization at all levels to bring equality in wages, opportunity in jobs, school admission, social security, health insurance and other such benefits.
6. Awareness programmes about nutritious food, health, hygiene, environmental sanitation, education etc. should be launched in the Tharu tribal area.
7. Compulsory education of Tharu children up to the age of fourteen for bridging the gender gap in literacy rates.
8. Establishing play houses for infants and stores for fuel and fodder at the village level by which children will not be unnecessarily engaged in looking infants and animals.

References

Anonymous. (2001a). *Census of India PCA totals, for the State of Uttarakhand, Jharkhand and Bihar in CD*. New Delhi: RGI of India.

Anonymous. (2011). Census of India Chapter-2 (*Scheduled Castes and Scheduled Tribes Population*). p. 36 and Website pca_state_dist_scheduled tribe 2011.

Anonymous. (1991). *Census of India, SC/ST population in CD*. New Delhi: RGI of India.

Anonymous. (2001). *A-11 State primary census abstract for individual scheduled tribe*. pp. 47–49.

Crook, W. (1896). *The tribes and castes of North Western Provinces and Oudh*. (in four volumes).

Gopalan, C., Ramashastry, B. V., & Subramanian, S. C. (2007). Revised and updated by Narsingha Rao, Y.G. Deosthale & K.C. Pant (2007) *Nutritive value of Indian foods* (p. 156). Hyderabad: National Institute of Nutrition, ICMR.

James, W. P. T., Anna Ferro, L., & Waterloo, J. C. (1988). Definition chronic energy deficiency in adults. *European Journal of Clinical Nutrition, 42*, 969.

Majumdar, D. N. (1942). Tharus and their blood groups. *Journal of Royal Asiatic Society of Bengal, 3*(1).

Majumdar, D. N., & Madan, T. N. (1970). *An introduction to social anthropology*. Bombay: Asia Publishing House.

Mohammad, A. (1977). *Food and nutrition in India* (p. 163). New Delhi: KB Publications.

Mohammad, N. (1978). *Agricultural landuse in India* (p. 231) Delhi: Inter-India Publications.

Surendra, P. (2010). Udham Singh nagar mai tharu janjati ke posan evam swasthya star ka bhaugolik adhayyan. Ph. D. Thesis, Nainital, Kumaun University, p 279 (in Hindi).

Pant, B R. (1988). Land utilization in Kotadun, Kumaun Himalaya, Ph. D. Thesis submitted in the Nainital, Kumaun University, p. 270.
Pant, B. R. (1994). *Food and nutrition: A study in Himalayan region* (p. 195). New Delhi: Anmol Publications Pvt.
Pant, B. R. (2004). *Women and nutrition in Himalaya* (p. 227). New Delhi: Anmol Publication Pvt Ltd.
Pant, B. R. (2006). Level. *Growth and gap in the literacy by sex in the Himalaya, India, man in India, 86*(1&2), 65–96.
Pant, B. R. (2010). *Tribal demography of India* (p. 288). New Delhi: Anamika Publication Pvt Ltd.
Pant, B. R., & Jalal, D. S. (1991). Nutrition deficiency diseases in Indian Central Himalaya. *The Geographer, 38*(1), 30–41.
Pant, B. R., Joshi, R. C., & Jalal, D. S. (1991). Agricultural landuse and nutrition in Kotadun Kumaun Himalaya. *The Geographical Review of India, 53*(4), 8–18.
Pant, B. R., Joshi, R. C., & Jalal, D. S. (1992). Agricultural panorama of Uttarakhand. In: Mohammad N (Ed.), *New dimensions in agricultural geography* (vol. 5, pp. 81–94).(Spatial Dimensions of Agriculture) New Delhi: Concept Publishing Company.
Pant, B. R., & Chand, R. (2009). Demographic behaviour and health status of Tharu Tribe in Uttarakhand. In MSS Rawat, & P. Dinesh (Eds.), *Management strategies for the Indian Himalaya: Development and conservation* (vol. II, pp. 104–134). Srinagar: Transmedia Publication Srinagar (Garhwal).
Srivastava, S. K. (1958). *The Tharus: A Study of Cultural Dynamics*. Agra: Agra University Press.
Shafi, M. (1967). Food production efficiency and nutrition in India. *The Geographer, 14,* 23–27.
Tiwari, P. D. (1988). *Agricultural development and nutrition: A case study of Rewa Plateau* (p. 110). New Delhi: Northern Book Center
Tiwari, P. D. (1991). Landuse, nutrition and deficiency diseases in Rajaura Village, District Sagar (M.P.). Geographical Review of India, *53*(1), 56–59.
Turner, A. C. (1931). Census report of the United Provinces. *18*, 599.

Part III
Migration and Depopulation

Chapter 15
Similarity of the Spatial Patterns of Net Migration and Unemployment in Finland

Olli Lehtonen and Markku Tykkyläinen

15.1 Introduction

Finland has faced the continuous migration of population and the associated concentration of jobs in the urban areas and the south of the country for decades. During the economic growth phase since 1993 and since Finland joined the EU in 1995, spatial restructuring has continued and indications show that economic growth and institutional changes has not changed these trends. High unemployment and out-migration concentrate mostly in the same peripheral areas. The purpose of this paper is to uncover this dependence between unemployment and migration. We investigate how systematically low unemployment increases in-migration and how high unemployment pushes people from the peripheries. A key question is whether this relationship is symmetrical in a core-periphery scenario. In this paper, the concept of spatial formation (cf. Boschma and Lambooy 1999; Fujita et al. 1999), defined empirically as the clustering of areas of similar development, is apposite as we seek to investigate uniform spatial structures brought about by economic development and its economic externalities. The paper reveals the patterns of winning and losing areas in socioeconomic development in Finland and their relations to migration and unemployment.

We focus in this paper on the years of 2004, 2006 and 2008 after the burst of the IT bubble revealing the regularity of the spatial coexistence of unemployment and net migration at a municipal level by means of spatial analysis. As the unbalance in the labour market and migration seems to concentrate geographically, we investigate the

O. Lehtonen (✉)
Natural Resources Institute Finland (Luke), Viikinkaari 4, FI-00790 Helsinki, Finland
e-mail: olli.lehtonen@luke.fi

M. Tykkyläinen
Department of Historical and Geographical Studies, University of Eastern Finland, PO Box 111, 80100 Joensuu, Finland
e-mail: markku.tykkylainen@uef.fi

R. Chand et al. (eds.), *Societies, Social Inequalities and Marginalization*, Perspectives on Geographical Marginality,
DOI 10.1007/978-3-319-50998-3_15

spatial formations of the different combinations of high and low unemployment and in-migrations and out-migrations occur and how they are linked to local socioeconomic conditions. Our first aim is to answer the question of how the spatial pattern of net migration are covariate with the spatial pattern of unemployment in Finland, and how regular the spatial coexistence between migration and unemployment was in the growth years of 2004 and 2006 and on the verge of the financial crisis of the late 2000s. Our second aim is to reveal the connections of these spatial formations of growth and decline to the socioeconomic conditions of particular areas. Thirdly, we measure similarity by presenting an index which is used in research.

15.2 From Migration and Unemployment to Spatial Formations

15.2.1 Emergence of the Spatial Patterns of Migration and Unemployment

The long-standing spatial formation of unemployment and migration reflects the imbalance in the labour market as a result of structural changes in the economy. Formerly having been dominated by the utilization of geographically scattered natural resources, the economy of Finland has grown in recent times towards a high technology-driven, urbanized economy with considerable research and experimental development (R&D) inputs (Leinbach and Brunn 2002). The growing demand for labour was met by means of migration flows that brought young adults to the metropolitan areas, while the middle-aged and elderly people mostly remained in the peripheries (Aro 2007a, p. 376–377; OECD 2008, p. 41; Heikkilä and Pikkarainen 2010). Selective migration of this kind is known to manifest itself in differences in well-being and human capital between localities.

Spatial equilibrium models interpret migration as the response of individuals to geographical differences in labour demand subsequent to their rational calculations of the costs and returns (Sjaastad 1962; Irvine et al. 2010). This implies that labour will move to where the wages are highest, and the theory postulates that the difference in income between the sending and receiving regions is the key attractive factor. It also argues that economic processes tend to even out disparities in unemployment between core areas and peripheries. Krugman's theory (1991) explains how a labour force will rationally move to areas of higher utility, which will lead to the concentrations of manufacturing due to the benefits brought about by migration-induced external economies. Nevertheless, this spatial process is sensitive to various conditions, such as transport costs (Krugman and Venables 1995), production costs and the dynamics in individual industries. The decision of an individual to move is dependent on several factors, which are increasingly coming to reflect social and environmental motives and the attraction of amenities (Lundholm 2007; Olfert and Partridge 2010), so that non-monetary factors are also part of the individual's search for higher utility. For settled adults, social insurance benefits hold back any immediate search for a job elsewhere, as migration is in essence a trade-off decision between expected costs and benefits. Education and career opportunities attract young adults from rural areas to the cities (Tuhkunen 2007, p. 98–104), while McCormick and Sheppard (1992) build their model of persistent inter-regional differences in unemployment rates on the premise 'that the willingness to migrate is an increasing function of the productivity of a worker'; regions with a large proportion of less-skilled workers may then have consistently the high rates of unemployment. Workers with higher education levels may also be better informed about job possibilities elsewhere.

Migration and unemployment are distinctive indicators used to show the degree of spatial economic equilibrium and regional socio-economic differences in the capacity to nurture industrial activities and to reflect regional competitiveness. The last-mentioned indicates the ability of the regional economy to attract firms and economic activity and to improve the standards of living (Storper 1997, p. 20). The spatial formations of migration are autocorrelative spatial structures of municipalities with a net migration surplus or deficit. The process of out-migration emerges from poor rural areas, mostly in the north-east of Finland, and to a few large centres and their surrounding areas (Lehtonen and Tykkyläinen 2010). Aro (2007b) and Gløersen et al. (2005) have shown that the main migration flows are from poorer regions to more prosperous regions, which has meant, since the early 1990s that they have focused mostly on larger urban areas and their surroundings. In this way, people are able to cope with unemployment, improve their wealth, and hence migration may be regarded as the response of individuals to unemployment and regional differences in economic development (Pekkala 2003, pp. 480). Peculiar to Finland is that these formations are relatively persistent as the regional differences in unemployment and migration has remained high (Lehtonen and Tykkyläinen 2010).

This paper investigates two kinds of spatial formation. The 'pull-win' formations are such where unemployment is low and in-migration high and the 'push-lose' formations consist of municipalities where both unemployment and out-migration are high. If these combinations were to lead to evening-out in the average values of unemployment and migration, the balancing mechanisms of the market would in theory lead towards a spatial equilibrium. This tendency can be revealed by the measuring of the similarity of the values of migration and unemployment. The areas where the values of similarity are low, indicate that spatial restructuring is at a lower level, or if either unemployment or migration only gets extreme values, spatial economic development is not on the way towards a spatial balance as the equilibrium theories assume (Irvin et al. 2010). The combination of high unemployment and high in-migration would be extremely exceptional, but relatively low values of out-migration combined with high unemployment would indicate that there are obstacles to move. Similarly, low unemployment combined with low in-migration would indicate obstacles to move in.

15.2.2 Why Spatial Patterns Should Be Studied?

The long-standing spatial formations of unemployment are symptoms of structural unemployment and hence the unbalance of the local labour market, which has a negative effect on the economic performance of an area as less value is produced (Irvin et al. 2010). In order to find solutions for a more efficient spatial use of human resources it is thus important to reveal what the reasons and conditions are for these vicious circles. As spatial formations exist, this leads to the question of what kinds of spatial conditions lead to different spatial formations and how these processes are linked to spatial proximity.

For individuals, persistent unemployment may indicate difficulties in searching for jobs, insufficient qualifications and obstacles to commuting or moving to the areas of high-employment. For the booming sectors and areas the shortage of labour causes a bottleneck for production and a rise in labour costs. While advanced industrial clusters in core areas are booming, economically weak areas are lagging behind and suffering from out-migration and a low employment rate. If migration doesn't balance the labor market geographically, human resources remain in part unused for some reasons in the formations of high unemployment.

Migration drains the younger generations and labour force from economically declining areas. In such a spatial economic system, this tendency leads to ageing and shrinking populations and a brain-drain from declining areas. When these characteristics of demography are combined with high unemployment in the same area, those areas face persistent development problems. In order to unravel what could be the reasons and mechanisms which maintain these spatial formations of (un)employment and migration we investigate them, unraveling how they are linked to the socio-economics of a municipality. Studying the reasons for the mismatch between migration and unemployment we get information of under which conditions some spatial economies are facing the persistent, unbalanced, labour market and what may hinder them from achieving a balance. Such results provide some conclusions regarding the strength and pace of spatial restructuring and what the available possibilities are to improve the situation.

15.3 Similarity Index Depicting Spatial Patterns of Two Variables

Spatial autocorrelation statistics measure and analyze the degree of dependency among observations in a geographic space and therefore offer a way to explore spatial formations. Classic global spatial autocorrelation statistics such as Moran's I and Geary's C provide summary statistics for overall spatial clustering (Moran 1948; Geary 1954; Cliff and Ord 1981; Griffith 1987; Odland 1988). They do not reveal the detailed spatial clusters involved in the spatial autocorrelation. Corresponding local spatial autocorrelation indices, such as LISA, give an indication of significant spatial clustering of similarly, dissimilarly and randomly deviating values around each observation and therefore allow researchers to explore local variations in spatial dependence and thus allows researchers to answer the question of where spatial autocorrelation exists in the dataset (Getis and Ord 1992; Ord and Getis 1995; Anselin 1995). These efforts are part of a broader endeavour to spatialize general statistics by recognizing that regular statistical assumptions seldom hold for spatial data (Lee 2001, p. 370).

15.3.1 Spatial Autocorrelation Indexes

Global spatial autocorrelation can be measured with Moran's I, which is an indicator of spatial proximity used in geography (Griffith 1987, p. 41). As Moran's I is applied here to two variables, additional indicators are defined based on its components. Moran's I usually varies from −1 to 1, with a value close to 0 indicating a random spatial pattern for the phenomenon in question and values close to −1 and 1 representing the highest geographical concentrations of dissimilar and similar values, respectively (Odland 1988). The expected Moran's I ($E(I)$) is $E(I) = -1/(m - 1)$ where m is the number of municipalities, and naturally values that are larger (or smaller) than the expected value indicate positive (negative) spatial autocorrelation (Odland 1988). In this case of the Finnish data $E(I) = -0.00293$.

Moran's I is a single global indicator which does not reveal the detailed clusters involved in the spatial autocorrelation, but the autocorrelation of individual spatial units can be investigated with local spatial autocorrelation indices such as LISA, which give an indication of significant spatial clustering of similarly, dissimilarly and randomly deviating values around each observation and therefore this allows us to answer the question of where spatial autocorrelation exists in the dataset (Anselin 1995, p. 98). In the randomization hypothesis the expected value for location i turns out to be $E(I_i) = -w_i/(m - 1)$ where w_i is the sum of the elements $\sum w_{ij}$ in a row of the spatial weight matrix W (Anselin 1995, p. 99). A positive value for I_i indicates spatial clustering of similar values for the region and its neighbours, whereas a negative value indicates spatial clustering of dissimilar values. An area is interpreted to be spatially autocorrelated if the indicators reveal statistically significant positive autocorrelation.

15.3.2 Analyzing the Similarity of the Spatial Patterns of the Migration and Unemployment

Most of the spatial autocorrelation statistics are univariate, and they investigate the spatial pattern of a single variable between spatial units. There is also much discussion about what is an appropriate measure of bivariate spatial association.

Watenberg (1985) formulated Cross-Moran which compares a data point in one variable and its neighbours in the other variable. Hubert et al. (1985) developed a nonparametric bivariate spatial association measure. Lee (2001) for example, developed an index that combines Pearson's bivariate correlation between G and H with Moran's spatial autocorrelation measures for G and H. However, bivariate spatial autocorrelation statistics do not compare the local spatial patterns of two variables and therefore some authors have used cross-tabulation in investigating the relationship between the spatial patterns of two variables.

In this article, we investigate the balance of the spatial formations of migration and unemployment with a similarity index which omits the information loss that happens in the cross-tabulation where data is classified in the nominal scale. We formalize this similarity index for two variables x_i and y_i, observed at location i, as a statistic:

$$L_i = f(x_{1,i}, x_{2,i}, y_{1,i}, y_{2,i}) \tag{15.1}$$

where *f* is a function, and the symbols $x_{1,i}$ and $y_{1,i}$ are the values of the variable 1 observed in the municipality i and its neighborhood, and the symbols $x_{2,i}$ and $y_{2,i}$ are values of the variable 2 observed in the municipality i and its neighborhood. The values of the variables 1 and 2 used in the computation of statistic should be standardized in order to avoid the scale dependence of the local indicators, similar to the practice often taken for the global and local indicators of spatial association. For example, in Moran's I the observations are taken as deviations from their mean (Anselin 1995).

The L_i should be such that it is possible to infer the statistical significance of the similarity of the spatial association of variables x and y in location i. More formally this requires the operationalization of a statement such as

$$\text{Prob}[L_i > \delta_i] \leq \alpha_i \tag{15.2}$$

where δ_i is a critical value, and α_i is a chosen pseudo significance level as the result of a randomization test.

The neighborhood values $y_{1,i}$ and $y_{2,i}$ for each municipality i's neighborhood are calculated by the means of a spatial weights or contiguity matrix, W. The exact operationalization of the matrix W is ultimately contractual, in that there are a variety of ways of applying the method. The values of spatially lagged variables depend on the definition of the spatial weight matrix $W = [w_{ij}]$. Here we are looking for continuous spatially lagged structures within socio-economic conditions and migration formatted for *adjacent*, that is neighbouring, municipalities, in order to reveal their spatial patterns and co-existence. The spatial weights matrix may be row-standardized to facilitate the interpretation of the statistics and when standardization is carried out, the values of the $y_{1,i}$ and $y_{2,i}$ corresponds to a form of weighted average of the values (Anselin 1995).

To test the similarity of the spatial patterns of two variables we define the index (L_i) as:

$$L_i = \mathbf{acos}\frac{\mathbf{x}_i \cdot \mathbf{y}_i}{|\mathbf{x}_i||\mathbf{y}_i|} \cdot |(|a| - |b|)| \quad (15.3)$$

where $\mathbf{acos}\frac{a \cdot b}{|a||b|}$ is the angle of two position vectors (law of cosines) a (here unemployment) and b (migration), and $||a| - |b||$ is the absolute value of the origin distance differences between position vectors a and b (Euclidean distance). The symbols $|a|$ and $|b|$ are norms of the position vectors in municipality i, and $a \cdot b$ is dot product of the vectors a and b in municipality i. The similarity measurement has maximum value 0 when $\cos\frac{a \cdot b}{|a||b|} = 0$ or when $||a| - |b|| = 0$ because in these cases positions vectors a and b would be same. By using definitions from the Eq. (15.1) and defining the position vectors $a = (x_{1,i}, y_{1,i})$ and $b = (x_{2,i}, y_{2,i})$, we can re-write index as

$$L_i = \mathbf{acos}\frac{x_{1,i}x_{2,i} + y_{1,i}y_{2,i}}{\sqrt{x_{1,i}^2 + y_{1,i}^2}\sqrt{x_{2,i}^2 + y_{2,i}^2}} \cdot \left|\sqrt{x_{1,i}^2 + y_{1,i}^2} - \sqrt{x_{2,i}^2 + y_{2,i}^2}\right| \quad (15.4)$$

The statistical significance of the index is assessed by the use of a conditional randomization or permutation approach to yield empirical so-called pseudo-significance levels. The randomization is conditional in the sense that the value of $x_{1,i}$ and $y_{1,i}$ is held fixed and the remaining values are randomly permuted over the location in the data set (Anselin 1995). For each of these re-sampled data sets, the value of L_i can be computed. The resulting empirical distribution function provides the basis for a statement about the extremeness (or lack of extremeness) of the observed statistic, relative to (and conditional on) the values computed under the null hypothesis (the randomly permuted values). In practice, this is straightforward to implement, since for each location only as many values as there are in the neighborhood set need to be re-sampled.

15.4 Data on Unemployment and Migration by Municipality in 2004, 2006 and 2008

A municipality is an essential local actor in the development and management of business environments and the provision of services in Finland. Statistics Finland (2011) provides a database of annual internal moves within the country, which are classified into moves between and within municipalities. For the present purpose we use country-internal net migration data by municipality, which is calculated as the difference between the sum of internal in-migration and internal out-migration for each municipality. Inter-municipal net migration expresses the direct impacts of migration on the distribution of population, and hence on regional structures.

The migration and unemployment data are from the three years of 2004, 2006, and 2008 according to the municipal division of the year of 2010. Net migration was growing in the 2000s and peaked in 2007 reaching 280,156 moves. The financial crisis in 2008 halted this growth and migration declined to 269,792 moves. Unemployment data was from the same years. Data was retrieved from the data bases of Statistics Finland (2011).

15.5 Similarity of the Spatial Patterns of Net Migration and Unemployment of the Finnish Municipalities

15.5.1 Spatial Autocorrelation of the Migration and Unemployment

When positive spatial autocorrelation exists, as indicated by statistically significant values for Moran's I and LISA, the scores on unemployment or net migration at a given location, tend to be similar to the scores in nearby locations. In other words, if the standardized component score or net migration rate is high in a given location, the presence of high positive spatial autocorrelation indicates that nearby scores will also be high. The LISA results are inferred at a pseudo-significance level of 1%, and the resulting local spatial autocorrelations are illustrated by means of choropleth maps which indicate those Finnish municipalities exhibiting significant local spatial autocorrelation. Such spatial formations are the result of economic development processes and migration, and the map shows their geographical morphology.

The Moran's I statistics for both variables imply that the spatial structures for the 342 municipalities of Finland are autocorrelated, and hence spatially clustered, over each of three years (Table 15.1). Both variables have a tendency to cluster and this indicate that it is possible that the spatial formations of net migration and unemployment emerge.

Each spatial cluster is defined as such when the value for a location (either high or low) is more similar to those of its neighbours than would be the case under the conditions of spatial randomness. The use of local spatial autocorrelative structures generates four LISA groups, which consist of the four types of local spatial association between a municipality and its neighbours in terms of the push component (or net migration rate): high-high municipalities, with a high value, i.e. above the

Table 15.1 Moran I values for different years

Weight matrix	Variable	Index	2004	2006	2008
Queen method	Net migration	LISA	0.294	0.290	0.185
		P	<0.001	<0.001	<0.001
	Unemployment	LISA	0.664	0.674	0.642
		P	<0.001	<0.001	<0.001

mean, surrounded by others with high values, low-high municipalities, with a low value, i.e. below the mean, surrounded by others with high values, and corresponding high-low and low-low municipalities. In high-high municipalities, with a high value, i.e. above the mean, surrounded by others with high values, low-low municipalities, with a low value, i.e. below the mean, surrounded by others with low values. The high-high and low-low LISA groups indicate the spatial clustering of similar positive or negative values of the unemployment and the net migration rate (Fig. 15.1), and refer to spatial autocorrelation. In addition to the similar autocorrelative structures shown by the LISA values, the remaining combinations, low-high and high-low, can be used to visualize hybrid clusters. Municipalities where spatial autocorrelation is not significant or autocorrelation is classified to low-high and high-low associations are left blank on the choropleth maps.

The local spatial associative patterns reveal the core-periphery structure of the vital urban regions and the poor capabilities of the distant inland areas (Fig. 15.1). The clustering of high unemployment is spatially concentrated in the regions of Kainuu, North Karelia, and the northern parts of Central Finland and North Ostrobothnia. These areas have had permanent settlements and relied on small-scale farming and trade for centuries, and they are perceived as the most poorly developed regional hinterlands in Finland. Their municipalities have disadvantageous socio-economic properties with regard to development.

The areas of attractive economic conditions revealed by low unemployment are assumed to pull migrants searching for higher utility to the urban municipalities that

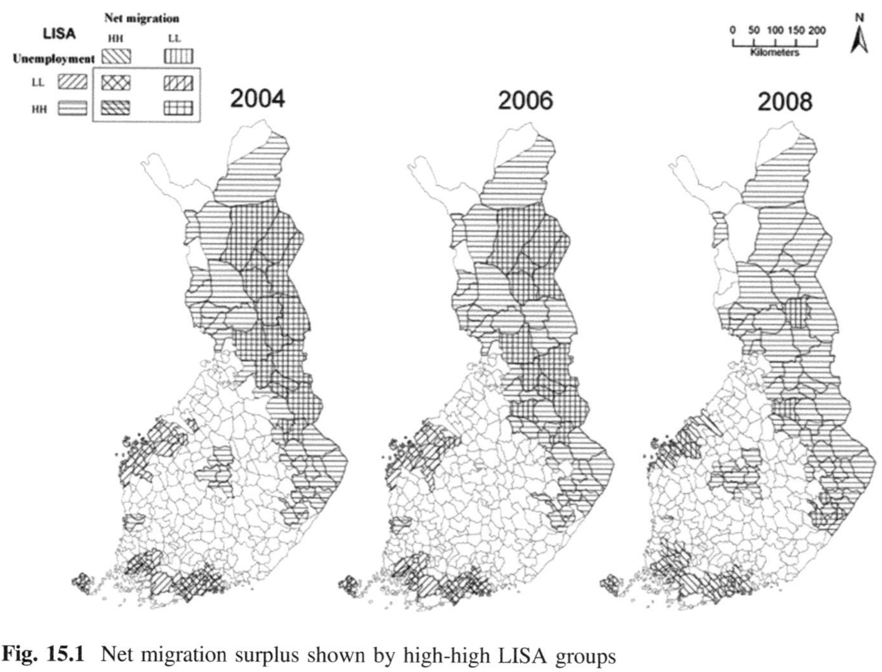

Fig. 15.1 Net migration surplus shown by high-high LISA groups

are located overwhelmingly in the largest concentration of population, in the south and south-west. A net migration surplus shown by high-high LISA groups is concentrated on the regions of Helsinki, and Turku and to some extent the Vaasa sub-regions (Fig. 15.1). The spatial accumulation of population has drawn human capital selectively to these regions of higher incomes as observed earlier (Ritsilä and Ovaskainen 2001, pp. 323–324; Pekkala 2003, p. 480). The most mobile groups, students and the unemployed, migrated to these growth centers, giving rise at first to in-migration into the cities and at a later stage in the family cycle to the development of their suburbs. Many of the fastest-growing and knowledge-intensive economic activities are located in advanced industrial metropolitan areas, and therefore students are likely to stay in these commuting areas after graduating.

The pull-win patterns consist of low-low (LL) unemployment municipalities and high-high (HH) migration municipalities where high in-migration prevails whereas the push-lose patterns consist of high-high (HH) unemployment municipalities and low-low (LL) migration where severe out-migration prevails. Other two combinations are rare.

The low-low clusters of net migration rate indicating high out-migration are highly uniformly distributed over northern and eastern Finland, and these uniform formations have spread spatially to wider and more northerly areas than are the high-high clusters of unemployment. Only certain small groups of municipalities in the local peripheries of southern Finland have confronted a clustered net migration deficit. Changes in the volume of net migration along with economic growth have an impact on the spatial extent of local spatial autocorrelation (Lehtonen and Tykkyläinen 2009).

15.5.2 Push-Lose and Pull-Win Municipalities in Finland

The theories of both migration and economic geography assume that migration is a response to differences in opportunities in geographical space, and thus the spatial patterns of migration and the push and pull factors should match. A contingency test performed on the LISA groups revealed the causality between the spatial clusters of net migration rate and the unemployment rate during the three years of 2004, 2006 and 2008 (Table 15.2). This showed that spatial net migration patterns in Finland are highly dependent on geographical unemployment differences, in that the spatially autocorrelated low unemployment areas identified here attract migrants for various reasons and, by contrast, the underdeveloped regions push migrants to move to wealthier, better developed municipalities. However, the cross-tabulation does not reveal the relationship between these two spatial patterns in detail but shows that these spatial formations have emerged. The pulses of migration waves are regulated by economic cycles (Lehtonen and Tykkyläinen 2009).

The long-term intra-regional outcome of the spatial flows of migrants relates to their path-forming impacts on regional competitiveness and performance, which also become spatially divergent, maintaining the process of cumulative causation as

Table 15.2 Cross-tabulation of the LISA groups of net migration and unemployment

			Net migration					
			0	HH (high in; win)	LL (severe out, lose)	LH	HL	Total
Unemployment	0	N	509	59	39	6	12	625
		%	49.6	5.8	3.8	0.6	1.2	60.9
	HH (high-high; push)	N	124	0	**33**	1	11	169
		%	12.1	0.0	**3.2**	0.1	1.1	16.5
	LL (low-low; pull)	N	103	**62**	8	9	1	183
		%	10.0	**6.0**	0.8	0.9	0.1	17.8
	LH	N	5	0	0	36	0	41
		%	0.5	0.0	0.0	3.5	0.0	4.0
	HL	N	4	2	1	1	0	8
		%	0.4	0.2	0.1	0.1	0.0	0.8
	Total	N	745	123	81	53	24	1026
		%	72.6	12.0	7.9	5.2	2.3	100.0

N refers to the numbers of LISA group associations. Data consist of the frequencies summed over the years of 2004, 2006 and 2008

observed in Finland for decades (Aro 2007b). It is therefore not only a question of how regions push and pull migrants, but of how migration creates spatial differences in the productive performance and competitiveness of each region and in all probability that the initial advantage that triggers regional growth will emerge over the course of time (Venables 2006). Owing to spatial selectivity, migration processes make for the formation of very competitive districts that are attractive to new industries, which in Finland prefer to locate their research and experimental development activities in metropolitan environments, and their routine production, if possible, in low-cost countries. As long as these push and pull forces prevail, self-reinforcing processes will continue, as the skilled laborer will concentrate on attractive districts and the peripheries will be caught up in a vicious circle of losing active cohorts of their human capital (Kitson et al. 2004). In total about 50.3% of the municipalities belongs to some spatial cluster of net migration or unemployment (Table 15.2).

The attractive districts are pull-win municipalities. The formations of competitive districts point to wealthy, predominantly suburbanized areas, named pull-win municipalities which have become real industrial and suburban districts and which are experiencing accelerated growth. These spatial clusters are rare, because only 6% of the municipalities belong to these clusters (Table 15.2). These formations concentrated geographically on the surroundings of Helsinki and Turku in the 2000s in the main (Fig. 15.2).

Push-lose formations strengthen, especially during an economic boom in peripheries. Unemployment and the meager opportunities there push migrants away from the spatial clusters of net migration deficit when an economic boom opens up

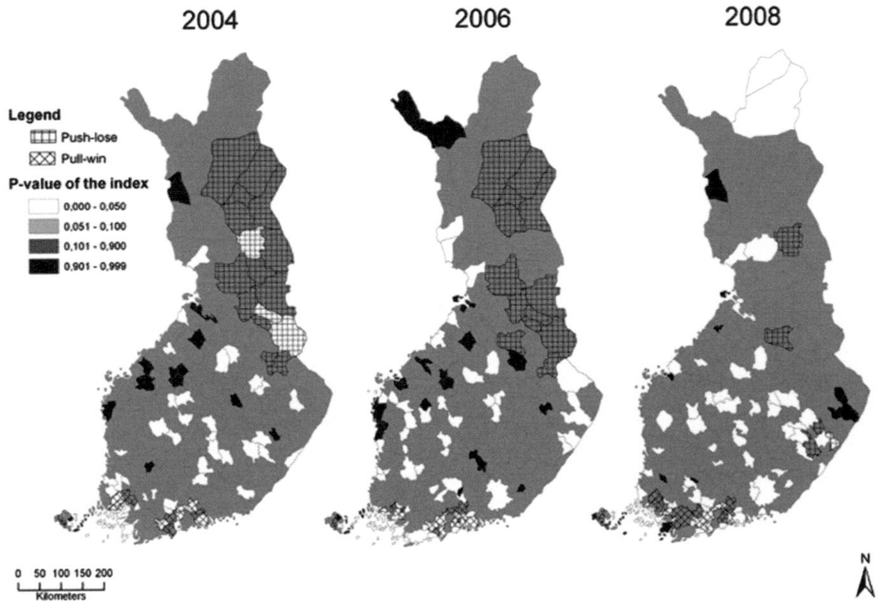

Fig. 15.2 Pull-win and push-lose formations and the p-values of the similarity index

new opportunities elsewhere. A high unemployment rate is a visible mark of this labour potential, but the underlying causes are the rationalization in the primary sector and, in mature industries, including the service sector, simply the fact that these areas do not have much to offer to advanced industries. Migrants leave these municipalities because this improves the likelihood of their becoming employed and enables them to achieve a better income level (Pekkala and Tervo 2002, p. 630; Pekkala 2002, p. 22). Compared with the pull-win municipalities the number of the push-lose municipalities is smaller than the congruent number of the pull-win municipalities which indicates that the emergence of the push-lose municipalities differs from the emergence of pull-win municipalities. Only 3.2% of the municipalities belong to push-lose formations (Table 15.2). These formations are geographically concentrated to the northern and eastern parts of Finland (Fig. 15.2).

The pull-win and push-lose patterns are not the only pairs of regular spatial autocorrelative structures. High positive net migration rate in a certain municipality and its surroundings is not always associated with the spatially clustered low values of the push component. As much as 48.0% of the winning municipalities summed over the years of 2004, 2006 and 2008 are not linked to any local spatial autocorrelative cluster of the unemployment (Table 15.2). This indicates that positive net migration can become concentrated for other reasons than spatial unemployment formations. The reasons can be many. Such cases may be referred to as non-pull-but-win spatial formations. Correspondingly, it is possible to find non-push-lose spatial formations, indicating the conditions of net migration deficit in an environment where no push factor is clustered. Even eight cases were in the spatial formations of LL unemployment but faced out-migration (Table 15.2).

15.5.3 The Similarity Index of the Spatial Patterns of Net Migration and Unemployment

The similarity statistics show the differences of the standardized values of net migration and unemployment in a municipality and its surroundings (defined by Queen method) and thus demonstrates the similarity of these two factors. The usefulness of the index comes from the fact that the probability of the existence of dependence can statistically be tested by a municipality and hence assessed. If the p-value of the similarity index in permutations is lower than 0.10 or bigger than 0.9, it shows that the relationship between net migration and unemployment is statistically similar or non-similar, which statistically confirms that the relationship between these variables is not spatially randomly formed. When the values in the geographical space are similar, the relationship between net migration and unemployment is most probably a result of economic development and the mobility of labor force. If the spatial patterns are not similar, municipalities have other reasons that retard migration from and to municipalities and to prohibit the emergence of spatial formations.

Figure 15.2 demonstrates that in most municipalities the relationship between net migration and unemployment is randomly formed, and the association between these spatial patterns remains unknown. Over 75% of the municipalities have a random association between spatial patterns which indicate weak associations between these two variables (Table 15.3; Fig. 15.2). This means that the spatial mobility of the labor force is not high, and thus retarding factors, such as social insurance benefits and property influence the lack of the similarity of spatial formations. The similarity between the spatial patterns of migration and unemployment is found in fewer than 20% of the municipalities. The number of similar associations was higher every year in pull-win municipalities than in push-lose municipalities (Table 15.3).

In the groups of push-lose municipalities, the number of significant municipalities varies between from 3 to zero annually indicating the existence of factors which prevent the emergence of a match. The results indicate that the balance of labour is not easily achieved as the coexistence of unemployment and out-migration does not match well. This characteristic associated to the push-lose municipalities is essential for the regional development, because it means that peripheries will e.g. ran out of human capital more slowly.

The differences of similarity between the pull-win and push-lose municipalities were tested through the use of the t-test and the difference was strongest in 2006. The values obtained from other years are not incongruent, but differences are not so significant statistically. Thus, the test results would seem to indicate that the

Table 15.3 Number of the significant values of similarity index in spatial formations

Weight	Year	Group		SIMI <0.100	0.101< SIMI <0.899	0.900< SIMI	Sum
Queen method	2004	All	N	57	268	17	342
			%	16.7	78.4	5.0	100.0
		Push-lose	N	3	13	0	16
			%	18.8	81.3	0.0	100.0
		Pull-win	N	9	9	0	18
			%	50.0	50.0	0.0	100.0
	2006	All	N	62	254	26	342
			%	18.1	74.3	7.6	100.0
		Push-lose	N	0	13	0	13
			%	0.0	100.0	0.0	100.0
		Pull-win	N	11	9	0	20
			%	55.0	45.0	0.0	100.0
	2008	All	N	66	263	13	342
			%	19.3	76.9	3.8	100.0
		Push-lose	N	1	3	0	4
			%	25.0	75.0	0.0	100.0
		Pull-win	N	8	16	0	24
			%	33.3	66.7	0.0	100.0

Table 15.4 Results from t-test between push-lose and pull-win municipalities

Weight	Year	Group	N	Mean	S.D	t	p-value
Queen method	2004	Push-lose	16	14.71	12.81	−1.77	0.088
		Pull-win	18	7.58	10.42		
	2006	Push-lose	13	17.76	13.52	−2.01	0.055
		Pull-win	20	8.38	12.24		
	2008	Push-lose	4	25.39	19.38	−1.39	0.174
		Pull-win	24	12.36	16.95		

patterns of similarity between the pull-win and push-lose municipalities are different and hence the dynamics between the core and periphery have their own, peculiar mechanisms which are not just the symmetric facets of the same general mechanism. The values of the similarity index are lower in pull-win municipalities than in push-lose municipalities (Table 15.4). The lower similarity value indicates that the relationship between the spatial formations of net migration and unemployment is more similar in pull-win formations than in push-lose formations which gives evidence that some factors retard migration in push-lose municipalities. These retarding factors in push-lose municipalities are related to the migration because the

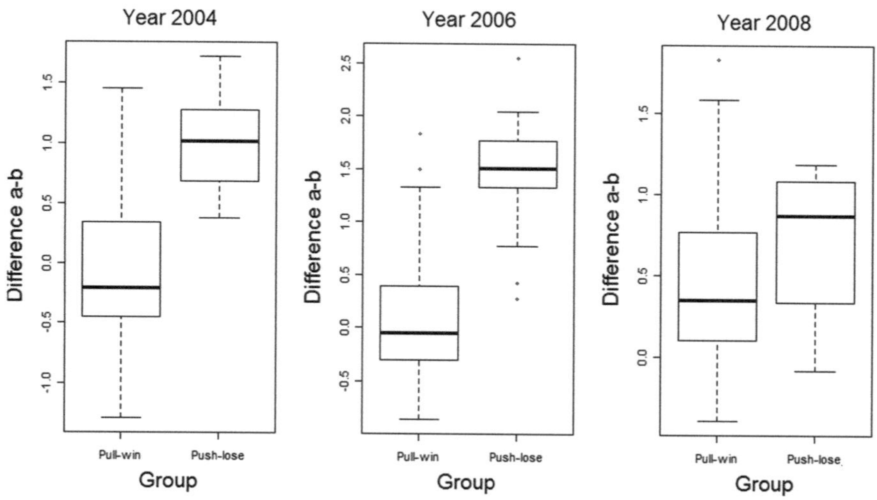

Fig. 15.3 Boxplots from the difference in origin distances between unemployment and migration

difference in the value of the origin distance differences between position vectors a (unemployment) and b (migration) in Eq. 15.3 is positive in all push-lose municipalities in 2004 and 2006. Only in 2008, in one municipality is the difference slightly negative (Fig. 15.3). The average differences are 1.02, 1.42 and 0.71 in the years of 2004, 2006 and 2008.

The relationship between net migration and unemployment match better in pull-win municipalities than in push-lose municipalities which indicate that job opportunities attract in-migration (Table 15.4). In these municipalities the retarding factors are mostly related to the migration but the average difference between origin distances are much lower. On average, the differences are −0.06, 0.15 and 0.52 in years 2004, 2006 and 2008 respectively (Fig. 15.3). From the viewpoint of the spatial equilibrium approach this balance is self-evident, but the questions are why the match is better in pull-win formations and especially why it so weak in peripheries.

The growth of the economy maintains this dynamics of pull and push, as in 2008 the differences between push-lose and pull-win municipalities are not evident as p-value is not statistically significant (Table 15.4) which indicates that during the economic crises the similarity of the spatial patterns declines when there is no pull of flourishing new economic sectors and the regional differences even out in attractiveness. At the same time migration fades.

15.5.4 *Factors that Explain the Differences of Similarity of Push-Lose and Pull-Win Municipalities*

Regression analysis is applied to explain differences in the values of the similarity index both in push-lose and pull-win municipalities. Demographic and some socio-economic characteristics in part explain the differences of the values of the similarity index of unemployment and migration in push-lose and pull-win municipalities (Table 15.5). Four variables are significant in the models and only two of the variables in both formations influence the match in the similar way. The higher the proportion of persons over 75 of age, the less is the match between migration and unemployment, and this association is stronger in the pull-win formations. The higher the proportion of people on low-incomes decreases so does the mismatch between employment and migration. At least in part this could be explained by the more flexible labour market of such groups as students and young employees who have relatively low incomes.

Naturally, agriculture and forestry do not play a significant role in pull-win formations. The more the proportion of those employed in agriculture and forestry in the push-lose formations decreases, the mismatch between unemployment and migration rises, which mirrors the situation where people move from areas where the rationalization of primary production is strong and other options than when migration are scarce. In the pull-win municipalities the increase of those young adults aged 17–24 who are not in education or training increases the match between employment and in-migration. It is explained that high employment means job opportunities which are responded to by young adult in pull-win municipalities. The results of the regression analyses thus explain a part of dynamics. However, a more robust interpretation of dynamics would require a more sophisticated spatio-temporal modeling.

15.6 Conclusions

Our first aim was to answer the question of how the spatial pattern of the net migration is covariate with the spatial pattern of unemployment in Finland, and how regular the spatial coexistence between migration and unemployment was during the years of economic growth. First, half of the municipalities are not linked to any autocorrelative structures. Second, although push-lose formations cover vast peripheral land masses in north-eastern Finland, only 3.2% of municipalities belong to these formations. The number of municipalities in pull-win formations is almost double the norm and their significance is much more important as they are populous places. We found very few municipalities where similarity was statistically significant and they were mostly attracting urban areas which pulled labour from all over the country. On the verge of the financial crisis of the late 2000s, spatial difference started to even-out as expected by previous research.

Table 15.5 The results of regression analysis explaining differences in the results of similarity index in push-lose and pull-win municipalities in Finland, 2006

Coefficient	Push-lose				Pull-win			
	Estimate	S.D	t	p-value	Estimate	S.D	t	p-value
Constant	82.929	46.528	1.782	0.125	2.321	22.664	0.102	0.919
Aged over 75 years	4.272	1.407	3.036	0.022	5.048	0.982	5.139	<0.001
Social assistance recipients during year, as % of total population	−4.577	2.753	−1.662	0.147	−1.676	1.091	−1.536	0.148
Highly educated as % of total population of over 15 years	−1.739	1.127	−1.543	0.173	−0.487	0.516	−0.943	0.362
Those aged 17–24 not in education or training, as % of total population of same age	2.711	2.277	1.190	0.278	1.713	0.362	4.733	<0.001
Employed in agriculture and forestry as % of economically active population	−2.117	0.943	−2.245	0.065	−1.605	1.180	−1.360	0.196
Low income rate, % of total population	−2.225	1.104	−2.014	0.090	−2.715	0.970	−2.715	0.015

These relations of pull-win formations and push-lose formations are not symmetrical in a core-periphery dimension. Our result indicated that the pattern between the pull-win and push-lose is different and hence the dynamics between the core and periphery have their own peculiar mechanisms which are not just the facets of the same general mechanism. Similarities in the spatial patterns of unemployment and migration are lower in pull-win municipalities than in push-lose municipalities. The lower similarity value indicates that the relationship between the spatial formations of net migration and unemployment is more similar in pull-win formations than in push-lose formations which gives evidence that some factors retard migration in push-lose municipalities. This result accorded with the recent findings on the role of other factors than income that determine the decisions of migration in peripheries.

Our second aim was to reveal the dependence of the spatial formations of growth and decline on the socioeconomic conditions of areas. We found that demographic and some socio-economic characteristics in part explain the differences in the spatial patterns of unemployment and migration in push-lose and pull-win municipalities. Parts of those factors promote the tendency toward the spatial equilibrium of employment and settlement systems and some of them increase the divergence and maintain spatial unbalance.

The third aim was to apply, for the first time, this bivariate index of similarity and to investigate its applicability to comparing spatial patterns and testing their differences. The index provided more information than traditional cross-tabulation and its tests on dependency and made it possible to statistically test the probability of the existence of the similarity of spatial patterns.

References

Aro, T. (2007a). Valikoiva muuttoliike osana pitkän aikavälin maassamuuttokehitystä. *Yhteiskuntapolitiikka, 72,* 371–379.

Aro, T. (2007b): "Julkinen valta ja maassamuuttoa edistävät ja rajoittavat tekijät Suomessa 1880-luvulta 2000-luvulle", *University of Turku. RUSE Research Reports* 69

Anselin, L. (1995). Local indicator of spatial association: LISA. *Geographical Analysis, 27,* 93–115.

Boschma, R. A., & Lambooy, J. G. (1999). Evolutionary economics and economic geography. *Journal of Evolutionary Economics, 9,* 411–429.

Cliff, A. D., & Ord, J. K. (1981). *Spatial processes: Models & applications*. London: Pion Limited.

Gløersen, E., Dubois, A., Copus, A., & Schürmann, C. (2005). Northern peripheral, sparsely populated regions in the European North. *Nordregio Report, 2005,* 4.

Geary, R. C. (1954). The contiguity ratio and statistical mapping. *Incorporated Statistician, 5,* 115–145.

Getis, A., & Ord, J. K. (1992). The analysis of spatial association by use of distance statistics. *Geographical Analysis, 24,* 189–206.

Griffith, D. (1987). *Spatial autocorrelation: A primer*. Washington DC: Association of American Geographers.

Fujita, M., Krugman, P., & Venables, A. J. (1999). *The Spatial Economy*. Cambridge (MA): The MIT Press.

Heikkilä, E., & Pikkarainen, M. (2010). Differential population development in the regions of Finland. *Population, Place and Space, 16,* 323–334.

Hubert, L. J., Golledge, R. G., Costanzo, C. M., & Gale, N. (1985). Measuring association between spatially defined variables: An alternative procedure. *Geographical Analysis, 17,* 36–46.

Irvin, E. G., Isserman, A. M., Kilkenny, M., & Partridge, M. D. (2010). A century of research on rural development and regional issues. *American Journal of Agricultural Economics, 92,* 522–553.

Kitson, M., Martin, R., & Tyler, P. (2004). Regional competitiveness: an elusive yet key concept? *Regional Studies, 38,* 991–999.

Krugman, P. (1991). Increasing returns and economic geography. *Journal of Political Economy, 99,* 183–199.

Krugman, P., & Venables, A. J. (1995). Globalization and the inequality of nations. *Quarterly Journal of Economics, 110,* 857–880.

Lee, S.-I. (2001). Developing a bivariate spatial association measure: An integration of Pearson's r and Moran's I. *Journal of Geographical Systems, 3,* 369–385.

Lehtonen, O., & Tykkyläinen, M. (2009). Muuttoliikkeen alueelliset muodostumat ja pulssi Suomessa 1980–2006 (Regional formations and pulse of migration in Finland in 1980–2006). *Terra, 121,* 119–137.

Lehtonen, O., & Tykkyläinen, M. (2010). Self-reinforcing spatial clusters of migration and socio-economic conditions in Finland in 1998–2006. *Journal of Rural Studies, 26,* 361–373.

Leinbach, T., & Brunn, S. (2002). National innovation systems, firm strategy, and enabling mobile communications: the case of Nokia. *Tijdschrift voor Economische en Sociale Geografie, 93,* 489–508.

Lundholm, E. (2007). New motives for migration? *Umeå Universitet Gerum kulturgeografi, 2007,* 2.

McCormick, B., & Sheppard, S. (1992). A model of regional contraction and unemployment. *Economic Journal, 102,* 366–377.

Moran, P. A. P. (1948). The interpretation of statistical maps. *Journal of the Royal Statistical Society Series B. (Methodological), 10*, 243–251

Odland, J. D. (1988). *Spatial autocorrelation*. Newbury Park (CA): Sage.

OECD. (2008). *OECD rural policy reviews Finland*. Paris: OECD.

Olfert, M. R., & Partridge, M. D. (2010). Best practices in twenty-first-century rural development policy. *Growth and Change, 41*(2), 147–164.

Ord, J. K., & Getis, A. (1995). Local spatial autocorrelation statistics: Distributional issues and an application. *Geographical Analysis, 27,* 286–306.

Pekkala, S. (2002). Migration and individual earnings in Finland: A regional perspective. *Regional Studies, 36,* 13–24.

Pekkala, S. (2003). Migration flows in Finland: Regional differences in migration determinants and migrant types. *International Regional Science Review, 26,* 466–482.

Pekkala, S., & Tervo, H. (2002). Unemployment and migration: Does moving help? *Scandinavian Journal of Economics, 104,* 621–639.

Ritsilä, J., & Ovaskainen, M. (2001). Migration and regional centralization of human capital. *Applied Economics, 33,* 317–325.

Sjaastad, L. A. (1962). The cost and returns of human migration. *Journal of Political Economy, 70,* 80–93.

Statistics Finland. (2011). *PX-Web data bases*. Statistics Finland, Helsinki. Downloaded on 14–25 March 2011 (http://www.stat.fi/til/muutl/index_en.html and http://www.stat.fi/tup/posnro/index.html)

Storper, M. (1997). *The regional world: Territorial development in a global economy*. New York: Guilford Press.

Tuhkunen, A. (2007). Between location and a sense of place. *Acta Universitatis Tamperensis*, 1207

Venables, A. (2006). Economic geography. In Barry, R., Weingast, R., & Wittman, D. (Eds.), *The Oxford handbook of political economy*. Oxford University Press, Oxford (UK), pp. 739–754.

Watenberg, D. (1985). Multivariate spatial correlation: A method for exploratory geographical analysis. *Geographical Analysis, 17,* 263–283.

Chapter 16
Regional Pattern of Migration Flows in Uttarakhand, India

R.C.S. Taragi and Raghubir Chand

16.1 Introduction

The Himalayan State of Uttarakhand is a newly created State in the Republic of India. It has a total geographic area of about 53,483 km^2. extending from 28° 44′ North to 31° 25′ North Latitudes and 77° 45′ to 81° East Longitudes. The State of Uttarakhand constitutes 13 districts with 95 development blocks. The state is bounded in the north by Tibet (China), in the east by Nepal and in the west and south by the states of Himanchal Pradesh and Uttar Pradesh respectively (Fig. 16.1). According to the Census 2011, it has a population of 10,086,292 with a density of 189 persons per km^2. There are 117 urban centers consisting of small towns to cities and urban agglomerations where 30.6% of the state's total population lives. Majority of the population is classified by the Census as rural living in over 16,000 villages.

Uttarakhand region presents very distinct demographic characteristics determined by the local relief and the harsh climate together with a host of other physical as well as cultural factors. There are areas with no population in the Trans Himalaya to densely populated areas along the foothills. The distribution pattern of population not only differs from rest of the country but greatly varies among districts belonging to one part to another part of the region. As one proceeds from southern Tarai–Bhabar tract to Shiwalik, Lesser Himalaya; and Greater Himalaya and beyond to Trans Himalayan region of the state, the population decreases markedly

R.C.S. Taragi (✉)
Ministry of Development Planning and Statistics, States of Qatar,
P.O. Box-1855, Doha, Qatar
e-mail: rcstaragi@gmail.com

R. Chand
Department of Geography, DSB Campus, Kumaun University,
Nainital 263002, Uttarakhand, India
e-mail: raghubir.seri@gmail.com

R. Chand et al. (eds.), *Societies, Social Inequalities and Marginalization*,
Perspectives on Geographical Marginality,
DOI 10.1007/978-3-319-50998-3_16

Fig. 16.1 Location map of Uttarakhand state in India

in accordance to the topography and environmental conditions (Chand and Thakur 1991). The slope gradient primarily controls the clustering of population and now the infrastructural development has altered and shifted the population nucleus from their original locations to the places receiving developmental inputs (Chand and Taragi 2014). About two thirds population of the state is concentrated in 4 districts belonging to the plains areas along foothills while remaining 40% population is distributed among 9 hilly districts. According to Census 2011, the overall increase in population in past decade is noted about 19% (1.5 million) with extreme variations if compared as districts of plains (31%) and hills (9%). Further highlighting fact to note from migration point of view is that on the one hand Almora (−1.64%) and Pauri Garhwal (−1.41%) districts of the state recorded a decrease in total population during the Census 2001–2011, the district of Dehradun, Nainital and UdhamSinghNagar belonging plains areas have shown a high growth (more than 30%). The growth trend clearly indicates the flow of people to these areas having big cities and centers of industrial and commercial activities, better infrastructure facilities and living opportunities.

16.2 Migration Issues

There is plenty of research work and literature available on migration issues relating to internal as well as international migrations. Rural to urban migration has dominated the academic discourse and policy making and planning strategies. The current state of knowledge on migration is well described by Lall et al. (2006) dealing with major concept and models. Some of the recent works by Lehtonen and Tykkyläinen (2009, 2010), Deshingkar (2006), Plane et al. (2005), Martin (2013), Chand (2009, 2013), Esipova et al. (2013), and Srivastava and Sasikumar (2003) portray many new dimensions in migration research. These broader concerns and renewed understanding is however restricted due to the nature of data availability in the context of present study area. There are some significant attempts in the last two decades on the migration issues of Uttaraklhand made by Mamgain (2003), Belwal (2007), Jain (2010), Grunawalt (2012) and Bahuguna and Belwal (2013). These attempts carry relevance in defining the need to conduct further research in Uttarakhand. The attempt here basically to add geographic interpretation and analyze movement of people and migration pattern taking place in the Uttarakhand region and its consequences marginalizing the hill society and its economy. The study is primarily based on Census results. Latest data on migration available for the State is according to Census 2001. The Census 2011 data on migration has not been released so far. Geographic Information System (GIS) techniques are used to make the analysis better and more presentable. The impact of ongoing process of globalization and the entry of market forces into these peripheral societies can be analyzed with the results obtained through this case study.

The main objectives of this study are:

1. To study the male-female division of migrants at four different levels-intra-district, inter-district, inter-state and; at international level.
2. To discover the major destination areas of out migration from Uttarakhand.
3. To look at the size, characteristics and composition of the inter-district migration within the state of Uttarakhand; and finally.
4. To study the regional pattern of migration flows from mountainous district to the districts located in the Tarai–Bhabar tract of the state.

16.3 Migration Streams

When a person is enumerated in census at a different place than his/her place of birth, she/he is considered to be a migrant. This may be due to marriage, which is the most common reason for migration among females. Employment/work and better living conditions which are in general the causes for migration among males. According to Census 2001, out of the 1.02 billion people in the country, 307 million (or 30%) were reported as migrants by place of birth. The distribution of migrants by main migration streams in the country is put under four important groups. Rural to rural migration within the country is maximum being 53.3 million followed by rural to urban migration recorded as 20.5 million. Other two categories are urban to urban migration with 14.3 million migrants and urban to rural migration which is 6.2 million (Census 2001).

As compared to the National average, Uttarakhand State reported about 3 million persons (or 29.84%) as migrants in 2001. Maximum share of these belong to the intra-district migrants constituting a little more than half of the total migrants (50.15%) in the State. The inter-state migration is the next with 28.8% share followed by 17.6% as inter-district migrants and 3.4% as international migrants. It is interesting to note that Uttarakhand has inter-state and international migrants more in ratio if compared to the National average. The migration streams in the State by Place of Birth and Place of Last Residence are given in the Table 16.1 as per the Census 2001.

In Uttarakhand, in comparison to place of birth, 60,490 people are more registered as migrants by place of last residence in 2001 census with a total of 3.07 million. By place of last residence also the highest 52.5% are the intra-district migrants in the State, followed by 27.98% as inter-state migrants and 16.60% as inter-district migrants. There are 2.8% classified as international migrants based on the place of last residence. It is important however that in both the situations, Uttarakhand has more inter-state and more international migrants as compared to the National average. The phenomenon of out migration is therefore the key feature of the State (Table 16.1).

Table 16.1 Migration streams in Uttarakhand state, 2001

Migrants by place of birth				Migrants by place of last residence		
Migrants	Persons	Males	Females	Persons	Males	Females
Intra-district migrants	1,510,000	306,194	1,203,806	1,613,899	349,944	1,263,955
Inter-district migrants	531,581	227,586	303,995	510,440	220,287	290,153
Inter-state migrants	866,894	390,547	476,347	859,598	388,335	471,263
International migrants	102,208	59,574	42,634	87,237	51,617	35,620
Total migrants	3,010,683	983,901	2,026,782	3,071,174	1,010,183	2,060,991

Source Census of India, 2001

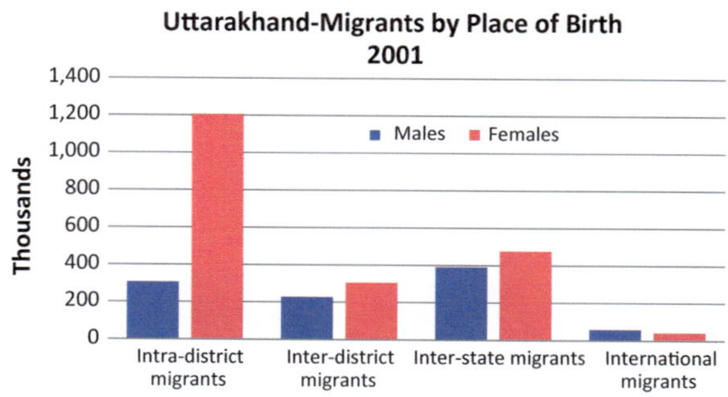

Fig. 16.2 Gender divide by migration streams in Uttarakhand, 2001

16.4 Gender Analysis of Migration Streams

Analyzing the migrations streams by gender, it is noted as female dominated in shorter distances that gradually decrease at larger distances. Figure 16.2 represents clearly that the overall share of female migrants is considerably very high (67.32%) in Uttarakhand although little less if compared with the national average. Intra-district migration contributes the largest proportion of female migrants (79.72%) in Uttarakhand which gradually decreases at inter-district level (57.18%), inter-state level (54.94%) and international level (41.71%) in the state. In all cases, female migrants dominate over male migrants except at international level. The reason behind the high female migration at the intra and inter-district level is due to marriages. In all the streams, female migrants dominate over male migrants except at international level.

16.5 Out Migration from Uttarakhand to Neighboring States

To examine the flow of out migration from Uttarakhand State to neighboring States/Union Territories in India, the top ten ranking states based on the Census 2001 are listed in Table 16.2.

The flow of out migration from Uttarakhand to any part of India is primarily influenced by the traditional linkages established during the past and also according to the availability of jobs and employment opportunities present in the destination areas. The bulk of out migration from Uttarakhand which constitutes 95% of total migrants is to these ten States/UTs listed in Table 16.2 and portrayed by flow chart map in Fig. 16.3.

According to Census 2001, about one third of the total migrants from Uttarakhand have their destination recorded as Delhi, the capital of the country. In one of the estimates, about 2% of Delhi population belongs to Uttarakhand. The State of Uttar Pradesh is next to Delhi with 30% share of migrants receiving from

Table 16.2 Top ten states/UTs ranked according to out migrants from Uttarakhand state, 2001

Sl. No	State	Migrants	% of migrants
1.	Delhi	306,503	33.1
2.	Uttar Pradesh	282,350	30.4
3.	Haryana	67,932	7.3
4.	Bihar	63,527	6.8
5.	Punjab	44,278	4.8
6.	Maharashtra	40,825	4.4
7.	Chandigarh	27,470	3.0
8.	Rajasthan	19,191	2.1
9.	Himachal Pradesh	17,754	1.9
10.	Madhya Pradesh	14,889	1.6

Source Census of India, 2001

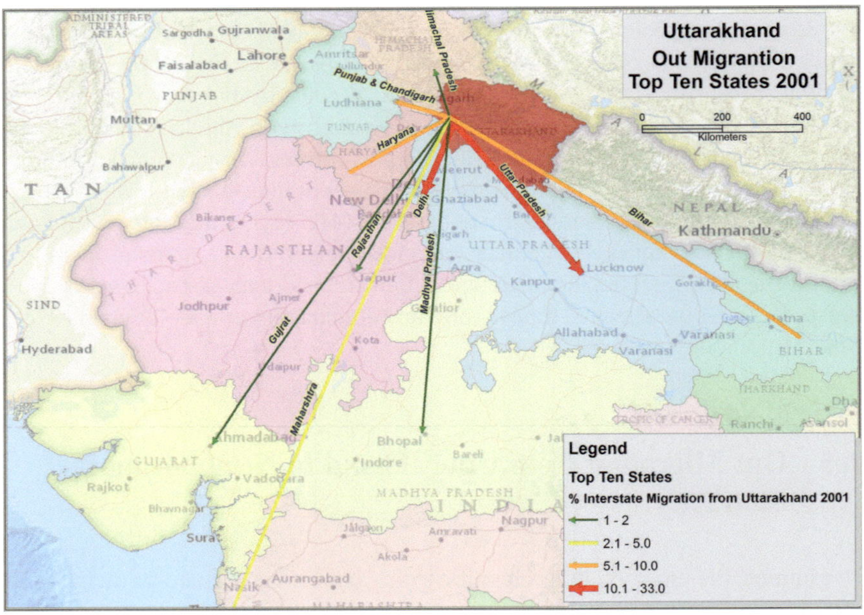

Fig. 16.3 Out migration from Uttarakhand to top ten ranked states in India, 2001

Uttarakhand. The main cities of concentration are Lucknow, Kanpur, Moradabad, Bareilly, Meerut etc. Next important state destinations are Haryana (7.3%) and Bihar (6.8%) ranking at the third and fourth place. Punjab (4.8%) and Maharashtra (4.4%) receive almost equal share of out migration from Uttarakhand. Chandigarh and Rajasthan are other two states receiving 3 and 2.1% migrants from Uttarakhand. The migrants from Uttarakhand to Himachal Pradesh (1.9%), Madhya Pradesh (1.6%) and Gujarat (1.1%) vary between 1 and 2%. The comparative picture is presented in Fig. 16.3.

16.6 Composition and Characteristics of Inter-district Migration in Uttarakhand

Inter-district analysis is important to understand population movement not only in terms of changing economic and political situations but also within the context of socio-cultural regional connections and existing traditional ecological systems. Census 2001 data on inter-district migration is the only data available to date for migration analysis. A total of 531,581 persons in the state were classified as inter-district migrants as per the Census 2001 which is 6.26% of the total 8,489,349 population of the State. The data presented in Table 16.3 at district level. It is a noticeable fact that the inter-district mobility among the females is appreciably higher being 3.58% compare to males (2.68%) in the State. While viewing the migrant population distribution at district level, Nainital received the maximum (15.22%) migrants followed by Dehradun (10.85%) (Table 16.3 and Fig. 16.4).

Relatively better infrastructural facilities and job opportunities available in these areas coupled with the tendency of continuous migration from hill areas towards the plains results into higher inter-district migration. Interestingly, the hill district Champawat with a partial population living in the foot hill areas has ranked third with 8.16% and Rudraprayag has occupied fifth place with 6.08% of the total district population. In this case, creation of new district seems to be a responsible

Table 16.3 Uttarakhand state-inter district migration (population classified by place of birth and sex), 2001

District	Total population	Place of enumeration—born in other districts of the state					
		Migrant population	%	Males	%	Females	%
Uttarkashi	295,013	21,123	7.16	8889	3.01	12,234	4.15
Chamoli	370,359	17,251	4.66	6860	1.85	10,391	2.81
Rudraprayag	227,439	13,828	6.08	2548	1.12	11,280	4.96
Tehri Garhwal	604,747	24,582	4.06	9261	1.53	15,321	2.53
Garhwal	697,078	25,596	3.67	11,342	1.63	14,254	2.04
Pithoragarh	462,289	17,949	3.88	6394	1.38	11,555	2.50
Bageshwar	249,462	14,147	5.67	2622	1.05	11,525	4.62
Almora	630,567	28,013	4.44	7306	1.16	20,707	3.28
Champawat	224,542	18,329	8.16	6236	2.78	12,093	5.39
Nainital	762,909	116,123	15.22	53,876	7.06	62,247	8.16
Udham Singh Nagar	1,235,614	65,308	5.29	30,404	2.46	34,904	2.82
Dehradun	1,282,143	139,111	10.85	69,428	5.41	69,683	5.43
Hardwar	1,447,187	30,221	2.09	12,420	0.86	17,801	1.23
Uttarakhand	8,489,349	531,581	6.26	227,586	2.68	303,995	3.58

Source Census of India, 2001

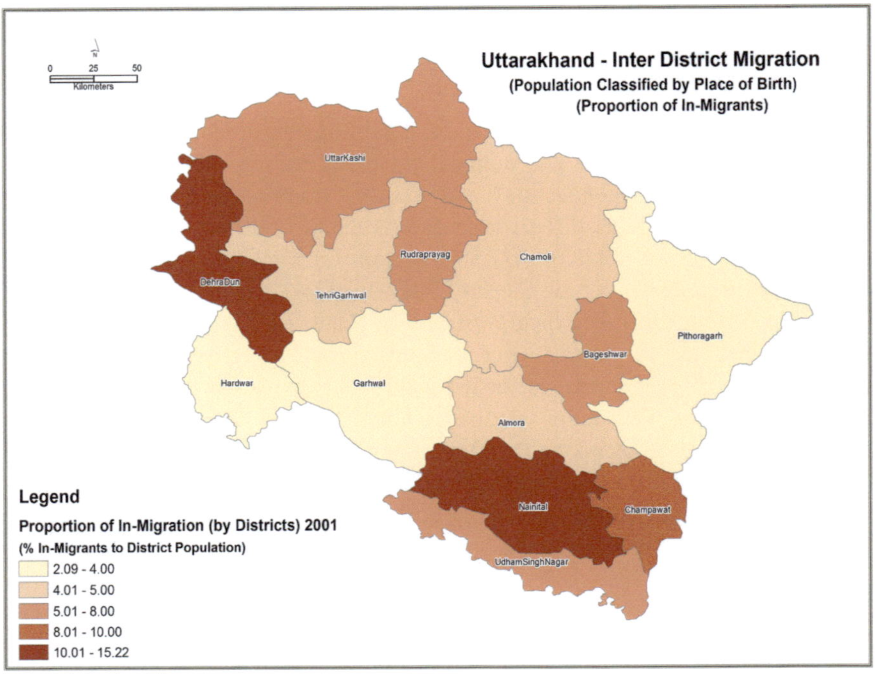

Fig. 16.4 Proportion of in-migrants by districts in Uttrakhand, 2001

factor since new infrastructural facilities have attracted people from neighboring districts.

Another hill district of Uttarakashi which ranked fourth with 7.16% inter-district migrants have local conditions and movement from and to neighboring districts is responsible. Further interesting fact is that district of Hardwar located in the plains has less inter-district migration population proportion compare to other plains districts. It seems that Hardwar even after being included in the state of Uttarakhand has not been able to attract in-migrants mainly because of a different socio-cultural setting than the hilly part of the state.

As noted before inter-district mobility among females is higher than males in almost all districts. Marriages are significantly performed at the inter-district level due to broader cultural interactions and mixed characteristics of socio-cultural groupings both in hills and plain districts. District Nainital ranks highest in both female (8.16%) and male (7.06%) inter-district migrant population in the state. Dehradun is at the second place following the same order with female (5.43%) and male (5.41%) migrants. The third rank is occupied by district Champawat which is a newly created district partly located in hills and partly in plains like that of district Nainital. It also has the highest female migrants (5.39%) followed by males (2.79%). The pattern is same in all other districts where a female dominated migrant population is common.

Males often move out of the district for employment and education. Inter-district mobility of males is not so pronounced as compared to females because the marriage is an associated feature with female migration. Apart from those strong economic reasons, social and cultural reasons are therefore also responsible for inter-district migration of population. However, there is a continuous migration of male and female active work force from hilly areas towards the plains and cities within and outside the State. There are immense local imbalances in the pattern of movement and inflow is concentrated in few areas for specific reasons. This is clearly reflected through the regional imbalances in the socio-economic development of the State.

The inter-district analysis of population movement in Uttarakhand shows a different picture when viewed from the regional perspective. The pull factor is dominant in district Nainital and Dehradun being infra-structurally and socio-economically more developed than their surrounding districts (Fig. 16.5). District Nainital pulls people mainly from Kumaun region while the latter does so from Garhwal region. The maximum number of in-migrant population is recorded in the State capital Dehradun with 1,39,111 persons consisting of a greater volume of in-migrants from Tehri (44,323) and Pauri Garhwal district (42,971). District Nainital with next highest in-migrants population of 1,16,123 persons draws a majority of its in-migrants from Almora (61,038), Pithoragarh (18,305) and Bageswar (12,161) districts. The remaining two districts located in plains in the state are Udhamsingnagar and Hardwar which fall in third and fourth place. Dominated by geographic and socio-cultural conditions, district Udhamsingnagar with 65,308 in-migrants receives maximum from Pithoragarh (24,681), Nainital (12,763) and Almora (11,427). Hardwar with 30,221 in-migrants receives a maximum from Dehradun (8877) and Pauri Garhwal (8373).

When the push factor is viewed, Almora ranks at the top in the volume of out-migration (99,279) from where a majority moves to Nainital district. The next ranking district in out-migration is Pauri Garhwal (76,403) followed closely by Tehri Garhwal (71,142) and Pithoragarh (70,685). The out migration from Pauri

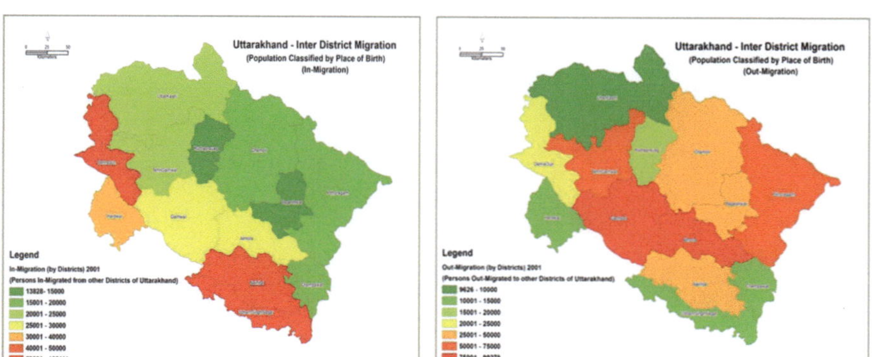

Fig. 16.5 Inter-district in and out migration by place of birth, 2001

Table 16.4 Inter-district movement of population by place of birth in Uttarakhand state, 2001

From	In-migrants													
To	Uttarkashi	Chamoli	Rudraprayag	Tehri Garhwal	Dehradun	Garhwal	Pithoragarh	Bageshwar	Almora	Champawat	Nainital	Us Nagar	Haridwar	Total
Uttarkash	0	284	95	4028	3715	687	98	9	42	16	210	93	349	9626
Chamoli	707	0	6194	1636	10,521	4722	327	988	1331	86	2732	1200	1138	3,1582
Rudrapray	362	5470	0	2806	3540	3471	77	22	36	16	122	62	363	16,347
Tehri Gar	13,104	1346	2563	0	44,323	6145	118	11	84	42	253	260	2893	71,142
Dehradun	2977	1835	254	4486	0	2681	245	32	328	121	1224	1240	8877	24,300
Garhwal	1811	2398	2127	7592	42,971	0	484	54	2314	273	5611	2395	8373	76,403
Pithoraga	185	523	42	284	3899	314	0	6237	5557	9399	18,305	24,681	1259	70,685
Bageshwar	56	1369	13	76	599	125	4708	0	6175	353	12,161	3042	238	28,915
Almora	264	2284	80	486	5445	2737	4791	5295	0	3551	61,038	11,427	1881	99,279
Champawat	21	72	6	54	251	67	3053	140	1262	0	3133	4742	321	13,122
Nainital	145	348	36	263	2392	863	1317	700	8540	2609	0	12,763	918	30,894
Udhamsing	55	117	2026	154	608	364	453	131	649	1443	8539	0	391	14,930
Hardwar	244	191	57	890	10,197	1583	67	14	129	80	623	662	0	14,737
Unclassif	1192	1014	335	1827	10,650	1837	2211	514	1566	340	2172	2741	3220	29,619
Total	21,123	17,251	13,828	24,582	139,111	25,596	17,949	14,147	28,013	18,329	116,123	65,308	30,221	531,581

Source Census of India, 2001

Note The total in-migrants are shown in the rows total and out migrants are in the columns total of the table

Garhwal and Tehri Garhwal districts takes mainly to Dehradun. The movement from Pithoragarh is mainly to Udamsinghnagar and Nainital. District Chamoli has recorded 31,582 out-migrants as per the Census 2001 (Table 16.4).

The Fig. 16.5 portrays the volume of inter-district movement of population highlighting the population moving in and out of the district. The dominant districts in movement are represented through darker shades and clearly reflect the regional pattern of movement and flow of out migration from Hill areas to the Tarai-Bahber and Plains region of the State.

It was found that the outflow is well pronounced in district Almora and Pauri Garhwal to the extent that these two districts have registered negative growth of population of −1.73 in Almora and −1.51 in Pauri Garhwal during 2001–2011. Other Hill districts with their poor infrastructural set-up and poor socio-cultural structure also represent push factor in dominance. The only exception in this case is district Uttarakashi with lowest number (9626) of out migrants compared to other hill districts in the State. There is an adequate reciprocity between the causes of both the 'in' as well as 'out' migration pattern. In general, the district of Dehradun, Nainital, Udhamsinghnagar and Hardwar have shown high inflow of population from the Hill district of the State, while outflow is a dominant feature of district Almora, Pauri Garhwal, Tehri Garhwal and Chamoli.

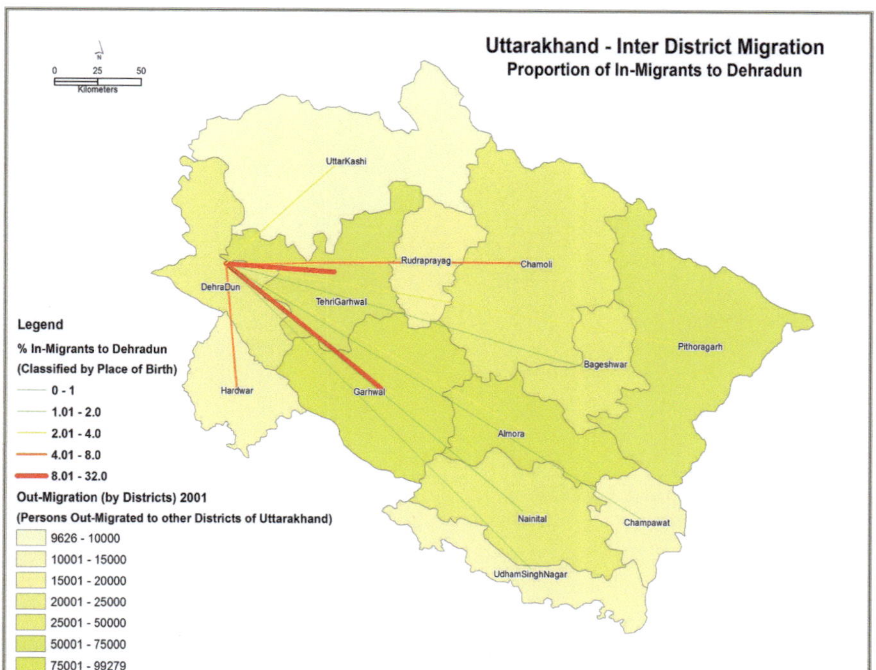

Fig. 16.6 District wise proportion of in-migrants in Dehradun, 2001

The facts of in and out movement of population from these districts is well reflected in the census 2011 with a marked increase in the population of Plain area districts which is about 31% more from the previous Census of 2001, while it has been just 3% in the Hill areas. The creation of new state more than a decade ago seems not been effective in checking this trend as appears from the results of Census 2011. Migration of people remains the biggest challenges faced by the State of Uttarakhand in its social and economic development. The population during 1981–2011 has continuously decreased in Hilly districts whereas increased throughout in case of Plains areas, especially in past two decades.

16.7 Regional Pattern of Movement

Kumaun and Garhwal are the two major regions for many well explained reasons which determine the regional flow of migration both in and out movement. The total in-migration from the Grahwal region is 51.11% followed closely by Kumaun region (48.88%) of the total migrant population of the state. The bulk of in-migration from Garhwal region is seen to Dehradun and Hardwar which together

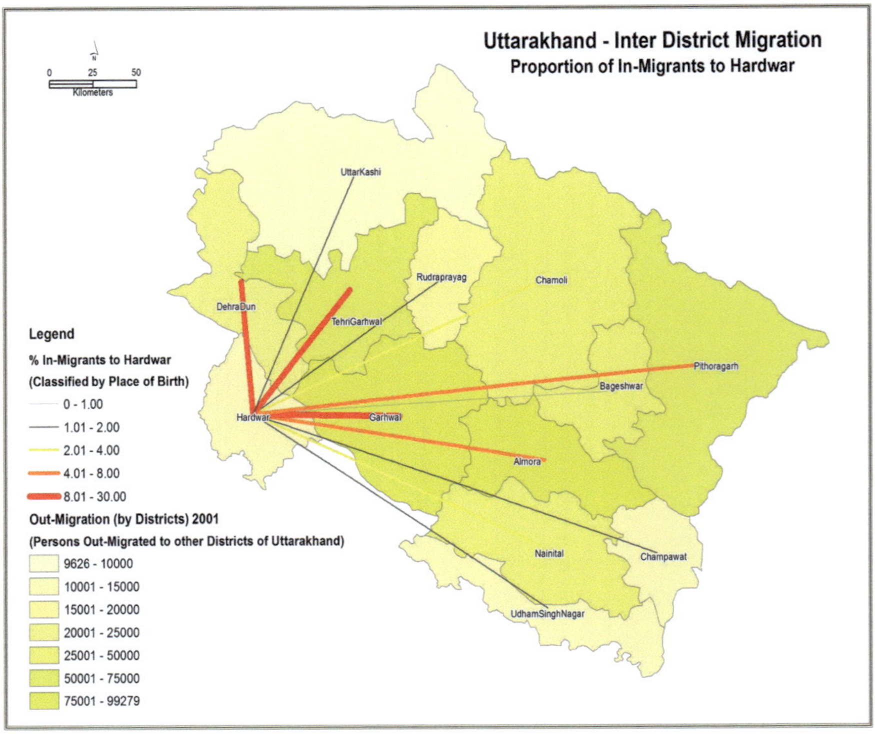

Fig. 16.7 District wise proportion of in-migrants in Hardwar, 2001

accounts for 62.32% of total Garhwal region. On the other hand, the share of in-migration to Nainital and Udhamsinghnagar from Kumaun region is little higher at 69.81%. Figures 16.6 and 16.7 represent the proportion of in-migrants to Dehradun and Hardwar. If we take the in-migrants of Dehradun that alone shares 26.17% of the total in-migrant population of the state. The bulk of these in-migrants come from Garhwal region (83.04%) and the rest 9.48 5% come from Kumaun region. In case of Hardwar, it shares just 5.68% of the total in-migrant population of the state. Hardwar draws 72.77% from Garhwal region and the rest 16.57% from Kumaun region.

The second most important gaining district of the state and first in Kumaun region in terms of receiving the in-migrant population is Nainital with a share of 21.84% of the total migrant population of the state (Fig. 16.8). It is worth mentioning that a majority of 89% in-migrants in Nainital district belongs to the Hill districts of Kumaun Region and less than 10% coming from Garhwal Region. The persons migrating from Kumaun to Dehardun and from Garhwal to Nainital are almost equal around 9%. Almost the similar pattern of in-migrant population is also observed in case of Udhamsinghnagar (Fig. 16.9). Udham Singh Nagar is second important pulling district in Kumaun region. A majority of in-migrants (86.75%)

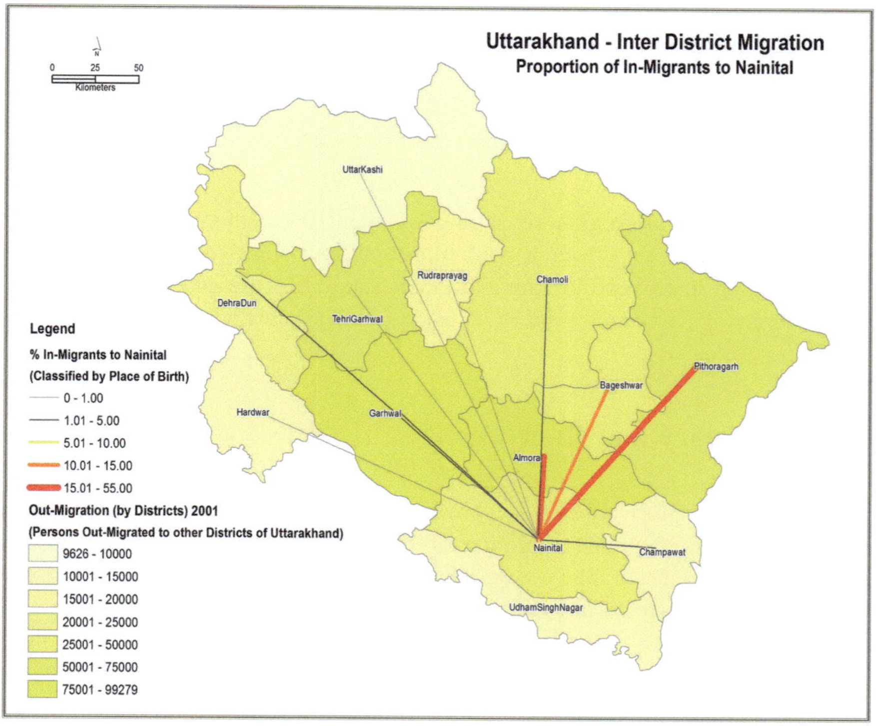

Fig. 16.8 District wise proportion in-migrants in Nainital, 2001

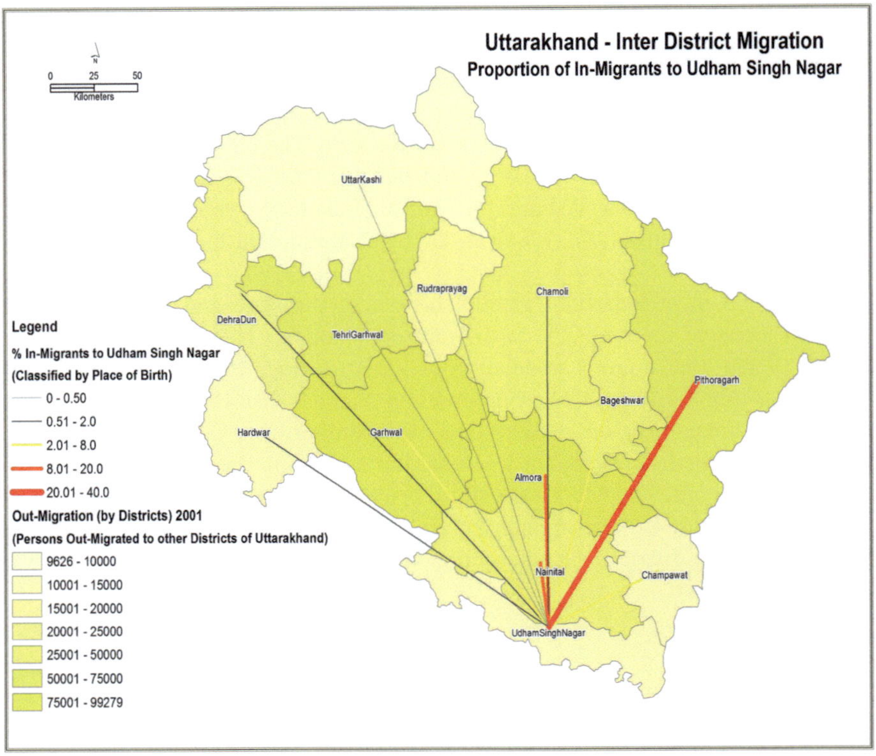

Fig. 16.9 District wise proportion in-migrants in Udham Singh Nagar, 2001

here are from Kumaun Region, remaining less than 10% coming from the Garhwal Region. Thus there has been a continuous outflow from the hilly districts of Garhwal region to its nearby Plain and Doon valley districts and an outflow from Hilly districts of Kumaun region to its adjoining Plain districts.

16.8 Summary and Conclusions

Rural-urban migration appears to be the predominant form of migration in Uttarakhand. The economic condition of migrants plays the main role whereas social and family aspects are more decisive in choice of destination. The majority of migrants are unmarried, young men in the age group 20–29 (about 42%) at the time of migration. Vast majority of migrants are literate, majority being in the educational level 10+2 or below. Most significantly, a good proportion of workforce goes to defence services constituting about one third of migrants from Uttarakhand to outside. In contrast to out migration from Uttarakhand there is noticeable in-migration of unskilled labourers from Nepal, and lately in-flow of semi-skilled and unskilled migrant workers, mainly as construction workers, is seen from UP and Bihar states.

The out-migration of active work force from hilly district of Uttarakhand state is the biggest socio-economic problem these areas face today. Not only the out migration is increasing year after year, but equally serious is the problem that the people who have once migrated are, most often reluctant to return to their origin; rather they become keen to stay and settle outside. The problem is not only an economic one; it imbibes sociological and psychological ingredients as well. In Uttarakhand, 29.84% of its total population is reported as migrants in 2001. The intra-district migrants are highest (50.15%) slightly below the country average, it is followed by inter-state migrants (28.8%) and inter-district migrants (17.66%). The State has 3.4% population classified as international migrants more than the National Average. The phenomenon of out migration is the key feature of the State. Such a persistent migration and its serious consequences are clearly seen in the prevailing socio-economic condition of the State. The flow of out migration from Uttarakhand to any part of India is primarily influenced by the traditional linkages established during the past and also according to the availability of jobs and employment opportunities present in the destination areas. The bulk of out migration from Uttarakhand which constitutes about 95% of total migrants is confined to ten Indian States only. Notably, majority one third is in Delhi alone followed by Uttar Pradesh (30%). It is assumed that more than 2% population of Delhi National Capital Region is from Uttarakhand.

The pull factor is dominant in district Nainital and Dehradun being infra-structurally and socio-economically more developed than their surrounding districts. At the regional level in the State, district Nainital pulls people mainly from Kumaun region while Dehradun does so from Garhwal region. The maximum number of in-migrant population is recorded in the state capital Dehradun dominated by in-migrants from Tehri and Pauri Garhwal districts. District Nainital next in rank draws a majority of its in-migrants from Almora, Pithoragarh and Bageswar. The other two districts located in plains areas of the State are Udhamsingnagar and Hardwar which fall in third and fourth in ranking by volume of in-migrants. District Udhamsingnagar with receives maximum in-migrants from Pithoragarh, Nainital and Almora whereas Hardwar receives a maximum from Dehradun and Pauri Garhwal. The one sided outflow and movement of people from hill areas in the State has marked negative impacts in the overall socio-cultural structure and economic conditions of many areas. The present trend of out-migration from Hills to Plans and Rural to Urban areas in the State is alarming and needs to be checked to save the socio-cultural fabric and base economy of the State. In near future, the main stay of the economy i.e. agriculture and allied activities are likely to come to a standstill. It has already affected agriculture productivity of the region which may result in food shortages, affect local cottage and small industries. There could be other very significant social and ecological issues affected badly due to migration of active force from the region although some researches also indicate that migration has indeed contributed towards better education and poverty eradication in Uttarakhand.

References

Census of India, 2001. D – Series: Migration Tables; Registrar General of India GOI, New Delhi.
Census of India, 2011. Primary Census Abstarct 2011, Registrar General of India, GOI, New Delhi http://www.censusindia.gov.in/
Bahuguna, P., & Belwal, O. K. (2013). Regression model approach for out-migration on demographic aspects of rural areas of Pauri Garhwal. *International Journal of Management and Social Sciences Research (IJMSSR), 2*(8), 175–182.
Belwal, O. K. (2007). Hill out-migration from Uttarakhand: Access to livelihood, education and other determinants of movement. *Demography India, 36*(1), 135–143.
Chand, R. (2009). *Population mobility behavior and external linkages of a village community in Eastern Bhutan* (pp. 117–138). Population Dynamics of Bhutan, An Occasional Publication of the Centre of Population and Development Studies, Sherubtse College, Royal University of Bhutan, Kanglung.
Chand, R. (2013). Labour migration as a livelihood strategy in Far East Bhutan: A case study of a marginal Bhutanese community. *Hrvatski Geografski Glasnik, 75*(2), 41–57, 9–23.
Chand, R., & Taragi, R. C. S. (2014). Population growth and distributional pattern in the Indian Himalayan region from 1901 to 2011. In M. S. S. Rawat & D. Pratap (Eds.), *Resource environment and development of the Indian Himalaya* (pp. 179–190). Srinagar (Garhwal), Uttarakhand, India: Transmedia Publication.
Chand, R., & Thakur, M. C. (1991). *Changing population profile* (pp. 19–23). Seminar, The Himalaya-378.
Deshingkar, P. (2006). Internal migration, poverty and development in Asia: Including the excluded. *Institute of Development Studies Bulletin, 37,* 88–100.
Esipova, N., Pugliese, A., & Ray, J. (2013). The demographics of global internal migration. *Migration Policy Practice, 3*(2), 3–6.
Grunawalt, P. (2012). *Why are cities the only place for dreams?* Outmigration of Youths From Rural Uttarakhand, Independent Study Project (ISP) Collection. Paper 1297, 44p. http://digitalcollections.sit.edu/isp_collection/1297
Jain, A. (2010). Labour migration and remittances in Uttarakhand (32p). Case Study Report, International Centre for Integrated Mountain Development(ICIMOD), Kathmandu, Nepal.
Lall, S. V., Selod, H., & Shalizi, Z. (2006). *Rural-urban migration in developing countries: A survey of theoretical prediction and empirical findings*. Development Research Group, The World Bank Policy Research Paper No. 3915, Vol. 1. Available at SSRN http://ssrn.com/abstract=920498
Lehtonen, O., & Tykkyläinen, M. (2009). Muuttoliikkeen alueelliset muodostumat ja pulssi Suomessa 1980–2006 (Regional formations and pulse of migration in Finland in 1980–2006). *Terra, 121*, 119–137.
Lehtonen, O., & Tykkylainen, M. (2010). Self-reinforcing spatial clusters of migration and socio-economic conditions in Finland in 1998–2006. *Journal of Rural Studies, 26*, 361–373.
Mamgain, R. P. (2003). Out-migration among rural households in Uttaranchal: Magnitude and characteristics. Labour and Development (special issue on migration), *9*, 259–287.
Martin, P. (2013). The global Challenge of managing migration. Population Bulletin, *68*(2), 2–16.
Plane, D. A., Henrie, C. J., & Perry, M. J. (2005). Migration up and down the urban hierarchy and across the life course. *Proceedings of the National Academy of Sciences, 102*(43), 15313–15318.
Srivastava, R., & Sasikumar, S. K. (2003). An overview of migration in India, its impacts and key issues. Regional Conference on Migration, Development and Pro-Poor Policy Choices in Asia. 22–24 June 2003, the Refugee and Migratory Movements Research Unit, Dhaka, Bangladesh.

Part IV
Policies and Strategies

Chapter 17
Tourism in the Swiss Alps: The Human Factor in Local Development

Walter Leimgruber

17.1 Introduction

Marginality is on the upsurge worldwide, both as a regional and a social phenomenon. Regions and people are vulnerable to multiple natural and societal processes and risk to become marginalized, i.e. overtaken by general mainstream development. As a result, regions may be forgotten and find themselves isolated, outside major communication flows. People will also become isolated and may end up in depression, thus putting additional stress on societies and their health system.

We contend that this is not necessarily the final state of the process of marginalization but that it can be reversed. However, it is a serious issue, in particular as it has been increasing over the past decades. Economic globalization is a major driving force, as it led to an increasing regional division of labour and fragmentation of the product cycles. It is accompanied by global political outlooks with less interest for local and regional issues, and both have consequences in the fields of society and culture, e.g. through the levelling of food patterns and cultural practices. In the field of languages, for example, the increasing use of English marginalizes other languages and dialects and thereby reduces cultural diversity. This can lead to the extinction of languages (see Kuntz 2009). The difficulty to meet the targets of the Millennium Development Goals (UN 2015) also shows that the social and economic mainstream passes over the needs of the less fortunate.

Marginality may be an undesired situation but it is not bad per se. It is part of the spatial and social differentiation that belongs to human societies and have been studied by geography for decades and from different perspectives (Leimgruber 1994, pp. 2 ff.). Marginality is a complex and scale-related quality (Leimgruber 2004, p. 37). Marginal regions may exist at one level and disappear at another, they

W. Leimgruber (✉)
Department of Geosciences, Geography, University of Fribourg,
chemin du Musée CH-1700, Fribourg, Switzerland
e-mail: walter.leimgruber@unifr.ch

R. Chand et al. (eds.), *Societies, Social Inequalities and Marginalization*,
Perspectives on Geographical Marginality,
DOI 10.1007/978-3-319-50998-3_17

may move from very marginal to less marginal, as Cravidão (1994) showed with respect to Portugal. The same holds good with mountain regions, in particular the Alps, which are the subject of this paper. It looks at the particular role individual actors play in helping small regions to get out of a marginal situation. Such efforts do, however, not guarantee that they are no longer vulnerable. In particular the emphasis put on tourism across the globe constitutes a risk as tourism is related to many factors, such as fashion, financial resources, overall economic situation, relative peace, weather etc.

17.2 Are Mountain Regions Marginal?

Mountain regions are often at risk to become forgotten corners of the world. Access is difficult, natural forces cause damage to human infrastructure and block access routes, the inhabitants lack contact with other people and work under difficult climatic and topographic conditions, etc. They were thus mostly considered marginal areas from the point of view of human occupancy (Gurung 1998, p. 36), an image they have largely retained despite modern technology (Leimgruber 1992, 1996). Living in and travelling through mountains still means to have to take nature into account—the many accidents and catastrophes in all mountain regions of the world bear testimony.

If we adopt a broader perspective, however, mountains are not necessarily marginal. They separate lowland regions from each other and thus invite people to overcome this obstacle and establish contact and exchange. Their geopolitical significance is highlighted through the wars that were fought to secure the pass routes and establish territorial control, and mountain people often rebelled against outside rulers who strived for the control of such routes and the income derived from transit traffic, fighting for their independence—the history of Switzerland is an example. The present-day conflicts in the Caucasus can also be seen in this light. To rashly define them as marginal because they are physical obstacles to land use and transit traffic is therefore to simplify a more complex reality.

One of the chief assets of mountains is doubtlessly their ecosystem, which is highly diverse according to the various levels of altitude (Gurung 1998, p. 35). It has generally suffered less from human activities than in lowland regions. The chief reason is the difficulty of access, but also that they are relatively hostile to man. As a consequence, mountain regions still maintain a high degree of biodiversity and are therefore attractive for nature lovers, recreationists and tourists (Leimgruber and Imhof 1998). In Switzerland, an estimated 19–20 million tourists stayed in accommodations in mountain regions (hotels, guest houses, youth hostels, campsites) in 2014, a figure which is likely to be substantially higher because second homes are not contained (BFS 2015, p. 51).

Since the 19th century mountains, in particular the Alps were opened up through improved transportation, emerging tourism, and the internationalization of the economy, which led to an increasing influx of people from the outside. The late 19th century saw the first tunnels dug through the Swiss Alps (the Gotthard railway tunnel, 1872–82); further tunnels for both road and rail were constructed during the 20th and early 21st century. The arrival of tourists triggered off another transport revolution: mountain railways (the first mountain cog railway of Europe reached the Rigi in 1871), cable cars, funiculars, chairlifts etc. were built—the mountains had definitely lost the image of terror. Tourists also appreciate nature reserves, for which mountain regions are particularly suited, given the generally low human impact, and it should be kept as low as possible (Leimgruber 2012).

However, tourism is not an option everywhere in the mountains, and its benefits are unevenly distributed. In most regions difficult living conditions persist. Mountain farming, the major activity before the arrival of tourism, is a very tough business, both physically and mentally: it includes hard work on harsh terrain for relatively low returns, and it operates under the constant threat of natural disasters such as floods, landslides, and avalanches. If the land is not properly managed, spontaneous vegetation will quickly establish itself and reduce potential agricultural surfaces. Working under such stress requires the mental strength to continue with a laborious life, a strength the younger generation does not necessarily possess. However, tourism has a positive side in that it enables people to continue living in their villages instead of becoming migrant workers; this, at least has been found in a remote Moroccon village (Metzger 2014, p. 54 f.).

Mountain regions, in Switzerland as elsewhere in Europe, have experienced substantial emigration since the 19th century. It was especially young people who left for a better life in the lowlands and in the cities. The exodus has led to considerable depopulation (Majoral 1984; Schuler 1984; Weatherley and Naylon 1984), with the remaining population consisting primarily of old people who did not want to move because of their attachment to the land and the difficulty to adjust to a new lifestyle away from home. Needless to say that the public infrastructure in such regions is at its minimum and not likely to be improved much. Rather have the populations to fight for the maintenance of public transport in order to ensure a minimum connection to the outside world. Survival has therefore become a major challenge, both to local populations and politicians on regional and national levels.

17.3 Survival in Remote Alpine Valleys

The Alps have been both an obstacle and an incentive to transport and conquest ever since pre-Roman times. Trade across them has been documented as far back as the Bronze Age (amber or fossilized tree resin from the Baltic has been used in the Mediterranean), and Hannibal's crossing in 218 BC is legendary. In the Roman period, but also during the Middle Ages, the alpine passes were important communication routes to maintain the unity of the various empires (Roman, German). However, settlement remained scarce, and people did not necessarily enjoy living in the mountain environment. Thomas Platter (1499–1582), humanist scholar, teacher and professor for Hebrew and Greek at Basle University, was born in the Valais and worked as a goatherd in his youth. In his autobiography, one of the few surviving literary witnesses that tell us about people's perceptions of the mountains in the 16th century, he complains of having had to walk up and down the high and cruel mountains (Leimgruber 1992).

It is therefore understandable that lowland areas, in particular cities, attracted people from mountain areas by offering better comfort and a regular life and salaries—Platter is no exception. This drain reduces their economic and development potential and increases their marginality. They usually became lost corners (*angle mort*, Reynaud 1981), which is one possible case for a marginal region (Leimgruber 2004, p. 42 ff.).

However, such a process is reversible if local initiatives coincide with lucky circumstances. It is often an individual agent that may help to put a place back on track in what seems to be a hopeless situation, compensating an isolated location by the development of a specific resource. Such occasions may be rare and not lend themselves for generalization, but each constitutes an example in itself that together show that ways out of seemingly desperate situations are possible.

Such is the case with the three examples to be discussed below (Fig. 17.1). They illustrate how local resources such as landscape and mineral springs have been and are put to human use, but also that expectations, aspirations, and sometimes excessive zeal are involved in such valorisations. While the impact of the first project remains to be seen, the second case demonstrates how even a remote village manages to survive. The third example illustrates the difficulties associated with exaggerated aspirations. They are all proof that humans may achieve something, but that we also have to calculate with human weaknesses.

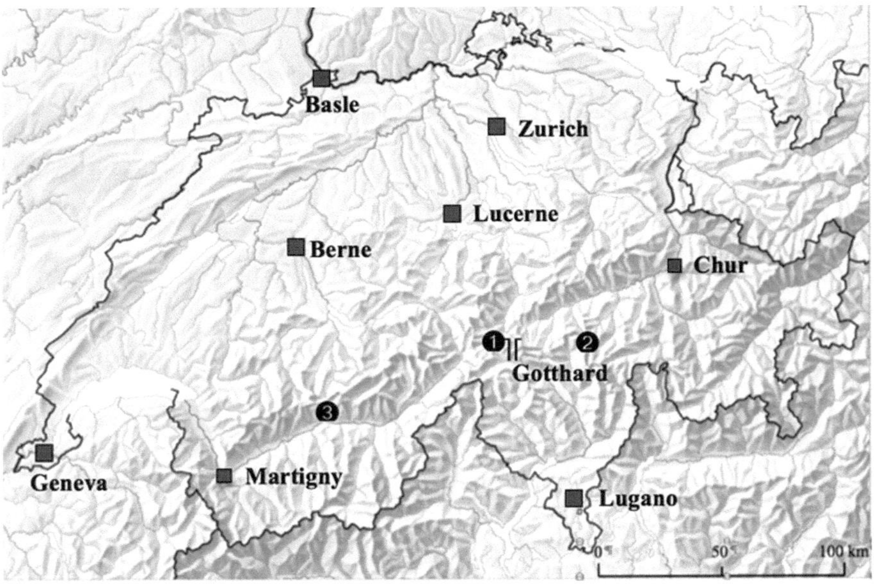

Fig. 17.1 Location of the three case studies (*1* Andermatt, *2* Vals, *3* Leukerbad) (Map base: Schweizer Weltatlas, EDK 2004)

The three case studies will then be put into the context of marginality theory to demonstrate that even individual cases are in principle not individual but part of a wider picture.

17.4 Andermatt: Transformation of the Cultural Landscape

The case of Andermatt in the canton of Uri offers a telling example of the ups and downs of a village, or how disinvestment leads to marginality, and investment can help to get out of it. The current situation is but a repetition of a process that has already taken place some time ago, but that then came to an end. An Egyptian investor is transforming the village into an upmarket tourist resort, benefitting from the availability of land and the beauty of the landscape. The story sounds like a fairy tale; however, the project is well on its way, the first buildings have been finished, and the resort has become operational with the opening of the central hotel in Andermatt in December 2013 (Orascom Development 2013).

The village of Andermatt (1447 m asl.) is in a crossroads situation, lying in the Urseren section of the alpine longitudinal valley that reaches from Chur in the Grisons to Martigny in the Valais, and on the European north-south axis between Basle and Lugano, the Gotthard pass route. When this pass was opened to transit traffic in the early 13th century, Andermatt became an important resting and animal changing place. The farmers also raised cattle and exported it to Northern Italy. Transit traffic offered various jobs to the local population and thus ensured permanent settlement, despite difficult living conditions in winter. However, the construction of the Gotthard railway tunnel in 1882 radically changed the situation: transit traffic disappeared almost overnight and the valley faced depopulation. When the motor-car became fashionable in the early decades of the 20th century, the pass route regained part of its former importance, and life continued to return to the village. The construction of a narrow gauge railway 1911–1925 that crosses from the Valais to the Grisons in an east-west direction also helped to stop decline, in particular as a rail link to the Gotthard railway in Göschenen had been constructed in 1917 (Zurfluh 2009).

Another important element was the arrival of the Swiss Army. From 1886 onwards, it started to construct the Gotthard fortress complex, and Andermatt became a cornerstone and chief garrison of this defence system. Soldiers were present all the year round, and the local economy boomed. The opening of the Gotthard motorway tunnel in 1982 did not affect the region very much, because the pass route still retained its attractiveness and has often to be used when the tunnel is temporarily closed. The presence of the army seemed to guarantee the long term survival of the valley.

However, the end of the Cold War in 1989 proved detrimental. The army lost its strategic role and has been in a constant reform process ever since. The fortification complex became superfluous, and the army began to withdraw from Andermatt. All of a sudden, the material basis for the survival of the population disappeared again. The situation began to resemble that a century before when transit traffic across the pass was abandoned because of the railway. The village and indeed the entire valley began to fear for its existence.

The Urseren valley comprises three municipalities and mirrors the demographic processes that can be encountered in the Alps (Table 17.1). Andermatt manages to retain or even increase its population, whereas the smaller and remoter villages are steadily losing it. Even if the situation does not look too dramatic, the inequality in the population distribution is evident.[1] Andermatt seems to have overcome the demographic shock caused by the departure of the army.

Who would come to the rescue of a community that again faces depopulation? Traditional mountain farming in a relatively dry valley with steep slopes is not an alternative, and tourism faces fierce competition from other resorts in the Swiss Alps which have much more of an international reputation than the Urseren valley and manage to attract people from around the globe. Andermatt was a spot on the map, with some skiing areas mainly of regional importance but little else to be a magnet for tourists.

Around 2005, the Egyptian investor Samih Sawiris discovered the Urseren valley more or less by chance. He had gained experience in building holiday resorts in his home country (e.g. El Gouna on the Red Sea) and was thrilled by Andermatt and its potential. He discussed his plan with local and cantonal authorities as well as with the local population. His situation was delicate: as an Egyptian he faced considerable distrust, partly dispelled by the fact that he is a Christian copt. His knowledge of German, acquired at a German school in Egypt and at university in Germany, enables him to talk to the people directly. In this way he could secure acceptance for this mega-project, which went under way in 2009. Luxury apartments instead of army barracks, a golf course in the place of rifle-ranges—what a particular way of demonstrating the change in the European political landscape.

In its current stage the project Andermatt Swiss Alps "is planned to offer 490 apartments and 25–30 private villas. In addition, 6 hotels classified as 4 and 5 star with a capacity of 844 rooms are planned." (Orascom Development 2008). This will lead to a radical transformation of the landscape, both in the physical and in the cultural sense. The project has been exempted from the so-called *Lex Koller* that restricts the purchase of land property by residents abroad, and thus an important legal obstacle has been lifted. The resort will increase the number of second homes, which runs contrary to a new article in the Swiss Constitution that limits the amount of second homes to 20% of the total dwelling stock (measured by gross surface). This article was introduced following a referendum in March 2012 (with a narrow outcome of 50.6% affirmative and 49.4% opposing votes), but the essential building licences had already been delivered.

[1]To maintain two small municipalities of less than 200 inhabitants looks like a luxury, but in a country where local autonomy is sacred, a merger looks highly unlikely.

Table 17.1 Population evolution of the Urseren valley, 1850–2013

Municipality	1850	1900	1950	2000	2013
Andermatt	677	818	1231	1335	1393
Hospenthal	424	290	282	233	202
Realp	203	208	186	177	148
Total valley	1304	1316	1699	1745	1743

Source Swiss statistics

Physically, there will be a substantial expansion of the built-up area and a reduction of agricultural activity. An 18-hole golf course has replaced the former army training grounds and partly surrounds the remaining agricultural zone on the valley floor. Culturally, the local population will be confronted with a massive influx of new (temporary) upper middle and upper class inhabitants, but the project will also create new jobs. It may well be that the demography of the entire valley will receive a boost in the long run.

In April 2011 it emerged that the Swedish firm Skistar (that runs several winter resorts in Scandinavia) wants to invest in the same region (including in neighbouring Grisons), cooperating also with the Sawiris project (NZZ 2011). In particular, it wants to join the Andermatt skiing area with the winter sports arena of Sedrun to the east, which would enhance the attraction of the new Andermatt resort. Major landscape transformation will occur as well as an increasing pressure on the ecosystem. First political conflicts emerged in the course of the year, and in March 2012 there were rumours that Skistar was going to withdraw from the project (Merki 2012). However, this engagement had been put on hold until the building permits had been delivered. This occurred in summer 2014, and work on the ski-arena Andermatt-Sedrun is about to begin (Andermatt Swiss Alps 2014).

17.5 Vals: Valorising a Local Resource or the Problems of Teamwork

The mineral source of Vals demonstrates how a remote village becomes a well-known place both as concerns bottled water and a spa. The village lies at the end of the road of a tributary valley to the Vorderrhein in the Grisons on 1254 m asl. The nearest regional centre is Ilanz in the Vorderrhein valley, 698 m asl. and 22 km away—but a proper road was only built in 1878–79 (Simonett 2006). Before that date, the village was oriented southwards, towards the Hinterrhein valley. Even nowadays access to the village is difficult because of the valley topography, and in winter, there is a risk of avalanches that may cut it off from the rest of the world.

Vals (971 inhabitants 2013) was probably settled in the middle of the 14th century (Kreis 1966, p. 82) by colonists who had originally emigrated from the upper Valais (Fig. 17.2). These people were part of an important south- and eastward oriented migration in the Middle Ages, known as the Walser migration and

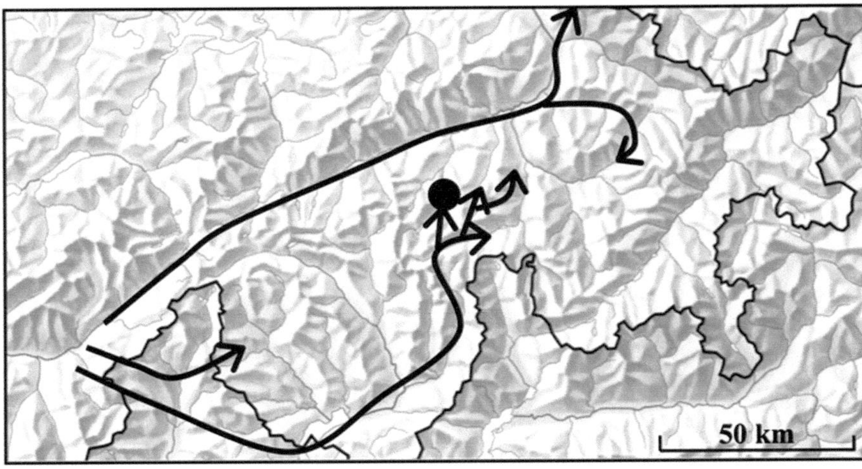

Fig. 17.2 The Walser migration and the colonisation of Vals

reaching as far as the Rhine valley and parts of western Austria. The newcomers, self-sustaining mountain farmers, settled especially in high areas that had not been taken over by the populations lower down in the valleys. The inhabitants of Vals also maintained an intense local and regional trade across the Valserberg pass into the Hinterrhein valley (Simonett 2006). Contrary to the lowland settlers who spoke Romansh, they have maintained their Germanic dialect until now.

The municipality's economy is based on three indigenous resources: stones, hydropower and mineral water (both as drinking water for export and in the spa, which gives rise to a modest year-round tourism). The mineral water is the best known product of the village and has been known since prehistoric (Bronze Age) times, probably mainly for drinking. The systematic marketing of the bottled mineral waters of the St. Peter's Spring began in 1961. A Spa has been documented as early as 1732, a large centre was opened in 1893 and was extended in 1970 (ibid.).

In 1983 the municipality acquired the spa complex in order to save it from bankruptcy and replaced it by a new building that was opened in 1996 (Krummenacher 2010). Designed by a well-know architect, Peter Zumthor, local stone was used for the construction, The collaboration between the architect and the owner of the local quarry became the cornerstone of the development of the entire municipality. More recently, fractures appeared in this cooperation, and the two protagonists stopped talking to each other. This is not unusual when long-term teamwork has to guarantee the survival of a community. Humans evolve over time and change their minds, and dissent may result from diverging opinions. If the future of the project is at stake, such disagreement will be counterproductive; in the present case, however, the future of the resort seems to be assured despite momentary divergences. Both chief protagonists have invested not only money but also enthusiasm and emotions into the project. Besides, the village has grown considerably around the water business with attached tourism, something that both the quarry and the power station alone could not have achieved. It is therefore a sign of hope that the spirit of teamwork has not completely disappeared, even if the village threatened to be split between two parties, the one siding with the architect, the other with the local protagonist (ibid.).

The story came to an end (hopefully) on March 9, 2012, when the municipal assembly decided to sell the spa to an outside investor (from the Grisons) and not to the architect who had constructed the modern complex. The municipality sold the complex because the renovation of the hotel would exceed its financial capacities (Krummenacher 2012). Further investment to enhance the village's tourism potential is not excluded.

17.6 Leukerbad: Individual Actions or How Far Should One Person Go?

In the late 1990s, the municipality of Leukerbad had been at the centre of a financial scandal that led to it being administered by the Canton between 1998 and 2004. Its president had overstretched his financial competences when he invested heavily into infrastructure. In order to keep these machinations secret, he manipulated the municipal accounts. Forced to obtain the necessary capital on the free market, the municipality eventually found itself with a debt of 346 million Swiss francs (Lötscher and Zulauf 2002, p. 5), this is roughly 200,000 francs per inhabitant. The former president was sentenced to 5 years in jail for his illegal manipulations.

Like Vals, the village lies at the end of a tributary valley, in this case of the River Rhone, and on 1402 m asl., 14 km from the railway station of Leuk (623 m asl.) and the main valley road. The nearest regional centres are French speaking Sierre (16,000 inhabitants, 20 km away) and German speaking Visp (7000 inhabitants, 33 km away). Leukerbad lies at the southern end of the Gemmi pass route, an important link (on foot) from the Valais to the Bernese Oberland. Much frequented during the Middle Ages, it lacked a well-defined transport organization, and the route was of local/regional importance only (Aerni 1999). Today, this pass is a popular hiking trek.

Leukerbad (1532 inhabitants 2013) had been a well-known spa since the late 15th century, but in the late 20th century it was at a comparative disadvantages compared to other spas that were better equipped. It was this deficit in infrastructure that triggered off the heavy investments and eventual near-bankruptcy of the municipality. The idea was good but the entire process obviously got out of control of the main protagonists. Local personal interconnections and related corruption played an important part in it. Again we are confronted with human weakness.

As a result of the heavy investment, Leukerbad became an important spa destination, and from this point of view, the policy certainly played in favour of the municipality. The number of inhabitants had begun to rise significantly after 1960,

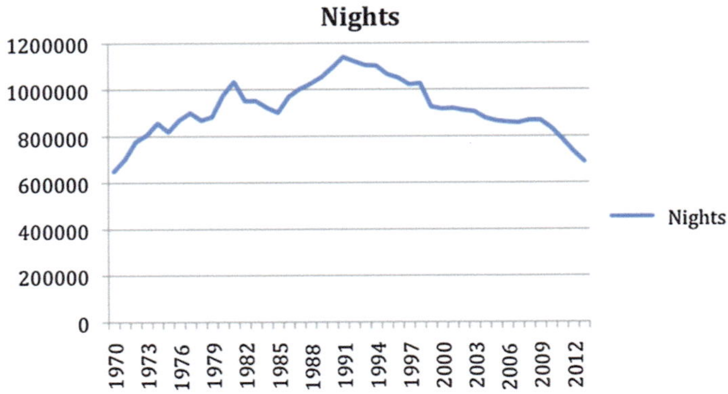

Fig. 17.3 Leukerbad: nights spent, 1970–2013 (Leukerbad Tourism Annual reports)

when the economy boomed, private transport became more common, and people increasingly drove to tourist destinations. Similarly, the arrival of visitors continued to rise, as the statistics of nights spent in hotels and other accommodation tell (Fig. 17.3). However, tourism remains a fragile industry, very much dependent on the economic situation at a given time. In Switzerland, there is a lot of competition between different spas when it comes to pleasure bathing as well as to medical treatments. In the case of Leukerbad, certain local spas are even in competition with each other, which may not be in the general interest. The peak number of guests was recorded in 1991 (1,139,594), but it was after 1998 that their number suddenly dropped sharply, and the drop has become dramatic after 2009. Whether this is related to the financial scandal or not remains an open question, given the uncertainty in the tourist business.

At the end of 2013, the municipality still had a debt of about 192 million Swiss francs or 116,696 francs per inhabitant (Munizipalgemeinde Leukerbad 2014), and it has been reducing it every year. However, this massive debt reduces the range of action for investments in the future. The canton imposes a limit of 900,000 francs on net investments, i.e. most investments have to be financed through income.

17.7 Comparison

Although the three villages are different from each other, there is at least one similarity. They all lie deeply inside the Alps, but whereas Andermatt is basically in a favourable situation as concerns road and rail traffic, the other two represent dead ends in valleys and can only be reached by road.[2] They all lie relatively high and

[2]Leukerbad had been connected to the main valley by a tramline: opened in 1915, it was scrapped in 1967 (Kalbermatter 2008).

access is difficult due to narrow and steep gorges that have to be overcome. Under these circumstances Vals and Leukerbad could be classed as marginal in the sense of Reynaud's (1981) lost corners in that local interactions would dominate over relations with the exterior, whereas Andermatt was an integrated periphery.

The valorisation of these villages has been based on several resources. In the case of Andermatt it was the location in the Gotthard region, first as an important resting place on the transit route, then as the pivot in the central defence system (the Gotthard fortress); nowadays it is the existence of large surfaces that have been abandoned by the army and could be re-allocated to a new role. They present an additional motive for development, because the valley was only marginally attractive for tourism; tourist activity was of little importance compared to the army. Prior to the arrival of Mr. Sawiris, nobody knew what would happen to the community and even the entire valley after the chief employer (the army) and the customers (the soldiers) had left.

In Leukerbad and Vals, the presence of mineral and thermal springs had given rise to a spa culture long before the 20th century. Both places, however, developed only in the second half of the last century, when private motor-cars had become the norm, given the difficulty of access. Leukerbad has only one resource (its thermal waters) and is therefore in a particularly vulnerable situation. Vals, on the other hand, has also quarries for building stones and a few jobs linked to the exploitation of its hydroelectric potential. The spa, important as it may be, is one of several sources of income. However, it is important to look well after it. The mineral water has a good position on the Swiss market, but since it has been sold to the Coca Cola Company in 2002, decisions are no longer taken on the local level.

The human factor, finally, varies from one village to the other. Andermatt is expecting wonders from a foreign investor, but external money is not necessarily better and safer than domestic one, and recent events demonstrate how dependent the village is on external decisions. In Vals and Leukerbad, local actors pushed change and made the two villages famous. However, human nature is not constant. Vals has been experiencing internal differences between two main protagonists, and only the sale of the spa could restore peace (at least temporarily). Leukerbad has become the victim of one person's exaggerated aspirations, but since the case has been resolved, the village is on a good way out of its desolate situation.

17.8 Conclusion

Let us confront these three examples with marginality theory. There are some points they have in common, but there are also differences from this perspective (Fig. 17.4).

The three villages are all in peripheral to marginal locations, characterized by difficult access, harsh physical environment, low economic potential, emigration of the young generation and an ageing population. They can be considered *geometrically* marginal, but this is only one side of the story. Andermatt lies in a key

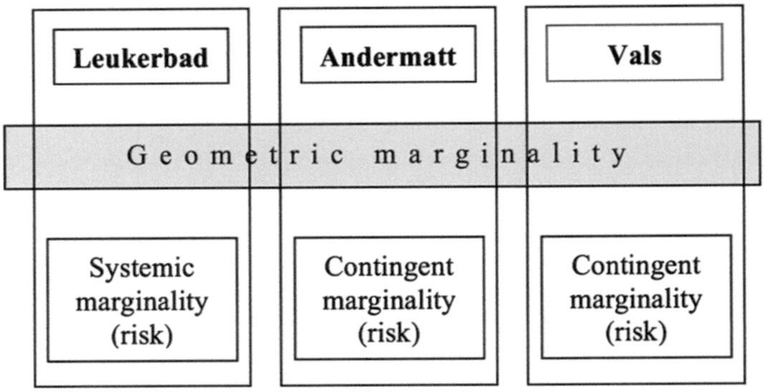

Fig. 17.4 Marginality types of the three case studies

region for transcontinental transit, yet modern technology has transformed this location from an asset into a handicap. Visiting Andermatt and the Urseren valley nowadays requires an extra effort, and the same holds good for Vals and Leukerbad, both in dead end locations.

This simple physical fact is a constant. However, the villages were not always marginalized if we adopt other theoretical perspectives. Marginalization is a process that can be reversed when circumstances change. It is therefore important to include the temporal dimension into our considerations. The look back has to be completed by a look into the future.

Historically, they all participated to a varying degree in transit trade, with Andermatt in a privileged position on a major European north-south route, whereas Vals and Leukerbad responded more to regional trade. It was the end of transport on foot, mules and (in the case of Andermatt) horse-drawn coaches that created situations of marginality. Vals and Leukerbad completely lost their significance for trade, whereas Andermatt could resume it to some extent thanks to the motor-car. When it opened in 1882, the Gotthard railway tunnel facilitated international transit transport, and the pass route lost its role. In the early 20th century, the motor-car put it back on the map, and the villages along the entire route (from Uri to Ticino) profited again from road transport, albeit mainly tourists passing through on their way to Ticino and Italy.

In the case of Andermatt, we are confronted with an example of *processual* marginality (Mehretu et al. 2002) due to two different processes. The first is technological progress, i.e. the replacement of roads across passes by the railway and (since 1982) the road tunnel, which facilitate and speed up transit transportation. In rare cases, when the Gotthard road tunnel is closed because of an accident or for maintenance, all road traffic will have to take the pass route again. However, unless drivers and passengers stop for a drink or petrol, the valley will only receive the nuisance of transit traffic, but hardly any benefit.

The second process is of a geopolitical nature. The collapse of the Soviet Union resulted in a new security situation across Europe and worldwide. The revaluation of the role of the army led to its departure from Andermatt. This is a case of *contingent* marginality: the village and the valley have no means to change the two basic conditions (technology and geopolitics), they are in a competitive disadvantage (ibid., p. 197). Andermatt's contingent marginality can only be reverted if Mr. Sawiri's entrepreneurial spirit will bring new life into the municipality. However, tourism at this scale is no longer a local or regional but an international business (with all the problems entailed), so there is still a risk that the village continues to be marginal.

Vals and Leukerbad are currently not marginal from this theoretical perspective, but they are in a risky situation. In both cases, the human factor (human weakness) is all decisive. If people become dependent on one or two leading figures who (overtly or indirectly) dictate the life of a community, there is the risk of *systemic* marginality. It occurs when dominant persons do not keep the common good in mind but distribute benefits unequally (ibid., p. 198). This type of marginality loomed over Leukerbad (and the drop in guests since 1998 may be an indicator), but has probably been avoided through a rigorous administration of the local accounts. Vals may stay away from a special form of *contingent* marginality if the protagonists and their partisans manage to overcome their divergences and continue to cooperate, but as long as the new owner of the spa makes clear statements the risk persists.

It has been said before: the three villages owe their relative prosperity to tourism, but this is a doubtful business, despite the fact that recreation occupies an important position in people's lives and budgets. Economic conditions and the potential guests' whims and fancies contribute to this uncertainty: leisure is also subject to fashion. Understandable, therefore, that competition is high in this field and marginalization a constant threat.

In all three cases the human factor plays a decisive role. We are confronted with individuals who venture into projects whose outcome is unknown. The public good is usually quoted as the chief goal, but behind it we often discover personal interests, be they financial, power, or simply prestige. This driving force had (or seems to have) positive impacts, but it can unexpectedly also lead to negative outcomes. Indeed, the examples also demonstrate the fragility of such individual initiatives. Two factors have to be considered: (1) Nobody can foretell the future of the economy (or anything else), and (2) humans are mortal beings and not necessarily consistent. For these two reasons the succession of projects in general is not automatically guaranteed. It is imperative to have a person or an organisation at hand if such a personal project is to thrive in the long run. De-marginalization may be the temporary result of such initiatives, but re-marginalization may result from a lack of long-term outlook.

The three examples demonstrate the many faces of the marginality concept. Geometric marginality need not be a handicap, it is what humans make out of a particular situation that counts. Factors contributing to marginality are both internal and external: internal if the social actors deviate from a common goal (resulting in systemic marginality), external if through technological innovations and/or the general economic situation the local framework is upset and contingent marginality will be the outcome. Both can be avoided, as the three examples have demonstrated.

References

Aerni, K. (1999). Die Gemmi—Von der Verbindung zum Weg. *Cartographica Helvetica, 19–20*, 3–15.

Andermatt Swiss Alps. (2014). SkiArena Andermatt-Sedrun: Auftragsvolumen von rund 100 Mio. ausgelöst. Press Release 13 August, www.andermatt-swissalps.ch/. Accessed September 23, 2014.

BFS. (2015). *Schweizer Tourismusstatistik 2014*. Neuchâtel, Bundesamt für Statistik (Swiss Federal Statistical Office).

Cravidão, F. D. (1994). A marginal country in the European Community or a European country with marginal regions? In C.-Y. D. Chang, S.-C. Jou, & Y.-Y. Lu (Eds.), *Marginality and development issues in marginal regions* (pp. 179–194). Taipei: National Taiwan University.

Gurung, H. (1998). Sustainability and development in mountain tourism. In P. East, K. Luger, & K. Inmann (Eds.), *Sustainability in mountain tourism. Perspectives for the Himalayan countries* (pp. 29–45). Innsbruck-Vienna: Book Faith, Delhi & Studienverlag.

Kuntz, L. I. (2009). A subtle interplay of forces. *UNESCO Courier* No. 2, p. 3.

Kalbermatter, P. (2008), Leukerbad, in Historisches Lexikon der Schweiz. http://www.hls-dhs-dss.ch/textes/d/D1461.php. Accessed April 02, 2011.

Kreis, H. (1966). *Die Walser. Ein Stück Siedlungsgeschichte der Zentralalpen* (2nd ed.). Bern: Francke.

Krummenacher, J. (2010). Valser Wirren. Mit oder ohne Peter Zumthor—der Tourismusort stellt die Weichen für seine Zukunft. *NZZ*, February 16, 2010.

Krummenacher, J. (2012). Vals bricht mit dem Architekten Peter Zumthor. *NZZ online*. http://www.nzz.ch/nachrichten/politik/schweiz/vals_bricht_mit_dem_architekten_peter_zumthor_1.15636092.html. Accessed March 10, 2012.

Leimgruber, W. (1992). Man and the mountains—The mountains and man: The changing role of mountain areas. *Occasional papers in geography and planning* (Vol. 4, pp. 121–135). Appalachian State University Boone N.C.

Leimgruber, W. (1994). Marginality and marginal regions: Problems of definition. In C.-Y. D. Chang, S.-C. Jou, & Y.-Y. Lu (Eds.), *Marginality and development issues in marginal regions* (pp. 1–18). Taipei: National Taiwan University.

Leimgruber, W. (1996). Telecom and marginality: Modern technology to prevent marginalization? In R. B. Sing & R. Majoral (Eds.), *Development issues in marginal regions. Processes, technological developments and societal reorganizations* (pp. 1–12). New Delhi: Oxford & IBH Publishers.

Leimgruber, W. (2004). *Between global and local. Marginality and marginal regions in the context of globalization and deregulation*. Aldershot: Ashgate.

Leimgruber, W. (2012). National parks and protected areas—policies and strategies in Switzerland. In W. Leimgruber, W. Zsilincsar, & E. Nel (Eds.), *Mountain regions in transformation. Global processes, regional and local impacts and responses* (pp. 223–238). Aachen: Shaker.

Leimgruber, W., & Imhof, G. (1998). Remote alpine valleys and the problem of sustainability. In L. Andersson & T. Blom (Eds.), *Sustainability and development. On the future of small society in a dynamic economy* (pp. 385–396). Sweden: University of Karlstad.

Leimgruber, W., Zsilincsar, W., & Nel, E. (Eds.). (2012). *Mountain regions in transformation. Global processes, regional and local impacts and responses*. Aachen: Shaker.

Lötscher, D., & Zulauf, B. (2002). *Die Finanzkrise in Leukerbad*, Seminararbeit in Finanzwissenschaft, Faculty of Economics and Social Sciences, University of Fribourg/CH (available online as pdf: http://www.unifr.ch/finwiss/assets/files/Diplom_Seminararbeiten/SeminarLoetscher.pdf

Majoral, R. (1984). Caractéristiques socio-structurales et modernisation de l'agriculture des Pyrénées occidentales catalanes. In R. Majoral & F. Lopez (Eds.), *Rural life and the exploitation of resources in Highlands and High-latitude zones* (pp. 113–122). Spain: Department of Geography, University of Barcelona.

Mehretu, A., Pigozzi, B. W., & Sommers, L. M. (2002). Spatial shifts in production and consumption: Marginality patterns in the new international division of labour. In H. Jussila, R. Majoral, & B. Cullen (Eds.), *Sustainable development and geographical space. Issues of population, environment, globalization and education in marginal regions* (pp. 195–208). Aldershot: Ashgate.

Merki, M. (2012). Verunsicherung in Andermatt. Der Ausbau des Skigebiets harzt, *NZZ online* 07.03.2012. http://www.nzz.ch/nachrichten/politik/schweiz/verunsicherung_in_andermatt_1.15520509.html. Accessed March 10, 2012.

Metzger, J. (2014). Arbeit ist nur das, was Geld bringt. Wandel der lokalen Ökonomie in Ameskar Fogani (Marokko) am Beispiel des Tourismus. *Geographica Helvetica, 69*(1), 49–58.

Munizipalgemeinde Leukerbad. (2014). *Finanzbericht & Rechnung 2013*. http://www.leukerbad.org/dl.php/de/536ce8c56f8a0/Jahresrechnung_2013.pdf. Accessed September 24, 2014.

NZZ. (2011). Über 140 Millionen für das Skigebiet Andermatt-Sedrun. *NZZ* 11.05.2011 (NZZ online). http://www.nzz.ch/nachrichten/wirtschaft/aktuell/ueber_140_millionen_fuer_das_skigebiet_andermatt-sedrun_1.10538769.html. Accessed May 18, 2011.

Orascom Development. (2008). *Andermatt—Switzerland*. http://www.orascomdh.com/en/destinations/operating-destinations/andermatt.html. Accessed September 23, 2014.

Orascom Development. (2013). The Chedi Andermatt: Super-class hotel begins operation. Press release 09.12. http://www.andermatt-swissalps.ch/en/news/pressreleases/2013/. Accessed September 23, 2014.

Reynaud, A. (1981). *Société, espace et justice. Inégalités régionales et justice socio-spatiale*. Paris: PUF.

Schuler, M. (1984). Migration patterns in the Swiss mountain areas. In E. A. Brugger, G. Furrer, B. Messerli, & P. Messerli (Eds.), *The transformation of Swiss mountain régions* (pp. 243–260). Bern: Haupt.

Simonett, J. (2006). Vals, in Historisches Lexikon der Schweiz. http://www.hls-dhs-dss.ch/textes/d/D1461.php. Accessed March 30, 2011.

UN. (2015). *The millennium development goals report 2015*. New York, United Nations.

Weatherley, R. D., & Naylon, J. (1984). Domestic tourism and second homes as motors of rural development in the Sierra Morena, Spain. In R. Majoral & F. Lopez (Eds.), *Rural life and the exploitation of resources in Highlands and High-latitude zones* (pp. 142–151). Spain: Department of Geography, University of Barcelona.

Zurfluh, A. (2009). Le cas d'Urseren. Note sur le role des expatriés dans le lancement d'un projet de développement d'une haute vallée alpestre. *Histoire des Alpes—Storia delle Alpi—Geschichte der Alpen, 14*, 121–128.

Chapter 18
Garhwal Himalaya—Potential of Cash Crops in Attaining Food Security and Enhancing Livelihoods—Khanda Gad Watershed Case Study

Vishwambhar Prasad Sati

18.1 Introduction

The Himalayan Mountain is geographically remote, economically underdeveloped and socially backward. These characteristics of the Himalaya have put people on the verge of economic and social marginality. Terraced agriculture is the main occupation, with limited arable land. The possibility for further expansion and modernization of agriculture is limited because of the fragility and steepness of the slope. Furthermore, the viability of the outcome from subsistence crops in the form of production and productivity is extremely low. The practice of cereal farming with rearing draught animals in mixed crop-livestock farming system has limited potential to maintain the food security, and for centuries the population was undernourished, struggling for livelihood sustainability. As the population grows, food insecurity increases. In the emerging trend of global change, the farmers are also fighting with market forces. Under such circumstances, enhancing and diversifying livelihood for attaining food security and poverty reduction are inevitable mainly in the Himalayan regions.

The main objective of our study was to evaluate the potential of cultivation of off-season vegetables for sustainable livelihood of the hill people. The main research question raised during the study was: What are the major cash crops grown and what is the production and productivity pattern of these crops?

The study area is located in the Khanda Gad sub-watershed, Garhwal Himalaya (Fig. 18.1). This perennial stream has its source in the Khirsu block of Pauri district and it joins the Alaknanda River at Kirtinagar, 3 km from Srinagar Garhwal towards Rishikesh. Total 12 villages lie along the Khanda Gad (both sides) where ample irrigation facilities are available. Farming system, in the watershed, characterizes cultivation of both cereals and cash crops. Native people grow cereals and

V.P. Sati (✉)
Department of Geography and Resource Management, Mizoram University, Aizawl, India
e-mail: sati.vp@gmail.com

R. Chand et al. (eds.), *Societies, Social Inequalities and Marginalization*,
Perspectives on Geographical Marginality,
DOI 10.1007/978-3-319-50998-3_18

cash crops are grown by the Nepali immigrants (total 65 households in number). A household level survey of all Nepali immigrants was carried out to reveal the cost-benefit analysis of cash generating and subsistence crops. We also interviewed the households about the output from vegetables and cereals.

In this chapter, we discussed mainly on the current patterns of cash and cereal crops, their economic valuation and case study of Khanda Gad sub watershed, and suggested policy measures for sustainable livelihoods.

18.2 Geographical and Socio-Economic Background

Garhwal Himalaya, lies in centre of the Indian Himalayan Region, forms an integral part of the Uttarakhand Himalaya. Various climate, numerous rivers, undulating terrain and topographies—valleys, mid-altitudes, highlands, alpine meadows and frigid high mountain ranges—make this part of the Himalaya so diverse in all respects. It is the home of the rivers—Bhagirathi, Alaknanda, Vishnu Ganga, Dauli Ganga, Nandakini, Pindar, Mandakini and their numerous tributaries—that form the Ganges system. The perpetual snow-clad mountain peaks—Chaukhamba, Kamet, Trishuli and Nandadevi—with their heights more than 7000 m, control the climatic conditions and make this part more panoramic. Altitude and slope aspects change

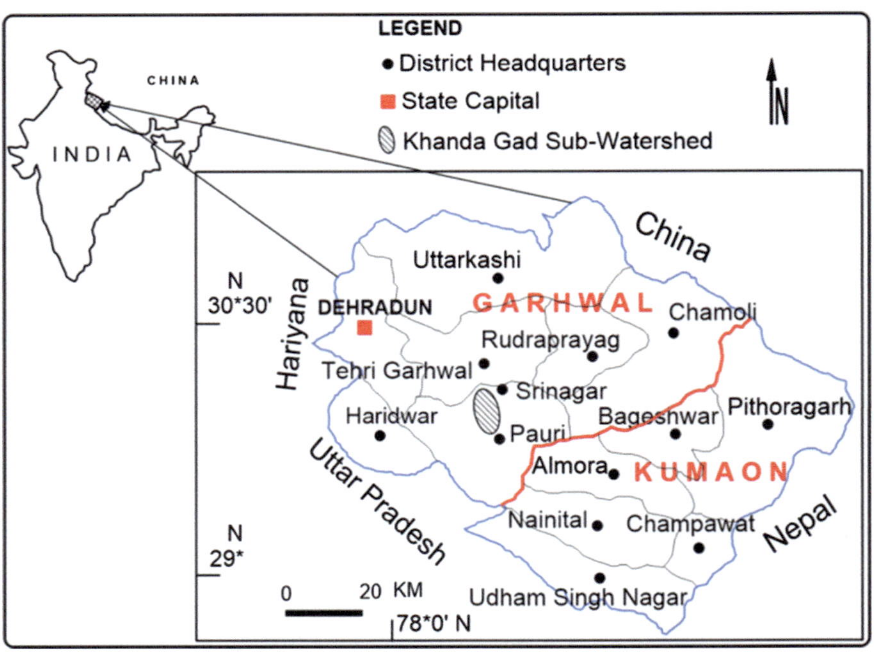

Fig. 18.1 Location map of the study area

climatic conditions and natural vegetations. Diversity in natural vegetations, from bushes and shrubs in the valleys to pine in the mid-altitudes, mixed oak forests in the temperate regions, coniferous forests in the higher altitudes above 2000 m and alpine meadows above 2800 m, is high although Garhwal region.

The Himalaya allows the cultivation of all kinds of vegetables, which grow in the different altitudinal zones and in all geographical locations. This diversity in vegetable crops can be observed everywhere. The whole region obtains suitable agro-ecological conditions for growing various types of vegetables. However, the production of vegetables is considerably low, mainly because of less proportion of arable land is devoted for its cultivation. Further, availability of infrastructural facilities such as cold storages, means of transportation and markets are lagging behind causing constraints to grow vegetables. Most of the vegetable products are consumed locally. Meanwhile, potatoes in the highlands and onions in the mid-altitudes and the valley regions grow largely and assist in food security.

India is the second largest producer of vegetables in the world. A report of the Indian Council of Agricultural Research (ICAR 2002) reveals that the present production of vegetables is 90.8 million tons and efforts are on full swing to raise its production to 250 million tons by 2024–2025. The Government undertakes significant efforts to produce substantial amount of vegetables in order to achieve the target. The most noticeable efforts are to increase the surfaces of vegetables by using hybrid seeds, and improved agro-techniques. Another potential approach is the perfection and promotion of protected vegetable cultivation (Singh 1998; Singh et al. 1999).

Like the other parts of the Himalaya, the Garhwal Himalaya practices subsistence cereal farming, which is insufficient for to carry livelihood sustainably. The terrain is not fit for the intensive cultivation of crops due to soil erosion, slope instability and the fragile ecosystem. Poverty and malnutrition have become common phenomena. This has led to high rates of out-migration to the foothills and plains. Under such circumstances, cultivation of cash crops—fruits, off-season vegetables, and medicinal plants—can help to enhance the livelihood of the population and can provide more employment to the unemployed youth. In our research presented in this chapter, we analyzed the economic viability of off-season vegetables.

The diversity in agro-ecological conditions provides suitability to grow variety of cash crops on the hill slopes with considerably high production. These crops can be a substantial help towards poverty alleviation and the reduction of malnutrition. In the Garhwal Himalaya, the production of potato, onion, and tomato is noteworthy as potato is exported to the regional markets on a large-scale. The other vegetables are pumpkin, cucumber, beans, reddish, carrot, coriander, and green leafs, but these are grown for domestic consumption. The commercial importance of these vegetables is also very high. Among spices, ginger, turmeric, chili, tejpat, coriander and garlic are common and grown extensively. Recently, the production of potato has assumed a remarkable position among crops, grow in high altitudes (1500–2200 m), and the cultivation of onion has become important in the valley regions and on the lower-middle slopes (800–1200 m).

18.3 Current Pattern of Cultivating Traditional Cereal Crops and Food Security

The Garhwal Himalaya characterises the dominance of traditional subsistence cereal farming, based upon century old practices; however, distribution of crops and their cultivating pattern are not uniform but vary between the valley regions and the highlands. Generally, millet grows in the highlands while the valleys characterises cultivating wheat and paddy crops. The economic viability of these crops in the valleys, the mid-slopes, and the highlands is insufficient in terms of meeting the food requirement of the population, due to low production. Several factors combine to propel changes in the regional agrarian system; uneconomic production on inconveniently located fields, growing off-farm opportunities to secure people's livelihood, the high population growth rate, which is exercising pressure on cultivable land, modern innovation in the agricultural fields and a high literacy rate. Together they promote the search for a possibility to adopt the cultivation of more economically viable crops. However, technological innovations such as chemical fertilizers, pesticides and high yield crop varieties, which transformed the valley's regions fields, did not succeed in changing the highland farming system because of the constraints specific to mountains. The dependence on forests for maintaining soil fertility in the croplands or the expansion of agricultural land was not substituted by the new technologies. As forests and livestock provide material and energy inputs in traditional mountain farming systems, the expansion of traditional agriculture did the risk of forest degradation. In order to deal with the present and future challenges of the sustainability criteria, the traditional systems need to be adapted in ways which enhance crop yields in a sustainable manner (Ramakrishnan et al. 1993).

The cropping pattern also varies with the variations in the climatic conditions and cropping seasons. The influence of the monsoon on the cropping pattern is very dominant; as a result, about 70.75% of the total cropped areas are under 'Kharif' (rainy season crops). In the region, whatever may be the type of soil or the amount of rainfall, the dominance of food grains in the cropping pattern is obvious everywhere (Sati and Rawat 1993). Presently, the cultivation of food grain is limited only on the terraced slopes in mid-altitudes (below 2000 m) or un-terraced gentle slopes in high altitudes (above 2000 m). In terms of, diversity in crops, it is high in the highlands in comparison with the crops of valley regions, whereas the valleys, throughout the area are much more intensively cropped than the slopes. Crop diversity is managed by mixed cropping with crop rotation (Sati 1993).

The Garhwal Himalaya provides a great opportunity for producing cash crops as it has wide range of agro-climatic conditions vary from sub-tropical to temperate and cold in different altitudes that suite to grow various types of cash crops—fruits and vegetables mainly. Similarly, cash crop farming varies from the valley regions to the mid-altitudes and the highlands according to terrain, slope, soil contents and availability of water. On the mid-altitudes and the highlands, potato grows extensively, while on the terraces of the valley regions, onion is the main vegetable.

Besides, almost all varieties of vegetables grow in the entire region, which has a high economic value. Vegetable farming can be divided into two vertical zones according to altitude.

Potato is the main commercial vegetable of the highlands and grows between 1500 and 2200 m. It is mostly found on the gentle slopes of the middle and highlands (10–15° slope angle). During the eighties, the farmers of the region started intensive cultivation of potato and they are now able to export them to the regional markets. The main areas where potatoes are grown on a large-scale are the highlands of the Garhwal Himalaya. Along potatoes, other vegetables and spices such as beans, ginger, cucumber, pumpkin, turmeric and chili are grown. These vegetables are mainly locally consumed.

Onions are grown in the lowlands, where water is amply available. They grow during summer in different places along the river terraces. Presently, the farmers are able to export onion on the regional markets (foothills of *Shivaliks*). Along with the production of vegetables, the region has a tremendous potential for the production of spices. The ecological conditions—climate, soil and landscape—of the region are highly suitable for the cultivation of spices such as ginger, turmeric, chili, tejpat, coriander, and garlic. Among them, ginger is the main cash crop supporting the livelihood and improving the economic level of many ginger growers in the basin.

Traditional cereal farming is the main occupation of the people of the mountain regions in general and the Garhwal Himalaya in particular (Sati 2005). The economic viability of these crops is not sufficient to guarantee daily meals. The nature of terrain, steep slopes, soil fertility, the availability of water and the uses of technology in the field of agriculture are the main limiting factors for food security and sustainable development. Under such circumstances, the marginal farmers of the region sought out the possibilities of cash crops, and they were highly successful.

18.4 Case Study of the Cluster of Villages in the Khanda Gad Watershed

We conducted a case study of 65 households of Nepali immigrants, spread in 12 villages of the Khanda Gad watershed in 2009. About 10 years ago, these Nepali inhabitants of the Rolpa District of Nepal visited the villages of Khanda Gad. Initially, they immigrated here in search of jobs. They met with the villagers of Khanda Gad (Margaon, Shrikot Khanda, Margadna, Bhitai Malli, Kaldung, Dhanak, Gaduwa Gad, Kamand, Dov, Shiyar Malla, Rawat Gaon, Bhitai Talli), mostly those who had permanently out-migrants. From them, they could lease small patches of land along the Khanda Gad (Rs. 4000 IC per family per crop season), which had been mostly abandoned. As a perennial stream, the Khanda Gad provides ample irrigation facility. The elders of the households are to a large extent illiterate, but recently their children have joined primary school. The overall literacy rate was 31.3% (among children 90%) with girls, however, 45.3%. Males and females are equally working on the farmlands. Table 18.1 shows demographic profile of Nepali immigrants. Total population of the 65 households was 409. Female population was higher (208) than male population (201).

The Nepali immigrants started cultivating cash generating crops such as potato, tomato, onion, cauliflower, capsicum, spinach, cucumber, pumpkin and beans, commercially. They use both chemical fertilizer (Rs. 1400/ha) and manure (Rs. 500/ha). Manure is easily available from the nearby villages. Every cluster of households has a pair of oxen, which they use to plow fields and get manure. The head of Nepali immigrants in Dhanak village informed us that there is no negative change in the production of off-season vegetables although they are using chemical fertilizers to boost the production of vegetables. However, subsistence agriculture still is done in the middle patches of the fields by the native farmers, but both production and per ha yields of the crops are considerably lower. Table 18.2 shows the production of off-season vegetables in 65 households of twelve villages of Khanda Gad in 2007–08. The highest yield among off-season vegetables was obtained from cauliflower (7.3 tones) followed by potato (4) and cucumber (3.1). Tomato's production was 1.8 tones. The figure was followed by beans (1.4) and capsicum (1.2) (Figs. 18.2 and 18.3).

Table 18.3 shows area and annual income of off-season vegetables and cereals in 2007–2008. The total area under off-season vegetables was 51.5 ha and annual income was Rs. 1,338,100 (Rs. 25,982 per ha) while under cereal crops, it is

Table 18.1 Demographic profile of Nepali immigrants (field survey 2009)

Village name	Total Nepali HHs	Total population	Male	Female	Total literacy (girls literacy)
Margaon (Khanda)	10	64	33	31	22 (9)
Koti	10	65	32	33	18 (7)
Dhamkeshwar	10	63	32	31	20 (11)
Khanda Shrikot	2	11	6	5	5 (2)
Margadana	2	13	7	6	4 (2)
Mitai	1	7	4	3	3 (1)
Kaldung	1	10	6	4	4 (2)
Dhanak	4	26	12	14	10 (6)
Gadowa Gad	2	14	7	7	5 (2)
Kamand	4	20	11	9	6 (2)
Dob Shrikot	4	22	12	10	7 (3)
Malli Sera	5	36	12	24	6 (3)
Rawat Gaon	4	24	11	13	7 (3)
Gadoli (tea state)	6	34	16	18	11 (5)
Total	65	409	201	208	128 (58)

Table 18.2 Production of off-season vegetables in Khanda Gad 2007–08

Village name	Area ha*	Production (in tones)					
		Cauliflower	Cucumber	Tomato	Capsicum	Beans	Potato
Margaon (Khanda)	5	0.7	0.3	0.4	0.1	0.2	Nil
Koti	5	0.7	0.3	0.4	0.1	0.2	1
Dhamkeshwar	5	0.7	0.3	0.4	0.1	0.2	2
Khanda Shrikot	2	0.3	0.1	0.1	0.04	0.04	1
Margadana	2	0.3	0.1	0.1	0.04	0.04	Nil
Mitai	1	0.1	0.06	0.06	0.02	0.02	
Kaldung	1.5	0.1	0.8	0.2	0.02	0.03	Nil
Dhanak	5	0.7	0.3	0.4	0.1	0.2	Nil
Gadowa Gad	2	0.3	0.1	0.1	0.04	0.04	Nil
Kamand	5	0.7	0.3	0.4	0.2	0.2	Nil
Dob Shrikot	5	0.7	0.3	0.4	0.1	0.2	Nil
Malli Sera	4	0.6	0.3	0.3	0.09	0.1	Nil
Rawat Gaon	4	0.6	0.3	0.3	0.09	0.1	Nil
Gadoli (tea state)	5	0.7	0.3	0.4	0.1	0.2	Nil
Total	51.5	7.3	3.1	1.8	1.2	1.4	4

Source Field survey, *Revenue department

330.4 ha land and Rs. 5,021,600 annual income, which is Rs. 15,198 per ha. Similarly, the numbers of households producing off-season vegetables was 65 while, 368 households cultivated cereals.

18.5 Discussion and Conclusions

The cultivation of subsistence cereal crops in the Garhwal Himalaya has not met the food requirement for centuries. The increase in population on limited agricultural surfaces further accelerated food scarcity and marginalized local population even more. The reaction to this was the transformation of subsistence crops to cash generating crops to achieve food security. Similarly, large-scale out-migration towards the metropolitan cities and the Ganges Plains in search of livelihood has taken place. The change in cultivation did not result in large-scale transformation because it was limited to certain valley regions where ample water supply was available, and to the uplands. Fruit cultivation also started during the 1980s but failed due to various reasons. The study area of the Khanda Gad sub-watershed is a unique example of food security through the cultivation of cash generating crops, but this is mainly done by the Nepali immigrants, not by the native people. They overlooked the success of the immigrants.

Fig. 18.2 Panoramic Khanda Gad watershed and paddy fields

The Garhwal Himalaya has suitable agro-ecological conditions for the cultivation of various crops, subsistence as well as cash generating. The emphasis is on land-based development, particularly on cultivation, rather than on looking for other options of livelihood, because the land provides the immediate base for food. Industrial development could not take place because of fragile terrain and the lack of infrastructure.

Subsistence cereal farming dominates in the cropping pattern of the Garhwal Himalaya; while their economic viability is below average. The reasons behind the low viability of cereal crops are:

- traditional methods of cultivation,
- seeds of low quality and poor land conditions,
- low soil fertility, and
- no use of modern innovation in the field of agriculture.

The amount of crops produced is significantly below the desired level. Furthermore, the fragmentation of the fields and fragility of the terrain prevent poor farmers to intensively cultivate their land. There is no use of innovation in the farmlands mainly because of undulating terrain and steep slope, which does not permit the introduction of innovative measures in the field of agriculture; therefore, agriculture is still traditional and heavily dependent on draught power. The use of chemical fertilizers is prohibited as past experiences show that the fertility of the

Fig. 18.3 Vegetables grow along the stream

Table 18.3 Area and annual income of off-season vegetables and cereal crops 2007–2008

Off-season vegetables	Area (ha)	Annual income (Rs.)	Cereal crops[a]	Area (ha)	Annual income (Rs.)
Cauliflower	15	508,900	Wheat	100	1,472,000
Cucumber	3	217,000	Rice	100	2,520,000
Tomato	20	179,000	Barley	50	760,000
Capsicum	3	118,000	Millets	50	121,600
Beans	3	115,200	Pulses	20	70,000
Potato	7.5	200,000	Oilseed	10.4	78,000
Total	51.5	1,338,100	Total	330.4	5,021,600

[a]Data on cereal crops was collected from village wise statistical diary of Pauri district, 2009–10
Source Field survey

soil diminished due to the excessive use of chemical fertilizers during the 1980s. Under the rain fed system of agriculture, these fertilizers burnt soil fertility. Already, agriculturists and horticulturists have initiated a great discussion about the viability of subsistence cereal crops, concluding that the traditional agricultural crops are more suited in the ecosystem of the mountain areas but that at the same time they are economically unviable. Great efforts are undertaken in the field of horticulture particularly in the cultivation of vegetables, and they yielded impressive success. Under such conditions, the efforts, which are already on the way, should be promoted more rigorously and systematically. The awareness programmes should be launched with full motivation and the farmers should be encouraged to devote the cereal cropped land into vegetable crops keeping all the positive factors in view. It will definitively improve the living standard of the people and the economy of the region.

Our study shows that the farming of off-season vegetables has the potential to enhance livelihood and food security as the agro-climatic conditions are suitable in this region. Potato and onion have the potential to ensure food security and sustainable development. The farmers of the regions have already started planting these two vegetables along with other spices and green leafy vegetables, and they have been very successful in their production and as well as sustainable livelihood. During the household level survey, the farmers responded that the main impediments on the way to production of vegetables are the lack of:

- cold storage,
- means of transportation,
- improved seeds, and
- fertilizer.

Because most of the vegetables are perishable in nature, cold storage is required for preservation.

A large proportion of arable land can be transformed to the cultivation of off-season vegetables to attain livelihood sustainability and food security. The cultivation of cereals should go parallel as it maintains ecology and diversity in the cropping pattern. The area has extensive alpine grasslands and feasible climatic conditions for rearing a high yield variety of indigenous livestock. This practice can lead to development of the dairy farming. Community participation, establishing cold storages, proper selection of land and imparting training to the farmers for scientific cultivation of vegetables are amongst the several measures that are required to attain food security and to enhance livelihoods.

References

ICAR. (2002). *Agricultural research data book*. New Delhi: Indian Council of Agricultural Research.

Ramakrishnan, P. S., Saxena, K. G., Swift, M. J., & Seward, P. D. (Eds.). (1993). *Tropical soil biology and fertility research: South Asian Context*. Dehra Dun: Oriental Enterprises.

Sati, V. P., & Rawat, M. S. S. (1993). Problems of agriculture ecology and their management in the Garhwal Himalaya. *National Geographer, 28*(2), 143–148.

Sati, V. P. (1993). Cropping pattern in the hill environment of the Garhwal Himalaya (A case study of a village). *National Geographer, 28*(1), 31–37.

Sati, V. P. (2005). Vertical zonation of horticultural farming in the Alaknanda basin of Garhwal Himalaya, India. *Journal of Mountain Science, 2*(4), 319–328.

Singh, B. (1998). Vegetable production under protected conditions: Problems and Prospects. Indian Soc. Veg. Sci. Souvenir: Silver Jubilee, National Symposium Dec 12–14, 1998, Varanasi, U.P. India pp. 90.

Singh, N., Diwedi, S. K., & Paljor, E. (1999). *Ladakh Mein Sabjion Kei Sanrakshit Kheti*. Regional Research Laboratory of DRDO, Leh. 56 A.P.O.

Index

R. Chand et al. (eds.), *Societies, Social Inequalities and Marginalization*,
Perspectives on Geographical Marginality,
DOI 10.1007/978-3-319-50998-3

MIX
Papier aus verantwortungsvollen Quellen
Paper from responsible sources
FSC® C105338

If you have any concerns about our products,
you can contact us on
ProductSafety@springernature.com

In case Publisher is established outside the EU,
the EU authorized representative is:
Springer Nature Customer Service Center GmbH
Europaplatz 3, 69115 Heidelberg, Germany

Printed by Libri Plureos GmbH
in Hamburg, Germany